Stochastic Optimal Control:
The Discrete-Time Case

Dimitri P. Bertsekas and Steven E. Shreve

WWW site for book information and orders

http://world.std.com/~athenasc/

Athena Scientific, Belmont, Massachusetts

Athena Scientific
Post Office Box 391
Belmont, Mass. 02178-9998
U.S.A.

Email: athenasc@world.std.com
WWW information and orders: http://world.std.com/~athenasc/

Cover Design: *Ann Gallager*

Originally published by Academic Press, Inc., in 1978

OPTIMIZATION AND NEURAL COMPUTATION SERIES

1. Dynamic Programming and Optimal Control, Vols. I and II, by
   Dimitri P. Bertsekas, 1995
2. Nonlinear Programming, by Dimitri P. Bertsekas, 1995
3. Neuro-Dynamic Programming, by Dimitri P. Bertsekas and John
   N. Tsitsiklis, 1996
4. Constrained Optimization and Lagrange Multiplier Methods, by
   Dimitri P. Bertsekas, 1996
5. Stochastic Optimal Control: The Discrete-Time Case by Dimitri
   P. Bertsekas and Steven E. Shreve, 1996

Publisher's Cataloging-in-Publication Data

Bertsekas, Dimitri P.
Stochastic Optimal Control: The Discrete-Time Case
Includes bibliographical references and index
1. Dynamic Programming. 2. Stochastic Processes. 3. Measure Theory. I.
Shreve, Steven E., joint author. II. Title.
T57.83.B49 1996   519.7'03   96-80191

ISBN 1-886529-03-5

To
Joanna
and
Steve's Mom and Dad

Contents

Preface   xi
Acknowledgments   xiii

Chapter 1 **Introduction**

1.1   Structure of Sequential Decision Models   1
1.2   Discrete-Time Stochastic Optimal Control Problems—Measurability
       Questions   5
1.3   The Present Work Related to the Literature   13

Part I ANALYSIS OF DYNAMIC
PROGRAMMING MODELS

Chapter 2 **Monotone Mappings Underlying Dynamic
Programming Models**

2.1   Notation and Assumptions   25
2.2   Problem Formulation   28
2.3   Application to Specific Models   29
       2.3.1   Deterministic Optimal Control   30
       2.3.2   Stochastic Optimal Control—Countable Disturbance Space   31
       2.3.3   Stochastic Optimal Control—Outer Integral Formulation   35
       2.3.4   Stochastic Optimal Control—Multiplicative Cost Functional   37
       2.3.5   Minimax Control   38

Chapter 3 **Finite Horizon Models**

3.1   General Remarks and Assumptions   39
3.2   Main Results   40
3.3   Application to Specific Models   47

Chapter 4 **Infinite Horizon Models under a Contraction Assumption**

4.1 General Remarks and Assumptions 52
4.2 Convergence and Existence Results 53
4.3 Computational Methods 58
    4.3.1 Successive Approximation 59
    4.3.2 Policy Iteration 63
    4.3.3 Mathematical Programming 67
4.4 Application to Specific Models 68

Chapter 5 **Infinite Horizon Models under Monotonicity Assumptions**

5.1 General Remarks and Assumptions 70
5.2 The Optimality Equation 71
5.3 Characterization of Optimal Policies 78
5.4 Convergence of the Dynamic Programming Algorithm—Existence of Stationary Optimal Policies 80
5.5 Application to Specific Models 88

Chapter 6 **A Generalized Abstract Dynamic Programming Model**

6.1 General Remarks and Assumptions 92
6.2 Analysis of Finite Horizon Models 94
6.3 Analysis of Infinite Horizon Models under a Contraction Assumption 96

Part II STOCHASTIC OPTIMAL CONTROL THEORY

Chapter 7 **Borel Spaces and Their Probability Measures**

7.1 Notation 102
7.2 Metrizable Spaces 104
7.3 Borel Spaces 117
7.4 Probability Measures on Borel Spaces 122
    7.4.1 Characterization of Probability Measures 122
    7.4.2 The Weak Topology 124
    7.4.3 Stochastic Kernels 134
    7.4.4 Integration 139
7.5 Semicontinuous Functions and Borel-Measurable Selection 145
7.6 Analytic Sets 156
    7.6.1 Equivalent Definitions of Analytic Sets 156
    7.6.2 Measurability Properties of Analytic Sets 166
    7.6.3 An Analytic Set of Probability Measures 169
7.7 Lower Semianalytic Functions and Universally Measurable Selection 171

Chapter 8 **The Finite Horizon Borel Model**

8.1 The Model 188

8.2   The Dynamic Programming Algorithm—Existence of Optimal
       and ϵ-Optimal Policies   194
8.3   The Semicontinuous Models   208

Chapter 9   **The Infinite Horizon Borel Models**

9.1   The Stochastic Model   213
9.2   The Deterministic Model   216
9.3   Relations between the Models   218
9.4   The Optimality Equation—Characterization of Optimal Policies   225
9.5   Convergence of the Dynamic Programming Algorithm—Existence
       of Stationary Optimal Policies   229
9.6   Existence of ϵ-Optimal Policies   237

Chapter 10   **The Imperfect State Information Model**

10.1 Reduction of the Nonstationary Model—State Augmentation   242
10.2 Reduction of the Imperfect State Information Model—Sufficient
       Statistics   246
10.3 Existence of Statistics Sufficient for Control   259
       10.3.1   Filtering and the Conditional Distributions of the States   260
       10.3.2   The Identity Mappings   264

Chapter 11   **Miscellaneous**

11.1 Limit-Measurable Policies   266
11.2 Analytically Measurable Policies   269
11.3 Models with Multiplicative Cost   271

Appendix A   **The Outer Integral**   273

Appendix B   **Additional Measurability Properties of Borel
                Spaces**

B.1   Proof of Proposition 7.35(e)   282
B.2   Proof of Proposition 7.16   285
B.3   An Analytic Set Which Is Not Borel-Measurable   290
B.4   The Limit σ-Algebra   292
B.5   Set Theoretic Aspects of Borel Spaces   301

Appendix C   **The Hausdorff Metric and the Exponential
                Topology**

References   312

Table of Propositions, Lemmas, Definitions, and Assumptions   317
Index   321

Preface

This monograph is the outgrowth of research carried out at the University of Illinois over a three-year period beginning in the latter half of 1974. The objective of the monograph is to provide a unifying and mathematically rigorous theory for a broad class of dynamic programming and discrete-time stochastic optimal control problems. It is divided into two parts, which can be read independently.

Part I provides an analysis of dynamic programming models in a unified framework applicable to deterministic optimal control, stochastic optimal control, minimax control, sequential games, and other areas. It resolves the *structural questions* associated with such problems, i.e., it provides results that draw their validity exclusively from the sequential nature of the problem. Such results hold for models where measurability of various objects is of no essential concern, for example, in deterministic problems and stochastic problems defined over a countable probability space. The starting point for the analysis is the mapping defining the dynamic programming algorithm. A single abstract problem is formulated in terms of this mapping and counterparts of nearly all results known for deterministic optimal control problems are derived. A new stochastic optimal control model based on outer integration is also introduced in this part. It is a broadly applicable model and requires no topological assumptions. We show that all the results of Part I hold for this model.

Part II resolves the *measurability questions* associated with stochastic optimal control problems with perfect and imperfect state information. These questions have been studied over the past fifteen years by several researchers in statistics and control theory. As we explain in Chapter 1, the approaches that have been used are either limited by restrictive assumptions such as compactness and continuity or else they are not sufficiently powerful to yield results that are as strong as their structural counterparts. These deficiencies can be traced to the fact that the class of policies considered is not sufficiently rich to ensure the existence of everywhere optimal or ϵ-optimal policies except under restrictive assumptions. In our work we have appropriately enlarged the space of admissible policies to include *universally measurable policies*. This guarantees the existence of ϵ-optimal policies and allows, for the first time, the development of a general and comprehensive theory which is as powerful as its deterministic counterpart.

We mention, however, that the class of universally measurable policies is not the smallest class of policies for which these results are valid. The smallest such class is the class of *limit measurable policies* discussed in Section 11.1. The σ-algebra of limit measurable sets (or C-sets) is defined in a constructive manner involving transfinite induction that, from a set theoretic point of view, is more satisfying than the definition of the universal σ-algebra. We believe, however, that the majority of readers will find the universal σ-algebra and the methods of proof associated with it more understandable, and so we devote the main body of Part II to models with universally measurable policies.

Parts I and II are related and complement each other. Part II makes extensive use of the results of Part I. However, the special forms in which these results are needed are also available in other sources (e.g., the textbook by Bertsekas [B4]). Each time we make use of such a result, we refer to both Part I and the Bertsekas textbook, so that Part II can be read independently of Part I. The developments in Part II show also that stochastic optimal control problems with measurability restrictions on the admissible policies can be embedded within the framework of Part I, thus demonstrating the broad scope of the formulation given there.

The monograph is intended for applied mathematicians, statisticians, and mathematically oriented analysts in engineering, operations research, and related fields. We have assumed throughout that the reader is familiar with the basic notions of measure theory and topology. In other respects, the monograph is self-contained. In particular, we have provided all necessary background related to Borel spaces and analytic sets.

Acknowledgments

This research was begun while we were with the Coordinated Science Laboratory of the University of Illinois and concluded while Shreve was with the Departments of Mathematics and Statistics of the University of California at Berkeley. We are grateful to these institutions for providing support and an atmosphere conducive to our work, and we are also grateful to the National Science Foundation for funding the research. We wish to acknowledge the aid of Joseph Doob, who guided us into the literature on analytic sets, and of John Addison, who pointed out the existing work on the limit σ-algebra. We are particularly indebted to David Blackwell, who inspired us by his pioneering work on dynamic programming in Borel spaces, who encouraged us as our own investigation was proceeding, and who showed us Example 9.2. Chapter 9 is an expanded version of our paper "Universally Measurable Policies in Dynamic Programming" published in *Mathematics of Operations Research*. The permission of The Institute of Management Sciences to include this material is gratefully acknowledged. Finally we wish to thank Rose Harris and Dee Wrather for their excellent typing of the manuscript.

Chapter 1

Introduction

1.1 Structure of Sequential Decision Models

Sequential decision models are mathematical abstractions of situations in which decisions must be made in several stages while incurring a certain cost at each stage. Each decision may influence the circumstances under which future decisions will be made, so that if total cost is to be minimized, one must balance his desire to minimize the cost of the present decision against his desire to avoid future situations where high cost is inevitable.

A classical example of this situation, in which we treat profit as negative cost, is portfolio management. An investor must balance his desire to achieve immediate return, possibly in the form of dividends, against a desire to avoid investments in areas where low long-run yield is probable. Other examples can be drawn from inventory management, reservoir control, sequential analysis, hypothesis testing, and, by discretizing a continuous problem, from control of a large variety of physical systems subject to random disturbances. For an extensive set of sequential decision models, see Bellman [B1], Bertsekas [B4], Dynkin and Juskevič [D8], Howard [H7], Wald [W2], and the references contained therein.

Dynamic programming (DP for short) has served for many years as the principal method for analysis of a large and diverse group of sequential

1

decision problems. Examples are deterministic and stochastic optimal control problems, Markov and semi-Markov decision problems, minimax control problems, and sequential games. While the nature of these problems may vary widely, their underlying structures turn out to be very similar. In all cases, the cost corresponding to a policy and the basic iteration of the DP algorithm may be described by means of a certain mapping which differs from one problem to another in details which to a large extent are inessential. Typically, this mapping summarizes all the data of the problem and determines all quantities of interest to the analyst. Thus, in problems with a finite number of stages, this mapping may be used to obtain the optimal cost function for the problem as well as to compute an optimal or ε-optimal policy through a finite number of steps of the DP algorithm. In problems with an infinite number of stages, one hopes that the sequence of functions generated by successive application of the DP iteration converges in some sense to the optimal cost function for the problem. Furthermore, all basic results of an analytical and computational nature can be expressed in terms of the underlying mapping defining the DP algorithm. Thus by taking this mapping as a starting point one can provide powerful analytical results which are applicable to a large collection of sequential decision problems.

To illustrate our viewpoint, let us consider formally a deterministic optimal control problem. We have a discrete-time system described by the system equation

$$x_{k+1} = f(x_k, u_k), \tag{1}$$

where x_k and x_{k+1} represent a state and its succeeding state and will be assumed to belong to some state space S; u_k represents a control variable chosen by the decisionmaker in some constraint set $U(x_k)$, which is in turn a subset of some control space C. The cost incurred at the kth stage is given by a function $g(x_k, u_k)$. We seek a finite sequence of control functions $\pi = (\mu_0, \mu_1, \ldots, \mu_{N-1})$ (also referred to as a *policy*) which minimizes the total cost over N stages. The functions μ_k map S into C and must satisfy $\mu_k(x) \in U(x)$ for all $x \in S$. Each function μ_k specifies the control $u_k = \mu_k(x_k)$ that will be chosen when at the kth stage the state is x_k. Thus the total cost corresponding to a policy $\pi = (\mu_0, \mu_1, \ldots, \mu_{N-1})$ and initial state x_0 is given by

$$J_{N,\pi}(x_0) = \sum_{k=0}^{N-1} g[x_k, \mu_k(x_k)], \tag{2}$$

where the states $x_1, x_2, \ldots, x_{N-1}$ are generated from x_0 and π via the system equation

$$x_{k+1} = f[x_k, \mu_k(x_k)], \qquad k = 0, \ldots, N-2. \tag{3}$$

Corresponding to each initial state x_0 and policy π, there is a sequence of control variables $u_0, u_1, \ldots, u_{N-1}$, where $u_k = \mu_k(x_k)$ and x_k is generated by

(3). Thus an alternative formulation of the problem would be to select a sequence of control variables minimizing $\sum_{k=0}^{N-1} g(x_k, u_k)$ rather than a policy π minimizing $J_{N,\pi}(x_0)$. The formulation we have given here, however, is more consistent with the DP framework we wish to adopt.

As is well known, the DP algorithm for the preceding problem is given by

$$J_0(x) = 0, \tag{4}$$

$$J_{k+1}(x) = \inf_{u \in U(x)} \{g(x, u) + J_k[f(x, u)]\}, \qquad k = 0, \ldots, N-1, \tag{5}$$

and the optimal cost $J^*(x_0)$ for the problem is obtained at the Nth step, i.e.,

$$J^*(x_0) = \inf_\pi J_{N,\pi}(x_0) = J_N(x_0).$$

One may also obtain the value $J_{N,\pi}(x_0)$ corresponding to any $\pi = (\mu_0, \mu_1, \ldots, \mu_{N-1})$ at the Nth step of the algorithm

$$J_{0,\pi}(x) = 0, \tag{6}$$

$$J_{k+1,\pi}(x) = g[x, \mu_{N-k-1}(x)] + J_{k,\pi}[f(x, \mu_{N-k-1}(x))], \qquad k = 0, \ldots, N-1. \tag{7}$$

Now it is possible to formulate the previous problem as well as to describe the DP algorithm (4)–(5) by means of the mapping H given by

$$H(x, u, J) = g(x, u) + J[f(x, u)]. \tag{8}$$

Let us define the mapping T by

$$T(J)(x) = \inf_{u \in U(x)} H(x, u, J) \tag{9}$$

and, for any function $\mu: S \to C$, define the mapping T_μ by

$$T_\mu(J)(x) = H[x, \mu(x), J]. \tag{10}$$

Both T and T_μ map the set of real-valued (or perhaps extended real-valued) functions on S into itself. Then in view of (6)–(7), we may write the cost functional $J_{N,\pi}(x_0)$ of (2) as

$$J_{N,\pi}(x_0) = (T_{\mu_0} T_{\mu_1} \cdots T_{\mu_{N-1}})(J_0)(x_0), \tag{11}$$

where J_0 is the zero function on S $[J_0(x) = 0 \ \forall x \in S]$ and $(T_{\mu_0} T_{\mu_1} \cdots T_{\mu_{N-1}})$ denotes the composition of the mappings $T_{\mu_0}, T_{\mu_1}, \ldots, T_{\mu_{N-1}}$. Similarly the DP algorithm (4)–(5) may be described by

$$J_{k+1}(x) = T(J_k)(x), \qquad k = 0, \ldots, N-1, \tag{12}$$

and we have

$$\inf_{\pi} J_{N,\pi}(x_0) = T^N(J_0)(x_0),$$

where T^N is the composition of T with itself N times. Thus *both the problem and the major algorithmic procedure relating to it can be expressed in terms of the mappings T and T_μ.*

One may also consider an infinite horizon version of the problem whereby we seek a sequence $\pi = (\mu_0, \mu_1, \ldots)$ that minimizes

$$J_\pi(x_0) = \lim_{N \to \infty} \sum_{k=0}^{N-1} g[x_k, \mu_k(x_k)] = \lim_{N \to \infty} (T_{\mu_0} T_{\mu_1} \cdots T_{\mu_{N-1}})(J_0)(x_0) \quad (13)$$

subject to the system equation constraint (3). In this case one needs, of course, to make assumptions which ensure that the limit in (13) is well defined for each π and x_0. Under appropriate assumptions, the optimal cost function defined by

$$J^*(x) = \inf_{\pi} J_\pi(x)$$

can be shown to satisfy Bellman's functional equation given by

$$J^*(x) = \inf_{u \in U(x)} \{g(x, u) + J^*[f(x, u)]\}.$$

Equivalently

$$J^*(x) = T(J^*)(x) \qquad \forall x \in S,$$

i.e., J^* is a fixed point of the mapping T. Most of the infinite horizon results of analytical interest center around this equation. Other questions relate to the existence and characterization of optimal policies or nearly optimal policies and to the validity of the equation

$$J^*(x) = \lim_{N \to \infty} T^N(J_0)(x) \qquad \forall x \in S, \tag{14}$$

which says that the DP algorithm yields in the limit the optimal cost function for the problem. Again the problem and the basic analytical and computational results relating to it can be expressed in terms of the mappings T and T_μ.

The deterministic optimal control problem just described is representative of a plethora of sequential optimization problems of practical interest which may be formulated in terms of mappings similar to the mapping H of (8). As shall be described in Chapter 2, one can formulate in the same manner stochastic optimal control problems, minimax control problems, and others. *The objective of Part I is to provide a common analytical frame-*

work for all these problems and derive in a broadly applicable form all the results which draw their validity exclusively from the basic sequential structure of the decision-making process. This is accomplished by taking as a starting point a mapping H such as the one of (8) and deriving all major analytical and computational results within a generalized setting. The results are subsequently specialized to five particular models described in Section 2.3: *deterministic optimal control problems, three types of stochastic optimal control problems (countable disturbance space, outer integral formulation, and multiplicative cost functional), and minimax control problems.*

1.2 Discrete-Time Stochastic Optimal Control Problems— Measurability Questions

The theory of Part I is not adequate by itself to provide a complete analysis of stochastic optimal control problems, the treatment of which is the major objective of this book. The reason is that when such problems are formulated over uncountable probability spaces nontrivial measurability restrictions must be placed on the admissible policies unless we resort to an outer integration framework.

A discrete-time stochastic optimal control problem is obtained from the deterministic problem of the previous section when the system includes a stochastic disturbance w_k in its description. Thus (1) is replaced by

$$x_{k+1} = f(x_k, u_k, w_k) \tag{15}$$

and the cost per stage becomes $g(x_k, u_k, w_k)$. The disturbance w_k is a member of some probability space $(W, \mathscr{F})$ and has distribution $p(dw_k | x_k, u_k)$. Thus the control variable u_k exercises influence over the transition from x_k to x_{k+1} in two places, once in the system equation (15) and again as a parameter in the distribution of the disturbance w_k. Likewise, the control u_k influences the cost at two points. This is a redundancy in the system equation model given above which will be eliminated in Chapter 8 when we introduce the transition kernel and reduced one-stage cost function and thereby convert to a model frequently adopted in the statistics literature (see, e.g., Blackwell [B9]; Strauch [S14]). The system equation model is more common in engineering literature and generally more convenient in applications, so we are taking it as our starting point. The transition kernel and reduced one-stage cost function are technical devices which eliminate the disturbance space $(W, \mathscr{F})$ from consideration and make the model more suitable for analysis. We take pains initially to point out how properties of the original system carry over into properties of the transition kernel and reduced one-stage cost function (see the remarks following Definitions 8.1 and 8.7).

Stochastic optimal control is distinguished from its deterministic counterpart by the concern with when information becomes available. In deterministic control, to each initial state and policy there corresponds a sequence of control variables $(u_0, \ldots, u_{N-1})$ which can be specified beforehand, and the resulting states of the system are determined by (1). In contrast, if the control variables are specified beforehand for a stochastic system, the decisionmaker may realize in the course of the system evolution that unexpected states have appeared and the specified control variables are no longer appropriate. Thus it is essential to consider *policies* $\pi = (\mu_0, \ldots, \mu_{N-1})$, where μ_k is a function from history to control. If x_0 is the initial state, $u_0 = \mu_0(x_0)$ is taken to be the first control. If the states and controls $(x_0, u_0, \ldots, u_{k-1}, x_k)$ have occurred, the control

$$u_k = \mu_k(x_0, u_0, \ldots, u_{k-1}, x_k) \tag{16}$$

is chosen. We require that the control constraint

$$\mu_k(x_0, u_0, \ldots, u_{k-1}, x_k) \in U(x_k)$$

be satisfied for every $(x_0, u_0, \ldots, u_{k-1}, x_k)$ and k. In this way the decisionmaker utilizes the full information available to him at each stage. Rather than choosing a sequence of control variables, the decisionmaker attempts to choose a policy which minimizes the total expected cost of the system operation. Actually, we will show that for most cases it is sufficient to consider only *Markov policies*, those for which the corresponding controls u_k depend only on the current state x_k rather than the entire history $(x_0, u_0, \ldots, u_{k-1}, x_k)$. This is the type of policy encountered in Section 1.1.

The analysis of the stochastic decision model outlined here can be fairly well divided into two categories—*structural considerations* and *measurability considerations*. Structural analysis consists of all those results which can be obtained if measurability of all functions and sets arising in the problem is of no real concern; for example, if the model is deterministic or, more generally, if the disturbance space W is countable. In Part I structural results are derived using mappings H, T_μ, and T of the kind considered in the previous section. Measurability analysis consists of showing that the structural results remain valid even when one places nontrivial measurability restrictions on the set of admissible policies. The work in Part II consists primarily of measurability analysis relying heavily on structural results developed in Part I as well as in other sources (e.g., Bertsekas [B4]).

One can best illustrate this dichotomy of analysis by the finite horizon DP algorithm considered by Bellman [B1]:

$$J_0(x) = 0, \tag{17}$$

$$J_{k+1}(x) = \inf_{u \in U(x)} E\{g(x, u, w) + J_k[f(x, u, w)]\}, \qquad k = 0, \ldots, N-1, \tag{18}$$

where the expectation is with respect to $p(dw|x, u)$. This is the stochastic counterpart of the deterministic DP algorithm (4)–(5).

It is reasonable to expect that $J_k(x)$ is the optimal cost of operating the system over k stages when the initial state is x, and that if $\mu_k(x)$ achieves the infimum in (18) for every x and $k = 0, \ldots, N - 1$, then $\pi = (\mu_0, \ldots, \mu_{N-1})$ is an optimal policy for every initial state x. If there are no measurability considerations, this is indeed the case under very mild assumptions, as shall be shown in Chapter 3. Yet it is a major task to properly formulate the stochastic control problem and demonstrate that the DP algorithm (17)–(18) makes sense in a measure-theoretic framework. One of the difficulties lies in showing that the expression in curly braces in (18) is measurable in some sense. Thus we must establish measurability properties for the functions J_k. Related to this is the need to balance the measurability of policies (necessary so the expected cost corresponding to a policy can be defined) against a desire to be able to select at or near the infimum in (18). We illustrate these difficulties by means of a simple two-stage example.

Two-Stage Problem   Consider the following sequence of events:

(a)   An initial state $x_0 \in R$ is generated (R is the real line).

(b)   Knowing x_0, the decisionmaker selects a control $u_0 \in R$.

(c)   A state $x_1 \in R$ is generated according to a known probability measure $p(dx_1|x_0, u_0)$ on $\mathscr{B}_R$, the Borel subsets of R, depending on x_0, u_0. [In terms of our earlier model, this corresponds to a system equation of the form $x_1 = w_0$ and $p(dw_0|x_0, u_0) = p(dx_1|x_0, u_0)$.]

(d)   Knowing x_1, the decisionmaker selects a control $u_1 \in R$.

Given $p(dx_1|x_0, u_0)$ for every $(x_0, u_0) \in R^2$ and a function $g: R^2 \to R$, the problem is to find a policy $\pi = (\mu_0, \mu_1)$ consisting of two functions $\mu_0: R \to R$ and $\mu_1: R \to R$ that minimizes

$$J_\pi(x_0) = \int g[x_1, \mu_1(x_1)] p(dx_1|x_0, \mu_0(x_0)). \tag{19}$$

We temporarily postpone a discussion of restrictions (if any) that must be placed on g, μ_0, and μ_1 in order for the integral in (19) to be well defined. In terms of our earlier model, the function g gives the cost for the second stage while we assume no cost for the first stage.

The DP algorithm associated with the problem is

$$J_1(x_1) = \inf_{u_1} g(x_1, u_1), \tag{20}$$

$$J_2(x_0) = \inf_{u_0} \int J_1(x_1) p(dx_1|x_0, u_0), \tag{21}$$

and, assuming that $J_2(x_0) > -\infty$, $J_1(x_1) > -\infty$ for all $x_0 \in R$, $x_1 \in R$, the results one expects to be true are:

R.1   There holds

$$J_2(x_0) = \inf_\pi J_\pi(x_0) \qquad \forall x_0 \in R.$$

R.2   Given $\varepsilon > 0$, there is an (everywhere) ε-optimal policy, i.e., a policy π_ε such that

$$J_{\pi_\varepsilon}(x_0) \le \inf_\pi J_\pi(x_0) + \varepsilon \qquad \forall x_0 \in R.$$

R.3   If the infimum in (20) and (21) is attained for all $x_1 \in R$ and $x_0 \in R$, then there exists a policy that is optimal for every $x_0 \in R$.

R.4   If $\mu_1^*(x_1)$ and $\mu_0^*(x_0)$, respectively, attain the infimum in (20) and (21) for all $x_1 \in R$ and $x_0 \in R$, then $\pi^* = (\mu_0^*, \mu_1^*)$ is optimal for every $x_0 \in R$, i.e.,

$$J_{\pi^*}(x_0) = \inf_\pi J_\pi(x_0) \qquad \forall x_0 \in R.$$

A formal derivation of R.1 consists of the following steps:

$$\inf_\pi J_\pi(x_0) = \inf_{\mu_0} \inf_{\mu_1} \int g[x_1, \mu_1(x_1)] p(dx_1 | x_0, \mu_0(x_0)) \qquad (22a)$$

$$= \inf_{\mu_0} \int \left\{ \inf_{u_1} g(x_1, u_1) \right\} p(dx_1 | x_0, \mu_0(x_0)) \qquad (22b)$$

$$= \inf_{\mu_0} \int J_1(x_1) p(dx_1 | x_0, \mu_0(x_0))$$

$$= \inf_{u_0} \int J_1(x_1) p(dx_1 | x_0, u_0) = J_2(x_0).$$

Similar formal derivations can be given for R.2, R.3, and R.4.

The following points need to be justified in order to make the preceding derivation meaningful and mathematically rigorous.

(a)   In (22a), g and μ_1 must be such that $g[x_1, \mu_1(x_1)]$ can be integrated in a well-defined manner.

(b)   In (22b), the interchange of infimization and integration must be legitimate. Furthermore g must be such that $J_1(x_1) [= \inf_{u_1} g(x_1, u_1)]$ can be integrated in a well-defined manner.

We first observe that if, for each (x_0, u_0), $p(dx_1 | x_0, u_0)$ has *countable support*, i.e., is concentrated on a countable number of points, then integration in (22a) and (22b) reduces to infinite summation. Thus there is no need to impose measurability restrictions on g, μ_0, and μ_1, and the interchange of infimization and integration in (22b) is justified in view of the assumption

$\inf_{u_1} g(x_1, u_1) > -\infty$ for all $x_1 \in R$. (For $\varepsilon > 0$, take $\mu_\varepsilon : R \to R$ such that

$$g[x_1, \mu_\varepsilon(x_1)] \le \inf_{u_1} g(x_1, u_1) + \varepsilon \qquad \forall x_1 \in R. \tag{23}$$

Then

$$\inf_{\mu_1} \int g[x_1, \mu_1(x_1)] p(dx_1|x_0, \mu_0(x_0)) \le \int g[x_1, \mu_\varepsilon(x_1)] p(dx_1|x_0, \mu_0(x_0))$$

$$\le \int \inf_{u_1} g(x_1, u_1) p(dx_1|x_0, \mu_0(x_0)) + \varepsilon. \tag{24}$$

Since $\varepsilon > 0$ is arbitrary, it follows that

$$\inf_{\mu_1} \int g[x_1, \mu_1(x_1)] p(dx_1|x_0, \mu_0(x_0)) \le \int \left\{ \inf_{u_1} g(x_1, u_1) \right\} p(dx_1|x_0, \mu_0(x_0)).$$

The reverse inequality is clear, and the result follows.) A similar argument proves R.2, while R.3 and R.4 are trivial in view of the fact that there are no measurability restrictions on μ_0 and μ_1.

If $p(dx_1|x_0, u_0)$ does not have countable support, there are two main approaches. The first is to *expand the notion of integration*, and the second is to *restrict g, μ_0, and μ_1 to be appropriately measurable.*

Expanding the notion of integration can be achieved by interpreting the integrals in (22a) and (22b) as *outer integrals* (see Appendix A). Since the outer integral can be defined for any function, measurable or not, there is no need to require that g, μ_0, and μ_1 are measurable in any sense. As a result, (22a) and (22b) make sense and an argument such as the one beginning with (23) goes through. This approach is discussed in detail in Part I, where we show that all the basic results for finite and infinite horizon problems of perfect state information carry through within an outer integration framework. However, there are inherent limitations in this approach centering around the pathologies of outer integration. Difficulties also occur in the treatment of imperfect information problems using sufficient statistics.

The major alternative approach was initiated in more general form by Blackwell [B9] in 1965. Here we assume at the outset that g is Borel- measurable, and furthermore, for each $B \in \mathcal{B}_R$ ($\mathcal{B}_R$ is the Borel σ-algebra on R), the function $p(B|x_0, u_0)$ is Borel-measurable in (x_0, u_0). In the initial treatment of the problem, the functions μ_0 and μ_1 were restricted to be Borel-measurable. With these assumptions, $g[x_1, \mu_1(x_1)]$ is Borel-measurable in x_1 when μ_1 is Borel-measurable, and the integral in (22a) is well defined.

A major difficulty occurs in (22b) since it is not necessarily true that $J_1(x_1) = \inf_{u_1} g(x_1, u_1)$ is Borel-measurable, even if g is. The reason can be traced to the fact that the orthogonal projection of a Borel set in R^2 on one of the axes need not be Borel-measurable (see Section 7.6). Since we have for $c \in R$

$$\{x_1 | J_1(x_1) < c\} = \underset{x_1}{\text{proj}}\{(x_1, u_1) | g(x_1, u_1) < c\},$$

where proj_{x_1} denotes projection on the x_1-axis, it can be seen that $\{x_1 | J_1(x_1) < c\}$ need not be Borel, even though $\{(x_1, u_1) | g(x_1, u_1) < c\}$ is. The difficulty can be overcome in part by showing that J_1 is a lower semi-analytic and hence also universally measurable function (see Section 7.7). Thus J_1 can be integrated with respect to any probability measure on $\mathcal{B}_R$.

Another difficulty stems from the fact that one cannot in general find a Borel-measurable ε-optimal selector μ_ε satisfying (23), although a weaker result is available whereby, given a probability measure p on $\mathcal{B}_R$, the existence of a Borel-measurable selector μ_ε satisfying

$$g[x_1, \mu_\varepsilon(x_1)] \leq \inf_{u_1} g(x_1, u_1) + \varepsilon$$

for p almost every $x_1 \in R$ can be ascertained. This result is sufficient to justify (24) and thus prove result R.1 ($J_2 = \inf_\pi J_\pi$). However, results R.2 and R.3 cannot be proved when μ_0 and μ_1 are restricted to be Borel-measurable except in a weaker form involving the notion of p-optimality (see [S14]; [H4]).

The objective of Part II is to resolve the measurability questions in stochastic optimal control in such a way that almost every result can be proved in a form as strong as its structural counterpart. This is accomplished by enlarging the set of admissible policies to include all *universally measurable policies*. In particular, we show the existence of policies within this class that are optimal or nearly optimal for *every* initial state.

A great many authors have dealt with measurability in stochastic optimal control theory. We describe three approaches taken and how their aims and results relate to our own. A fourth approach, due to Blackwell *et al.* [B12] and based on analytically measurable policies, is discussed in the next section and in Section 11.2.

I   The General Model

If the state, control, and disturbance spaces are arbitrary measure spaces, very little can be done. One attempt in this direction is the work of Striebel [S16] involving p-essential infima. Geared toward giving meaning to the dynamic programming algorithm, this work replaces (18) by

$$J_{k+1}(x) = p_k\text{-essential} \inf_\mu E\{g[x, \mu(x), w] + J_k[f(x, \mu(x), w)]\}, \quad (25)$$

$k = 0, \ldots, N - 1$, where the p-essential infimum is over all measurable μ from state space S to control space C satisfying any constraints which may have been imposed. The functions J_k are measurable, and if the probability measures $p_0, \ldots, p_{N-1}$ are properly chosen and the so-called countable ε-lattice property holds, this modified dynamic programming algorithm generates the optimal cost function and can be used to obtain policies which are optimal or nearly optimal for p_{N-1} almost all initial states. The selection of the proper probability measures $p_0, \ldots, p_{N-1}$, however, is at least as difficult as executing the dynamic programming algorithm, and the verification of the countable ε-lattice property is equivalent to proving the existence of an ε-optimal policy.

II The Semicontinuous Models

Considerable attention has been directed toward models in which the state and control spaces are Borel spaces or even R^n and the reduced cost function

$$h(x, u) = \int g(x, u, w) p(dw | x, u)$$

has semicontinuity and/or convexity properties. A companion assumption is that the mapping

$$x \to U(x)$$

is a measurable closed-valued multifunction [R2]. In the latter case there exists a Borel-measurable selector $\mu : S \to C$ such that $\mu(x) \in U(x)$ for every state x (Kuratowski and Ryll–Nardzewski [K5]). This is of course necessary if any Borel-measurable policy is to exist at all.

The main fact regarding models of this type is that under various combinations of semicontinuity and compactness assumptions, the functions J_k defined by (17) and (18) are semicontinuous. In addition, it is often possible to show that the infimum in (18) is achieved for every x and k, and there are Borel-measurable selectors $\mu_0, \ldots, \mu_{N-1}$ such that $\mu_k(x)$ achieves this infimum (see Freedman [F1], Furukawa [F3], Himmelberg, *et al.* [H3], Maitra [M2], Schäl [S3], and the references contained therein). Such a policy $(\mu_0, \ldots, \mu_{N-1})$ is optimal, and the existence of this optimal policy is an additional benefit of imposing topological conditions to ensure that the problem is well defined. In Section 9.5 we show that lower semicontinuity and compactness conditions guarantee convergence of the dynamic programming algorithm over an infinite horizon to the optimal cost function, and that this algorithm can be used to generate an optimal stationary policy.

Continuity and compactness assumptions are integral to much of the work that has been done in stochastic programming. This work differs from our own in both its aims and its framework. First, in the usual stochastic programming model, the controls cannot influence the distribution of future states (see Olsen [O1–O3], Rockafellar and Wets [R3–R4], and the references contained therein). As a result, the model does not include as special cases many important problems such as, for example, the classical linear quadratic stochastic control problem [B4, Section 3.1]. Second, assumptions of convexity, lower semicontinuity, or both are made on the cost function, the model is designed for the Kuratowski–Ryll–Nardzewski selection theorem, and the analysis is carried out in a finite-dimensional Euclidean state space. All of this is for the purpose of overcoming measurability problems. Results are not readily generalizable beyond Euclidean spaces (Rockafellar [R2]). The thrust of the work is toward convex programming type results, i.e., duality and Kuhn–Tucker conditions for optimality, and so a narrow class of problems is considered and powerful results are obtained.

III The Borel Models

The Borel space framework was introduced by Blackwell [B9] and further refined by Strauch, Dynkin, Juskevič, Hinderer, and others. The state and control spaces S and C were assumed to be Borel spaces, and the functions defining the model were assumed to be Borel-measurable. Initial efforts were directed toward proving the existence of "nice" optimal or nearly optimal policies in this framework. Policies were required to be Borel-measurable. For this model it is possible to prove the universal measurability of the optimal cost function and the existence for every $\varepsilon > 0$ and probability measure p on S of a p–ε-optimal policy (Strauch [S14, Theorems 7.1 and 8.1]). A p–ε-optimal policy is one which leads to a cost differing from the optimal cost by less than ε for p almost every initial state. As discussed earlier, even over a finite horizon the optimal cost function need not be Borel-measurable and there need not exist an everywhere ε-optimal policy (Blackwell [B9, Example 2]). The difficulty arises from the inability to choose a Borel-measurable function $\mu_k : S \to C$ which nearly achieves the infimum in (18) uniformly in x. The nonexistence of such a function interferes with the construction of optimal policies via the dynamic programming algorithm (17) and (18), since one must first determine at each stage the measure p with respect to which it is satisfactory to nearly achieve the infimum in (18) for p almost every x. This is essentially the same problem encountered with (25). The difficulties in constructing nearly optimal policies over an infinite horizon are more acute. Furthermore, from an applications point of view, a p–ε-optimal policy, even if it can be constructed, is a much less appealing object than an everywhere ε-optimal policy, since in many situations the distribution p is unknown or may change when the system is operated repetitively, in which case a new p–ε-optimal policy must be computed.

In our formulation, the class of admissible policies in the Borel model is enlarged to include all universally measurable policies. We show in Part II that this class is sufficiently rich to ensure that *there exist everywhere ε-optimal policies and, if the infimum in the DP algorithm (18) is attained for every x and k, then an everywhere optimal policy exists.* Thus the notion of p-optimality can be dispensed with. The basic reason why optimal and nearly optimal policies can be found within the class of universally measurable policies may be traced to the selection theorem of Section 7.7. Another advantage of working with the class of universally measurable functions is that this class is closed under certain basic operations such as integration with respect to a universally measurable stochastic kernel and composition.

Our method of proof of infinite horizon results is based on an equivalence of stochastic and deterministic decision models which is worked out in Sections 9.1–9.3. The conversion is carried through only for the infinite horizon model, as it is not necessary for the development in Chapter 8. It is also done only under assumptions (P), (N), or (D) of Definition 9.1, although the models make sense under conditions similar to the (F^+) and (F^-) assumptions of Section 8.1. The relationship between the stochastic and the deterministic models is utilized extensively in Sections 9.4–9.6, where structural results proved in Part I are applied to the deterministic model and then transferred to the stochastic model. The analysis shows how results for stochastic models with measurability restrictions on the set of admissible policies can be obtained from the general results on abstract dynamic programming models given in Part I and provides the connecting link between the two parts of this work.

1.3 The Present Work Related to the Literature

This section summarizes briefly the contents of each chapter and points out relations with existing literature. During the course of our research, many of our results were reported in various forms (Bertsekas [B3–B5]; Shreve [S7–S8]; Shreve and Bertsekas [S9–S12]). Since the present monograph is the culmination of our joint work, we report particular results as being new even though they may be contained in one or more of the preceding references.

Part I

The objective of Part I is to provide a unifying framework for finite and infinite horizon dynamic programming models. We restrict our attention to three types of infinite horizon models, which are patterned after the discounted and positive models of Blackwell [B8–B9] and the negative model of Strauch [S14]. It is an open question whether the framework of Part I can be effectively extended to cover other types of infinite horizon models such as the average cost model of Howard [H7] or convergent dynamic programming models of the type considered by Dynkin and Juskevič [D8] and Hordijk [H6].

The problem formulation of Part I is new. The work that is most closely related to our framework is the one by Denardo [D2], who considered an abstract dynamic programming model under contraction assumptions. Most of Denardo's results have been incorporated in slightly modified form in Chapter 4. Denardo's problem formulation is predicated on his contraction assumptions and is thus unsuitable for finite horizon models such as the one in Chapter 3 and infinite horizon models such as the ones in Chapter 5. This fact provided the impetus for our different formulation.

Most of the results of Part I constitute generalizations of results known for specific classes of problems such as, for example, deterministic and stochastic optimal control problems. We make an effort to identify the original sources, even though in some cases this is quite difficult. Some of the results of Part I have not been reported earlier even for a specific class of problems, and they will be indicated as new.

Chapter 2   Here we formulate the basic abstract sequential optimization problem which is the subject of Part I. Several classes of problems of practical interest are described in Section 2.3 and are shown to be special cases of the abstract problem. All these problems have received a great deal of attention in the literature with the exception of the stochastic optimal control model based on outer integration (Section 2.3.3). This model, as well as the results in subsequent chapters relating to it, is new. A stochastic model based on outer integration has also been considered by Denardo [D2], who used a different definition of outer integration. His definition works well under contraction assumptions such as the one in Chapter 4. However, many of the results of Chapters 3 and 5 do not hold if Denardo's definition of outer integral is adopted. By contrast, all the basic results of Part I are valid when specialized to the model of Section 2.3.3.

Chapter 3   This chapter deals with the finite horizon version of our abstract problem. The central results here relate to the validity of the dynamic programming algorithm, i.e., the equation $J_N^* = T^N(J_0)$. The validity of this equation is often accepted without scrutiny in the engineering literature, while in mathematical works it is usually proved under assumptions that are stronger than necessary. While we have been unable to locate an appropriate source, we feel certain that the results of Proposition 3.1 are known for stochastic optimal control problems. The notion of a sequence of policies exhibiting $\{\varepsilon_n\}$-dominated convergence to optimality and the corresponding existence result (Proposition 3.2) are new.

Chapter 4   Here we treat the infinite horizon version of our abstract problem under a contraction assumption. The developments in this chapter overlap considerably with Denardo's work [D2]. Our contraction assumption C is only slightly different from the one of Denardo. Propositions 4.1, 4.2, 4.3 (a), and 4.3 (c) are due to Denardo [D2], while Proposition 4.3 (b) has been shown by Blackwell [B9] for stochastic optimal control problems. Proposition 4.4 is new. Related compactness conditions for existence of a stationary optimal policy in stochastic optimal control problems were given by Maitra [M2], Kushner [K6], and Schäl [S5]. Propositions 4.6 and 4.7 improve on corresponding results by Denardo [D2] and McQueen [M3]. The modified policy iteration algorithm and the corresponding convergence result (Proposition 4.9) are new in the form given here. Denardo [D2] gives a somewhat less general form of policy iteration. The idea of policy iteration for deterministic and stochastic optimal control problems dates, of course, to the early days of dynamic programming (Bellman [B1]; Howard [H7]). The mathematical programming formulation of Section 4.3.3 is due to Denardo [D2].

Chapter 5   Here we consider infinite horizon versions of our abstract model patterned after the positive and negative models of Blackwell [B8, B9] and Strauch [S14]. When specialized to stochastic optimal control problems, most of the results of this chapter have either been shown by these authors or can be trivially deduced from their work. The part of Proposition 5.1 dealing with existence of an ε-optimal stationary policy is new, as is the last part of Proposition 5.2. Forms of Propositions 5.3 and 5.5 specialized to certain gambling problems have been shown by Dubins and Savage [D6], whose monograph provided the impetus for much of the subsequent work on dynamic programming. Propositions 5.9–5.11 are new. Results similar to those of Proposition 5.10 have been given by Schäl [S5] for stochastic optimal control problems under semicontinuity and compactness assumptions.

Chapter 6   The analysis in this chapter is new. It is motivated by the fact that the framework and the results of Chapters 2–5 are primarily applicable to problems where measurability issues are of no essential concern. While it is possible to apply the results to problems where policies are subject to measurability restrictions, this can be done only after a fairly elaborate reformulation (see Chapter 9). Here we generalize our framework so that problems in which measurability issues introduce genuine complications can be dealt with directly. However, only a portion of our earlier results carry through within the generalized framework—primarily those associated with finite horizon models and infinite horizon models under contraction assumptions.

Part II

The objective of Part II is to develop in some detail the discrete-time stochastic optimal control problem (additive cost) in Borel spaces. The measurability questions are addressed explicitly. This model was selected from among the specialized models of Part I because it is often encountered and also because it can serve as a guide in the resolution of measurability difficulties in a great many other decision models.

In Chapter 7 we present the relevant topological properties of Borel spaces and their probability measures. In particular, the properties of analytic sets are developed. Chapter 8 treats the finite horizon stochastic optimal control problem, and Chapter 9 is devoted to the infinite horizon version. Chapter 10 deals with the stochastic optimal control problem when only a "noisy" measurement of the state of the system is possible. Various extensions of the theory of Chapters 8 and 9 are given in Chapter 11.

Chapter 7   The properties presented for metrizable spaces are well known. The material on Borel spaces can be found in Chapter 1 of Partha-sarathy [P1] and is also available in Kuratowski [K2–K3]. A discussion of the weak topology can be found in Parthasarathy [P1]. Propositions 7.20, 7.21, and 7.23 are due to Prohorov [P2], but their presentation here follows Varadarajan [V1]. Part of Proposition 7.21 also appears in Billingsley [B7]. Proposition 7.25 is an extension of a result for compact X found in Dubins and Freedman [D5]. Versions of Proposition 7.25 have been used in the literature for noncompact X (Strauch [S14]; Blackwell *et al.* [B12]), the authors evidently intending an extension of the compact result by using Urysohn's theorem to embed X in a compact metric space. Proposition 7.27 is reported by Rhenius [R1], Juskevič [J3] and Striebel [S16]. We give Striebel's proof. Propositions 7.28 and 7.29 appear in some form in several texts on probability theory. A frequently cited reference is Loève [L1]. Propositions 7.30 and 7.31 are easily deduced from Maitra [M2] or Schäl [S4], and much of the rest of the discussion of semicontinuous functions is found in Hausdorff [H2]. Proposition 7.33 is due to Dubins and Savage [D6]. Proposition 7.34 is taken from Freedman [F1].

The investigation of analytic sets in Borel spaces began several years ago, but has been given additional impetus recently by the discovery of their applications to stochastic processes. Suslin schemes and analytic sets first appear in a paper by M. Suslin (or Souslin) in 1917 [S17], although the idea is generally attributed to Alexandroff. Suslin pointed out that every Borel subset of the real line could be obtained as the nucleus of a Suslin scheme for the closed intervals, and non-Borel sets could be obtained this way as well. He also noted that the analytic subsets of R were just the projections on an axis of the Borel subsets of R^2. The universal measurability of analytic sets (Corollary 7.42.1) was proved by Lusin and Sierpinski [L3] in 1918. (See also Lusin [L2].) Our proof of this fact is taken from Saks [S1]. We have also taken material on analytic sets from Kuratowski [K2], Dellacherie [D1], Meyer [M4], Bourbaki [B13], Parthasarathy [P1], and Bressler and Sion [B14]. Proposition 7.43 is due to Meyer and Traki [M5], but our proof is original. The proofs given here of Propositions 7.47 and 7.49 are very similar to those found in Blackwell et al. [B12]. The basic result of Proposition 7.49 is due to Jankov [J1], but was also worked out about the same time and published later by von Neumann [N1, Lemma 5, p. 448]. The Jankov–von Neumann result was strengthened by Mackey [M1, Theorem 6.3]. The history of this theorem is related by Wagner [W1, pp. 900–901]. Proposition 7.50(a) is due to Blackwell et al. [B12]. Proposition 7.50(b) together with its strengthened version Proposition 11.4 generalize a result by Brown and Purves [B15], who proved existence of a universally measurable φ for the case where f is Borel measurable.

Chapter 8   The finite horizon stochastic optimal control model of Chapter 8 is essentially a finite horizon version of the models considered by Blackwell [B8, B9], Strauch [S14], Hinderer [H4], Dynkin and Juskevič [D8], Blackwell et al. [B12], and others. With the exception of [B12], all these works consider Borel-measurable policies and obtain existence results of a p–ε-optimal nature (see the discussion of the previous section). We allow universally measurable policies and thereby obtain everywhere ε-optimal existence results. While in Chapters 8 and 9 we concentrate on proving results that hold everywhere, the previously available results which allow only Borel-measurable policies and hold p almost everywhere can be readily obtained as corollaries. This follows from the following fact, whose proof we sketch shortly:

(F)   *If X and Y are Borel spaces, $p_0, p_1, \ldots$ is a sequence of probability measures on X, and μ is a universally measurable map from X to Y, then there is a Borel measurable map μ' from X to Y such that*

$$\mu(x) = \mu'(x)$$

for p_k almost every x, $k = 0, 1, \ldots$.

As an example of how this observation can be used to obtain p almost everywhere existence results from ours, consider Proposition 9.19. It states in part that if $\varepsilon > 0$ and the discount factor α is less than one, then an ε-optimal nonrandomized stationary policy exists, i.e., a policy $\pi = (\mu, \mu, \ldots)$, where μ is a universally measurable mapping from S to C. Given p_0 on S, this policy generates a sequence of measures $p_0, p_1, \ldots$ on S, where p_k is the distribution of the kth state when the initial state has distribution p_0 and the policy π is used. Let $\mu': S \to C$ be Borel-measurable and equal to μ for p_k almost every x, $k = 0, 1, \ldots$. Let $\pi' = (\mu', \mu', \ldots)$. Then it can be shown that for p_0 almost every initial state, the cost corresponding to π' equals the cost corresponding to π, so π' is a p_0–ε-optimal nonrandomized stationary *Borel*-measurable policy. The existence of such a π' is a new result. This type of argument can be applied to all the existence results of Chapters 8 and 9.

We now sketch a proof of (F). Assume first that Y is a Borel subset of $[0, 1]$. Then for $r \in [0, 1]$, r rational, the set

$$U(r) = \{x | \mu(x) \le r\}$$

is universally measurable. For every k, let $p_k^*[U(r)]$ be the outer measure of $U(r)$ with respect to p_k and let $B_{k1}, B_{k2}, \ldots$ be a decreasing sequence of Borel sets containing $U(r)$ such that

$$p_k^*[U(r)] = p_k\left[\bigcap_{j=1}^{\infty} B_{kj}\right].$$

Let $B(r) = \bigcap_{k=1}^{\infty} \bigcap_{j=1}^{\infty} B_{kj}$. Then

$$p_k^*[U(r)] = p_k[B(r)], \qquad k = 0, 1, \ldots,$$

and the argument of Lemma 7.27 applies. If Y is an arbitrary Borel space, it is Borel isomorphic to a Borel subset of $[0, 1]$ (Corollary 7.16.1), and (F) follows.

Proposition 8.1 is due to Strauch [S14], and Proposition 8.2 is contained in Theorem 14.4 of Hinderer [H4]. Example 8.1 is taken from Blackwell [B9]. Proposition 8.3 is new, the strongest previous result along these lines being the existence of an analytically measurable ε-optimal policy when the one-stage cost function is nonpositive [B12]. Propositions 8.4 and 8.5 are new, as are the corollaries to Proposition 8.5. Lower semicontinuous models have received much attention in the literature (Maitra [M2]; Furukawa [F3]; Schäl [S3–S5]; Freedman [F1]; Himmelberg et al. [H3]). Our lower semicontinuous model differs somewhat from those in the literature, primarily in the form of the control constraint. Proposition 8.6 is closely related to the analysis in several of the previously mentioned references. Proposition 8.7 is due to Freedman [F1].

Chapter 9 Example 9.1 is a modification of Example 6.1 of Strauch [S14], and Proposition 9.1 is taken from Strauch [S14]. The conversion of the stochastic optimal control problem to the deterministic one was suggested by Witsenhausen [W3] in a different context and carried out systematically for the first time here. This results in a simple proof of the lower semianalyticity of the infinite horizon optimal cost function (cf. Corollary 9.4.1 and Strauch [S14, Theorem 7.1]). Propositions 9.8 and 9.9 are due to Strauch [S14], as are the (D) and (N) parts of Proposition 9.10. The (P) part of Proposition 9.10 is new. Proposition 9.12 appears as Theorem 5.2.2 of Schäl [S5], but Corollary 9.12.1 is new. Proposition 9.14 is a special case of Theorem 14.5 of Hinderer [H4]. Propositions 9.15–9.17 and the corollaries to Proposition 9.17 are new, although Corollary 9.17.2 is very close to Theorem 13.3 of Schäl [S5]. Propositions 9.18–9.20 are new. Proposition 9.21 is an infinite horizon version of a finite horizon result due to Freedman [F1], except that the nonrandomized ε-optimal policy Freedman constructs may not be semi-Markov.

   Chapter 10 The use of the conditional distribution of the state given the available information as a basis for controlling systems with imperfect state information has been explored by several authors under various assumptions (see, for example, Åström [A2], Striebel [S15], and Sawaragi and Yoshikawa [S2]). The treatment of imperfect state information models with uncountable Borel state and action spaces, however, requires the existence of a regular conditional distribution with a measurable dependence on a parameter (Proposition 7.27), and this result is quite recent (Rhenius [R1]; Juskevič [J3]; Striebel [S16]). Chapter 10 is related to Chapter 3 of Striebel [S16] in that the general concept of a statistic sufficient for control is defined. We use such a statistic to construct a perfect state information model which is equivalent in the sense of Propositions 10.2 and 10.3 to the original imperfect state information model. From this equivalence the validity of the dynamic programming algorithm and the existence of ε-optimal policies under the mild conditions of Chapters 8 and 9 follow. Striebel justifies use of a statistic sufficient for control by showing that under a very strong hypothesis [S16, Theorem 5.5.1] the dynamic programming algorithm is valid and an ε-optimal policy can be based on the sufficient statistic. The strong hypothesis arises from the need to specify the null sets in the range spaces of the statistic in such a way that this specification is independent of the policy employed. This need results from the inability to deal with the pointwise partial infima of multivariate functions without the machinery of universally measurable policies and lower semianalytic functions. Like Striebel, we show that the conditional distributions of the states based on the available information constitute a statistic sufficient for control (Proposition 10.5), as do the vectors of available information themselves (Proposition 10.6).

   The treatments of Rhenius [R1] and Juskevič [J3] are like our own in that perfect state information models which are equivalent to the original one are defined. In his perfect state information model, Rhenius bases control on the observations and conditional distributions of the states, i.e., these objects are the states of his perfect state information model. It is necessary in Rhenius' framework for the controller to know the most recent observation, since this tells him which controls are admissible. We show in Proposition 10.5 that if there are no control constraints, then there is nothing to be gained by remembering the observations. In the model of Juskevič [J3], there are no control constraints and control is based on the past controls and conditional distributions. In this case, ε-optimal control is possible without reference to the past controls (Propositions 10.5, 8.3, 9.19, and 9.20), so our formulation is somewhat simpler and just as effective.

Chapter 10 differs from all the previously mentioned works in that simple conditions which guarantee the existence of a statistic sufficient for control are given, and once this existence is established, all the results of Chapters 8 and 9 can be brought to bear on the imperfect state information model.

Chapter 11   The use in Section 11.1 of limit measurability in dynamic programming is new. In particular, Proposition 11.3 is new, and as discussed earlier in regard to Proposition 7.50(b), a result by Brown and Purves [B15] is generalized in Proposition 11.4. Analytically measurable policies were introduced by Blackwell *et al.* [B12], whose work is referenced in Section 11.2. Borel space models with multiplicative cost fall within the framework of Furukawa and Iwamoto [F4–F5], and in [F5] the dynamic programming algorithm and a characterization of uniformly N-stage optimal policies are given. The remainder of Proposition 11.7 is new.

Appendix A   Outer integration has been used by several authors, but we have been unable to find a systematic development.

Appendix B   Proposition B.6 was first reported by Suslin [S17], but the proof given here is taken from Kuratowski [K2, Section 38VI]. According to Kuratowski and Mostowski [K4, p. 455], the limit σ-algebra $\mathscr{L}_X$ was introduced by Lusin, who called its members the "C-sets." A detailed discussion of the σ-algebra was given by Selivanovskij [S6] in 1928. Propositions B.9 and B.10 are fairly well known among set theorists, but we have been unable to find an accessible treatment. Proposition B.11 is new. Cenzer and Mauldin [C1] have also shown independently that $\mathscr{L}_X$ is closed under composition of functions, which is part of the result of Proposition B.11. Proposition B.12 is new.

It seems plausible that there are an infinity of distinct σ-algebras between the limit σ-algebra and the universal σ-algebra that are suitable for dynamic programming. One promising method of constructing such σ-algebras involves the R-operator of descriptive set theory (see Kantorovitch and

Livenson [K1]). In a recent paper [B11], Blackwell has employed a different method to define the "Borel-programmable" σ-algebra and has shown it to have many of the same properties we establish in Appendix B for the limit σ-algebra. It is not known, however, whether the Borel-programmable σ-algebra satisfies a condition like Proposition B.12 and is thereby suitable for dynamic programming. It is easily seen that the limit σ-algebra is contained in Blackwell's Borel-programmable σ-algebra, but whether the two coincide is also unknown.

Appendix C A detailed discussion of the exponential topology on the set of closed subsets of a topological space can be found in Kuratowski [K2–K3]. Properties of semicontinuous (K) functions are also proved there, primarily in Section 43 of [K3]. The Hausdorff metric is discussed in Section 38 of [H2].

Part I

Analysis of Dynamic Programming Models

Chapter 2

Monotone Mappings Underlying Dynamic Programming Models[†]

This chapter formulates the basic abstract sequential optimization problem which is the subject of Part I. It also provides examples of special cases which include wide classes of problems of practical interest.

2.1 Notation and Assumptions

Our usage of mathematical notation is fairly standard. For the reader's convenience we mention here that we use R to denote the real line and R^* to denote the extended real line, i.e., $R^* = R \cup \{-\infty, \infty\}$. The sets $(-\infty, \infty] = R \cup \{\infty\}$ and $[-\infty, \infty) = R \cup \{-\infty\}$ will be written out explicitly. We will assume throughout that R is equipped with the usual topology generated by the open intervals (α, β), $\alpha, \beta \in R$, and with the (Borel) σ-algebra generated by this topology. Similarly R^* is equipped with the topology generated by the open intervals (α, β), $\alpha, \beta \in R$, together with the sets $(\gamma, \infty]$, $[-\infty, \gamma)$, $\gamma \in R$, and with the σ-algebra generated by this topology. The Cartesian product of sets $X_1, X_2, \ldots, X_n$ is denoted $X_1 X_2 \cdots X_n$.

[†] Parts I and II can be read independently. The reader may proceed directly to Part II if he so wishes.

The following definitions and conventions will apply throughout Part I.

(1) S and C are two given sets referred to as the *state space* and *control space*, respectively.

(2)   For each $x \in S$, there is given a nonempty subset $U(x)$ of C referred to as the *control constraint set at* x.

(3)   We denote by M the set of all functions $\mu : S \to C$ such that $\mu(x) \in U(x)$ for all $x \in S$. We denote by Π the set of all sequences $\pi = (\mu_0, \mu_1, \ldots)$ such that $\mu_k \in M$ for all k. Elements of Π are referred to as *policies*. Elements of Π of the form $\pi = (\mu, \mu, \ldots)$, where $\mu \in M$, are referred to as *stationary policies*.

(4)   We denote:

   F the set of all extended real-valued functions $J : S \to R^*$;

   B the Banach space of all bounded real-valued functions $J : S \to R$ with the supremum norm $\|\cdot\|$ defined by

$$\|J\| = \sup_{x \in S} |J(x)| \qquad \forall J \in B.$$

(5)   For all $J, J' \in F$ we write

$$J = J' \quad \text{if} \quad J(x) = J'(x) \qquad \forall x \in S,$$
$$J \leq J' \quad \text{if} \quad J(x) \leq J'(x) \qquad \forall x \in S.$$

For all $J \in F$ and $\varepsilon \in R$, we denote by $J + \varepsilon$ the function taking the value $J(x) + \varepsilon$ at each $x \in S$, i.e.,

$$(J + \varepsilon)(x) = J(x) + \varepsilon \qquad \forall x \in S.$$

(6)   Throughout Part I the analysis is carried out within the set of extended real numbers R^*. We adopt the usual conventions regarding ordering, addition, and multiplication in R^* except that we take

$$\infty - \infty = -\infty + \infty = \infty,$$

and we take the product of zero and infinity to be zero. In this way the sum and the product of any two extended real numbers is well defined. Division by zero or ∞ does not appear in our analysis. In particular, we adopt the following rules in calculations involving ∞ and $-\infty$:

$$\alpha + \infty = \infty + \alpha = \infty \qquad \text{for} \quad -\infty \leq \alpha \leq \infty,$$
$$\alpha - \infty = -\infty + \alpha = -\infty \qquad \text{for} \quad -\infty \leq \alpha < \infty;$$
$$\alpha\infty = \infty\alpha = \infty, \qquad \alpha(-\infty) = (-\infty)\alpha = -\infty \qquad \text{for} \quad 0 < \alpha \leq \infty,$$
$$\alpha\infty = \infty\alpha = -\infty, \qquad \alpha(-\infty) = (-\infty)\alpha = \infty \qquad \text{for} \quad -\infty \leq \alpha < 0;$$
$$0\infty = \infty 0 = 0 = 0(-\infty) = (-\infty)0, \qquad -(-\infty) = \infty;$$
$$\inf \varnothing = +\infty, \qquad \sup \varnothing = -\infty,$$

where $\varnothing$ is the empty set.

Under these rules the following laws of arithmetic are still valid:

$$\alpha_1 + \alpha_2 = \alpha_2 + \alpha_1, \qquad (\alpha_1 + \alpha_2) + \alpha_3 = \alpha_1 + (\alpha_2 + \alpha_3),$$
$$\alpha_1\alpha_2 = \alpha_2\alpha_1, \qquad (\alpha_1\alpha_2)\alpha_3 = \alpha_1(\alpha_2\alpha_3).$$

We also have

$$\alpha(\alpha_1 + \alpha_2) = \alpha\alpha_1 + \alpha\alpha_2$$

if either $\alpha \geq 0$ or else $(\alpha_1 + \alpha_2)$ is not of the form $+\infty - \infty$.

(7) For any sequence $\{J_k\}$ with $J_k \in F$ for all k, we denote by $\lim_{k \to \infty} J_k$ the pointwise limit of $\{J_k\}$ (assuming it is well defined as an extended real-valued function) and by $\lim\sup_{k \to \infty} J_k$ ($\lim\inf_{k \to \infty} J_k$) the pointwise limit superior (inferior) of $\{J_k\}$. For any collection $\{J_\alpha | \alpha \in A\} \subset F$ parameterized by the elements of a set A, we denote by $\inf_{\alpha \in A} J_\alpha$ the function taking the value $\inf_{\alpha \in A} J_\alpha(x)$ at each $x \in S$.

The Basic Mapping

We are given a function H which maps SCF (Cartesian product of S, C, and F) into R^*, and we define for each $\mu \in M$ the mapping $T_\mu : F \to F$ by

$$T_\mu(J)(x) = H[x, \mu(x), J] \qquad \forall x \in S. \tag{1}$$

We define also the mapping $T : F \to F$ by

$$T(J)(x) = \inf_{u \in U(x)} H(x, u, J) \qquad \forall x \in S. \tag{2}$$

We denote by T^k, $k = 1, 2, \ldots$, the composition of T with itself k times. For convenience we also define $T^0(J) = J$ for all $J \in F$. For any $\pi = (\mu_0, \mu_1, \ldots) \in \Pi$ we denote by $(T_{\mu_0} T_{\mu_1} \cdots T_{\mu_k})$ the composition of the mappings $T_{\mu_0}, \ldots, T_{\mu_k}$, $k = 0, 1, \ldots$.

The following assumption will be in effect throughout Part I.

Monotonicity Assumption For every $x \in S$, $u \in U(x)$, $J, J' \in F$, we have

$$H(x, u, J) \leq H(x, u, J') \qquad \text{if} \quad J \leq J'. \tag{3}$$

The monotonicity assumption implies the following relations:

$$J \leq J' \Rightarrow T(J) \leq T(J') \qquad \forall J, J' \in F,$$
$$J \leq J' \Rightarrow T_\mu(J) \leq T_\mu(J') \qquad \forall J, J' \in F, \quad \mu \in M.$$

These relations in turn imply the following facts for all $J \in F$:

$$J \leq T(J) \Rightarrow T^k(J) \leq T^{k+1}(J), \qquad k = 0, 1, \ldots,$$
$$J \geq T(J) \Rightarrow T^k(J) \geq T^{k+1}(J), \qquad k = 0, 1, \ldots,$$
$$J \leq T_\mu(J) \qquad \forall \mu \in M \Rightarrow (T_{\mu_0} \cdots T_{\mu_k})(J) \leq (T_{\mu_0} \cdots T_{\mu_{k+1}})(J),$$
$$\qquad\qquad k = 0, 1, \ldots, \quad \pi = (\mu_0, \mu_1, \ldots) \in \Pi,$$
$$J \geq T_\mu(J) \qquad \forall \mu \in M \Rightarrow (T_{\mu_0} \cdots T_{\mu_k})(J) \geq (T_{\mu_0} \cdots T_{\mu_{k+1}})(J),$$
$$\qquad\qquad k = 0, 1, \ldots, \quad \pi = (\mu_0, \mu_1, \ldots) \in \Pi.$$

Another fact that we shall be using frequently is that for each $J \in F$ and $\varepsilon > 0$, there exists a $\mu_\varepsilon \in M$ such that

$$T_{\mu_\varepsilon}(J)(x) \leq \begin{cases} T(J)(x) + \varepsilon & \text{if} \quad T(J)(x) > -\infty, \\ -1/\varepsilon & \text{if} \quad T(J)(x) = -\infty. \end{cases}$$

In particular, if J is such that $T(J)(x) > -\infty$ for $\forall x \in S$, then for each $\varepsilon > 0$, there exists a $\mu_\varepsilon \in M$ such that

$$T_{\mu_\varepsilon}(J) \leq T(J) + \varepsilon.$$

2.2 Problem Formulation

We are given a function $J_0 \in F$ satisfying

$$J_0(x) > -\infty \qquad \forall x \in S, \tag{4}$$

and we consider for every policy $\pi = (\mu_0, \mu_1, \ldots) \in \Pi$ and positive integer N the functions $J_{N,\pi} \in F$ and $J_\pi \in F$ defined by

$$J_{N,\pi}(x) = (T_{\mu_0} T_{\mu_1} \cdots T_{\mu_{N-1}})(J_0)(x) \qquad \forall x \in S, \tag{5}$$

$$J_\pi(x) = \lim_{N \to \infty} (T_{\mu_0} T_{\mu_1} \cdots T_{\mu_{N-1}})(J_0)(x) \qquad \forall x \in S. \tag{6}$$

For every result to be shown, appropriate assumptions will be in effect which guarantee that the function J_π is well defined (i.e., the limit in (6) exists for all $x \in S$). We refer to $J_{N,\pi}$ as the *N-stage cost function for π* and to J_π as the *cost function for π*. Note that $J_{N,\pi}$ depends only on the first N functions in π while the remaining functions are superfluous. Thus we could have considered policies consisting of finite sequences of functions in connection with the N-stage problem, and this is in fact done in Chapter 8. However, there are notational advantages in using a common type of policy in finite and infinite horizon problems, and for this reason we have adopted such a notation for Part I.

Throughout Part I we will be concerned with the N-stage optimization problem

$$\begin{aligned} &\text{minimize} \quad J_{N,\pi}(x) \\ &\text{subject to} \quad \pi \in \Pi, \end{aligned} \tag{F}$$

and its infinite horizon version

$$\begin{aligned} &\text{minimize} \quad J_\pi(x) \\ &\text{subject to} \quad \pi \in \Pi. \end{aligned} \tag{I}$$

We refer to problem (F) as the *N-stage finite horizon problem* and to problem (I) as the *infinite horizon problem*.

For a· fixed $x \in S$, we denote by $J_N^*(x)$ and $J^*(x)$ the optimal costs for these problems, i.e.,

$$J_N^*(x) = \inf_{\pi \in \Pi} J_{N,\pi}(x) \qquad \forall x \in S, \tag{7}$$

$$J^*(x) = \inf_{\pi \in \Pi} J_\pi(x) \qquad \forall x \in S. \tag{8}$$

We refer to the function J_N^* as the *N-stage optimal cost function* and to the function J^* as the *optimal cost function*.

We say that a policy $\pi^* \in \Pi$ is *N-stage optimal at* $x \in S$ if $J_{N,\pi^*}(x) = J_N^*(x)$ and *optimal at* $x \in S$ if $J_{\pi^*}(x) = J^*(x)$. We say that $\pi^* \in \Pi$ is *N-stage optimal* (respectively *optimal*) if $J_{N,\pi^*} = J_N^*$ (respectively $J_{\pi^*} = J^*$). A policy $\pi^* = (\mu_0^*, \mu_1^*, \ldots)$ will be called *uniformly N-stage optimal* if the policy $(\mu_i^*, \mu_{i+1}^*, \ldots)$ is $(N - i)$-stage optimal for all $i = 0, 1, \ldots, N - 1$. Thus if a policy is uniformly N-stage optimal, it is also N-stage optimal, but not conversely. *For a stationary policy* $\pi = (\mu, \mu, \ldots) \in \Pi$, *we write* $J_\pi = J_\mu$. Thus a stationary policy $\pi^* = (\mu^*, \mu^*, \ldots)$ is optimal if $J^* = J_{\mu^*}$.

Given $\varepsilon > 0$, we say that a policy $\pi_\varepsilon \in \Pi$ is *N-stage ε-optimal* if

$$J_{N,\pi_\varepsilon}(x) \leq \begin{cases} J_N^*(x) + \varepsilon & \text{if } J_N^*(x) > -\infty, \\ -1/\varepsilon & \text{if } J_N^*(x) = -\infty. \end{cases}$$

We say that $\pi_\varepsilon \in \Pi$ is *ε-optimal* if

$$J_{\pi_\varepsilon}(x) \leq \begin{cases} J^*(x) + \varepsilon & \text{if } J^*(x) > -\infty, \\ -1/\varepsilon & \text{if } J^*(x) = -\infty. \end{cases}$$

If $\{\varepsilon_n\}$ is a sequence of positive numbers with $\varepsilon_n \downarrow 0$, we say that a sequence of policies $\{\pi_n\}$ exhibits *$\{\varepsilon_n\}$-dominated convergence to optimality* if

$$\lim_{n \to \infty} J_{N,\pi_n} = J_N^*,$$

and, for $n = 2, 3, \ldots,$

$$J_{N,\pi_n}(x) \leq \begin{cases} J_N^*(x) + \varepsilon_n & \text{if } J_N^*(x) > -\infty, \\ J_{N,\pi_{n-1}}(x) + \varepsilon_n & \text{if } J_N^*(x) = -\infty. \end{cases}$$

2.3   Application to Specific Models

A large number of sequential optimization problems of practical interest may be viewed as special cases of the abstract problems (F) and (I). In this section we shall describe several such problems that will be of continuing interest to us throughout Part I. Detailed treatments of some of these problems can be found in DPSC.[†]

[†] We denote by DPSC the textbook by Bertsekas, "Dynamic Programming and Stochastic Control." Academic Press. New York. 1976.

2.3.1 Deterministic Optimal Control

Consider the mapping $H : SCF \to R^*$ defined by

$$H(x, u, J) = g(x, u) + \alpha J[f(x, u)] \qquad \forall x \in S, \quad u \in C, \quad J \in F. \qquad (9)$$

Our standing assumptions throughout Part I relating to this mapping are:

(1)   The functions g and f map SC into $[-\infty, \infty]$ and S, respectively.
(2)   The scalar α is positive.

The mapping H clearly satisfies the monotonicity assumption. Let J_0 be identically zero, i.e.,

$$J_0(x) = 0 \qquad \forall x \in S.$$

Then the corresponding N-stage optimization problem (F) can be written as

$$\text{minimize} \quad J_{N,\pi}(x_0) = \sum_{k=0}^{N-1} \alpha^k g[x_k, \mu_k(x_k)]$$

$$\text{subject to} \quad x_{k+1} = f[x_k, \mu_k(x_k)], \qquad \mu_k \in M, \quad k = 0, \dots, N-1. \qquad (10)$$

This is a finite horizon deterministic optimal control problem. The scalar α is known as the *discount factor*. The infinite horizon problem (I) can be written as

$$\text{minimize} \quad J_{\pi}(x_0) = \lim_{N \to \infty} \sum_{k=0}^{N-1} \alpha^k g[x_k, \mu_k(x_k)]$$

$$\text{subject to} \quad x_{k+1} = f[x_k, \mu_k(x_k)], \qquad \mu_k \in M, \quad k = 0, 1, \dots. \qquad (11)$$

This limit exists if any one of the following three conditions is satisfied:

$$g(x, u) \geq 0 \qquad \forall x \in S, \quad u \in U(x), \qquad (12)$$

$$g(x, u) \leq 0 \qquad \forall x \in S, \quad u \in U(x), \qquad (13)$$

$$\alpha < 1, \quad 0 \leq g(x, u) \leq b \qquad \text{for some } b \in (0, \infty) \text{ and all } x \in S, u \in U(x). \qquad (14)$$

Every result to be shown for problem (11) will explicitly assume one of these three conditions. Note that the requirement $0 \leq g(x, u) \leq b$ in (14) is no more strict than the usual requirement $|g(x, u)| \leq b/2$. This is true because adding the constant $b/2$ to g increases the cost corresponding to every policy by $b/2(1 - \alpha)$ and the problem remains essentially unaffected.

Deterministic optimal control problems such as (10) and (11) and their stochastic counterparts under the countability assumption of the next subsection have been studied extensively in DPSC (Chapters 2, 6, and 7). They are given here in their stationary form in the sense that the state and control spaces S and C, the control constraint $U(\cdot)$, the system function f, and the cost per stage g do not change from one stage to the next. When this is not the case, we are faced with a *nonstationary problem*. Such a problem, however, may be converted to a stationary problem by using a procedure described in Section 10.1 and in DPSC (Section 6.7). For this reason, we will not consider further nonstationary problems in Part I. Notice that within our formulation it is possible to handle state constraints of the form $x_k \in X$, $k = 0, 1, \ldots$, by defining $g(x, u) = \infty$ whenever $x \notin X$. This is our reason for allowing g to take the value ∞. Generalized versions of problems (10) and (11) are obtained if the scalar α is replaced by a function $\alpha: SC \to R^*$ with $0 \leq \alpha(x, u)$ for all $x \in S$, $u \in U(x)$, so that the discount factor depends on the current state and control. It will become evident to the reader that our general results for problems (F) and (I) are applicable to these more general deterministic problems.

2.3.2   Stochastic Optimal Control—Countable Disturbance Space

Consider the mapping $H: SCF \to R^*$ defined by

$$H(x, u, J) = E\{g(x, u, w) + \alpha J[f(x, u, w)]\,|\,x, u\}, \tag{15}$$

where the following are assumed:

(1)   The parameter w takes values in a *countable* set W with given probability distribution $p(dw|x, u)$ depending on x and u, and $E\{\cdot\,|x, u\}$ denotes expected value with respect to this distribution. (See a detailed definition below.)

(2)   The functions g and f map SCW into $[-\infty, \infty]$ and S, respectively.

(3)   The scalar α is positive.

Our usage of expected value in (15) is consistent with the definition of the usual integral (Section 7.4.4) and the outer integral (Appendix A), where the σ-algebra on W is taken to be the set of all subsets of W. Thus if w^i, $i = 1, 2, \ldots$, are the elements of W, $(p^1, p^2, \ldots)$ any probability distribution on W, and $z: W \to R^*$ a function, we define

$$E\{z(w)\} = \sum_{i=1}^{\infty} p^i z^+(w_i) - \sum_{i=1}^{\infty} p^i z^-(w_i),$$

where

$$z^+(w_i) = \max\{0, z(w_i)\}, \qquad i = 1, 2, \ldots,$$
$$z^-(w_i) = \max\{0, -z(w_i)\}, \qquad i = 1, 2, \ldots.$$

In view of our convention $\infty - \infty = \infty$, the expected value $E\{z(w)\}$ is well defined for every function $z: W \to R^*$ and every probability distribution $(p^1, p^2, \ldots)$ on W. In particular, if we denote by $(p^1(x, u), p^2(x, u), \ldots)$ the probability distribution $p(dw|x, u)$ on $W = \{w^1, w^2, \ldots\}$, then (15) can be written as

$$H(x, u, J) = \sum_{i=1}^{\infty} p^i(x, u) \max\{0, g(x, u, w^i) + \alpha J[f(x, u, w^i)]\}$$

$$- \sum_{i=1}^{\infty} p^i(x, u) \max\{0, -[g(x, u, w^i) + \alpha J[f(x, u, w^i)]]\}.$$

A point where caution is necessary in the use of expected value defined this way is that for two functions $z_1 : W \to R^*$ and $z_2 : W \to R^*$, the equality

$$E\{z_1(w) + z_2(w)\} = E\{z_1(w)\} + E\{z_2(w)\} \qquad (16)$$

need not always hold. It is guaranteed to hold if (a) $E\{z_1^+(w)\} < \infty$ and $E\{z_2^+(w)\} < \infty$, or (b) $E\{z_1^-(w)\} < \infty$ and $E\{z_2^-(w)\} < \infty$, or (c) $E\{z_1^+(w)\} < \infty$ and $E\{z_1^-(w)\} < \infty$ (see Lemma 7.11). We always have, however,

$$E\{z_1(w) + z_2(w)\} \leq E\{z_1(w)\} + E\{z_2(w)\}.$$

It is clear that the mapping H of (15) satisfies the monotonicity assumption. Let J_0 be identically zero, i.e.,

$$J_0(x) = 0 \qquad \forall x \in S.$$

Then if $g(x, u, w) > -\infty$ for all x, u, w, the N-stage cost function can be written as

$$J_{N, \pi}(x_0) = E_{w_0}\{g[x_0, \mu_0(x_0), w_0] + E_{w_1}\{\alpha g[x_1, \mu_1(x_1), w_1] + E_{w_2}\{\cdots$$

$$+ E_{w_{N-1}}\{\alpha^{N-1} g[x_{N-1}, \mu_{N-1}(x_{N-1}), w_{N-1}]|x_{N-1},$$

$$\mu_{N-1}(x_{N-1})\}|\cdots\}|x_0, \mu_0(x_0)\}$$

$$= E_{w_0}\{E_{w_1}\{\cdots E_{w_{N-1}}\{\sum_{k=0}^{N-1} \alpha^k g[x_k, \mu_k(x_k), w_k]|x_{N-1},$$

$$\mu_{N-1}(x_{N-1})\}|\cdots\}|x_0, \mu_0(x_0)\}, \qquad (17)$$

where the states $x_1, x_2, \ldots, x_{N-1}$ satisfy

$$x_{k+1} = f[x_k, \mu_k(x_k), w_k], \qquad k = 0, \ldots, N - 2. \qquad (18)$$

The interchange of expectation and summation in (17) is valid, since $g(x, u, w) > -\infty$ for all x, u, w, and we have for any measure space $(\Omega, \mathscr{F}, \nu)$, measurable $h: \Omega \to R^*$, and $\lambda \in (-\infty, +\infty]$,

$$\lambda + \int h \, dv = \int (\lambda + h) \, dv.$$

When Eq. (18) is used successively to express the states $x_1, x_2, \ldots, x_{N-1}$ exclusively in terms of $w_0, w_1, \ldots, w_{N-1}$ and x_0, one can see from (17) that $J_{N,\pi}(x_0)$ is given in terms of successive iterated integration over $w_{N-1}, \ldots, w_0$. For each $x_0 \in S$ and $\pi \in \Pi$ the probability distributions $p^i(x_0, \mu_0(x_0)), \ldots, p^i(x_{N-1}, \mu_{N-1}(x_{N-1}))$, $i = 1, 2, \ldots$, over W specify, by the product measure theorem [A1, Theorem 2.6.2], a unique product measure on the cross product W^N of N copies of W. If Fubini's theorem [A1, Theorem 2.6.4] is applicable, then from (17) the N-stage cost function $J_{N,\pi}(x_0)$ can be alternatively expressed as

$$J_{N,\pi}(x_0) = E\left\{ \sum_{k=0}^{N-1} \alpha^k g[x_k, \mu_k(x_k), w_k] \right\}, \tag{19}$$

where this expectation is taken with respect to the product measure on W^N and the states $x_1, x_2, \ldots, x_{N-1}$ are expressed in terms of $w_0, w_1, \ldots, w_{N-1}$ and x_0 via (18). Fubini's theorem can be applied if the expected value in (19) is not of the form $\infty - \infty$, i.e., if either

$$E\left\{ \max\left\{ 0, \sum_{k=0}^{N-1} \alpha^k g[x_k, \mu_k(x_k), w_k] \right\} \right\} < \infty$$

or

$$E\left\{ \max\left\{ 0, -\sum_{k=0}^{N-1} \alpha^k g[x_k, \mu_k(x_k), w_k] \right\} \right\} < \infty.$$

In particular, this is true if either

$$E\{\max\{0, g[x_k, \mu_k(x_k), w_k]\}\} < \infty, \qquad k = 0, \ldots, N-1,$$

or

$$E\{\max\{0, -g[x_k, \mu_k(x_k), w_k]\}\} < \infty, \qquad k = 0, \ldots, N-1$$

or if g is uniformly bounded above or below by a real number. If $J_{N,\pi}(x_0)$ can be expressed as in (19) for each $x_0 \in S$ and $\pi \in \Pi$, then the N-stage problem can be written as

$$\text{minimize} \quad J_{N,\pi}(x_0) = E\left\{ \sum_{k=0}^{N-1} \alpha^k g[x_k, \mu_k(x_k), w_k] \right\}$$

$$\text{subject to} \quad x_{k+1} = f[x_k, \mu_k(x_k), w_k], \qquad \mu_k \in M, \quad k = 0, \ldots, N-1,$$

which is the traditional form of an N-stage stochastic optimal control problem and is also the starting point for the N-stage model of Part II (Definition 8.3).

The corresponding infinite horizon problem is (cf. Definition 9.3)

$$\text{minimize} \quad J_\pi(x_0) = \lim_{N \to \infty} E \left\{ \sum_{k=0}^{N-1} \alpha^k g[x_k, \mu_k(x_k), w_k] \right\} \tag{20}$$

$$\text{subject to} \quad x_{k+1} = f[x_k, \mu_k(x_k), w_k], \qquad \mu_k \in M, \quad k = 0, 1, \dots .$$

This limit exists under any one of the conditions:

$$g(x, u, w) \geq 0 \qquad \forall x \in S, \quad u \in U(x), \quad w \in W, \tag{21}$$

$$g(x, u, w) \leq 0 \qquad \forall x \in S, \quad u \in U(x), \quad w \in W, \tag{22}$$

$$\alpha < 1, \qquad 0 \leq g(x, u, w) \leq b \qquad \text{for some } b \in (0, \infty)$$

$$\text{and all } x \in S, u \in U(x), w \in W. \tag{23}$$

Every result to be shown for problem (20) will explicitly assume one of these three conditions.

Similarly as for the deterministic problem, a generalized version of the stochastic problem is obtained if the scalar α is replaced by a function $\alpha : SCW \to R^*$ satisfying $0 \leq \alpha(x, u, w)$ for all (x, u, w). The mapping H takes the form

$$H(x, u, J) = E\{g(x, u, w) + \alpha(x, u, w)J[f(x, u, w)] | x, u\}.$$

This case covers certain semi-Markov decision problems (see [J2]). We will not be further concerned with this mapping and will leave it to the interested reader to obtain specific results relating to the corresponding problems (F) and (I) by specializing abstract results obtained subsequently in Part I. Also, nonstationary versions of the problem may be treated by reduction to the stationary case (see Section 10.1 or DPSC, Section 6.7).

The countability assumption on W is satisfied for many problems of interest. For example, it is satisfied in stochastic control problems involving Markov chains with a finite or countable number of states (see, e.g., [D3], [K6]). When the set W is not countable, then matters are complicated by the need to define the expected value

$$E\{g[x, \mu(x), w] + \alpha J[f(x, \mu(x), w)] | x, u\}$$

for every $\mu \in M$. There are two approaches that one can employ to overcome this difficulty. One possibility is to define the expected value as an outer integral, as we do in the next subsection. The other approach is the subject of Part II where we impose an appropriate measurable space structure on S, C, and W and require that the functions $\mu \in M$ be measurable. Under these circumstances a reformulation of the stochastic optimal control problem into the form of the abstract problems (F) or (I) is not straightforward. Nonetheless, such a reformulation is possible as well as useful as we will demonstrate in Chapter 9.

2.3.3 Stochastic Optimal Control—Outer Integral Formulation

Consider the mapping $H: SCF \to R^*$ defined by

$$H(x, u, J) = E^*\{g(x, u, w) + \alpha J[f(x, u, w)] | x, u\}, \tag{24}$$

where the following are assumed:

(1) The parameter w takes values in a measurable space $(W, \mathscr{F})$. For each fixed $(x, u) \in SC$, a probability measure $p(dw|x, u)$ on $(W, \mathscr{F})$ is given and $E^*\{\cdot | x, u\}$ in (24) denotes the outer integral (see Appendix A) with respect to that measure. Thus we may write, in the notation of Appendix A,

$$H(x, u, J) = \int^* \{g(x, u, w) + \alpha J[f(x, u, w)]\} p(dw|x, u).$$

(2) The functions g and f map SCW into $[-\infty, \infty]$ and S, respectively.
(3) The scalar α is positive.

We note that mappings (9) and (15) of the previous two subsections are special cases of the mapping H of (24). The mapping (9) (deterministic problem) is obtained from (24) when the set W consists of a single element. The mapping (15) (stochastic problem with countable disturbance space) is the special case of (24) where W is a countable set and $\mathscr{F}$ is the σ-algebra consisting of all subsets of W. For this reason, in our subsequent analysis we will not further consider the mappings (9) and (15), but will focus attention on the mapping (24).

Clearly H as defined by (24) satisfies the monotonicity assumption. Just as for the models of the previous two sections, we take

$$J_0(x) = 0 \qquad \forall x \in S$$

and consider the corresponding N-stage and infinite horizon problems (F) and (I).

If appropriate measurability assumptions are placed on S, C, f, g, and p, then the N-stage cost

$$J_{N, \pi}(x) = (T_{\mu_0} \cdots T_{\mu_{N-1}})(J_0)(x)$$

can be rewritten in terms of ordinary integration for every policy $\pi = (\mu_0, \mu_1, \ldots)$ for which μ_k, $k = 0, 1, \ldots$, is appropriately measurable. To see this, suppose that S has a σ-algebra $\mathscr{S}$, C has a σ-algebra $\mathscr{C}$, and $\mathscr{B}$ is the Borel σ-algebra on R^*. Suppose f is $(\mathscr{SCF}, \mathscr{S})$-measurable and g is $(\mathscr{SCF}, \mathscr{B})$-measurable, where $\mathscr{SCF}$ denotes the product σ-algebra on SCW. Assume that for each fixed $B \in \mathscr{F}$, $p(B|x, u)$ is $\mathscr{SC}$-measurable in (x, u) and consider a policy $\pi = (\mu_0, \mu_1, \ldots)$, where μ_k is $(\mathscr{S}, \mathscr{C})$-measurable for all k. These conditions guarantee that $T_{\mu_k}(J)$ given by

$$T_{\mu_k}(J)(x) = \int \{g[x, \mu_k(x), w] + \alpha J[f(x, \mu_k(x), w)]\} p(dw|x, u)$$

is $\mathscr{S}$-measurable for all k and $J \in F$ that are $\mathscr{S}$-measurable. Just as in the previous section, for a fixed $x_0 \in S$ and $\pi = (\mu_0, \mu_1, \ldots) \in \Pi$, the probability measures $p(\cdot \,|x_0, \mu_0(x_0)), \ldots, p(\cdot \,|x_{N-1}, \mu_{N-1}(x_{N-1}))$ together with the system equation

$$x_{k+1} = f[x_k, \mu_k(x_k), w_k], \qquad k = 0, \ldots, N-2, \tag{25}$$

define a unique product measure $p(d(w_0, \ldots, w_{N-1})|x_0, \pi)$ on the cross product W^N of N copies of W. [Note that x_k, $k = 0, 1, \ldots, N-1$, can be expressed as a measurable function of $(w_0, \ldots, w_{N-1})$ via (25)]. Using the calculation of the previous section, we have that if $g(x, u, w) > -\infty$ for all x, u, w, and Fubini's theorem is applicable, then

$$J_{N,\pi}(x_0) = E\left\{ \sum_{k=0}^{N-1} \alpha^k g[x_k, \mu_k(x_k), w_k] \right\}$$

$$= \int_{W^N} \left\{ \sum_{k=0}^{N-1} \alpha^k g[x_k, \mu_k(x_k), w_k] \right\} p(d(w_0, \ldots, w_{N-1})|x_0, \pi),$$

where $x_1, x_2, \ldots, x_{N-1}$ are expressed in terms of $w_0, w_1, \ldots, w_{N-1}$ and x_0 via (25). Also, as in the previous section, Fubini's theorem applies if either

$$E\left\{ \max\left\{ 0, \sum_{k=0}^{N-1} \alpha^k g[x_k, \mu_k(x_k), w_k] \right\} \right\} < \infty$$

or

$$E\left\{ \max\left\{ 0, -\sum_{k=0}^{N-1} \alpha^k g[x_k, \mu_k(x_k), w_k] \right\} \right\} < \infty.$$

Thus if appropriate measurability conditions are placed on S, C, W, f, g, and $p(dw|x, u)$ and Fubini's theorem applies, then the N-stage cost $J_{N,\pi}$ corresponding to measurable π reduces to the traditional form

$$J_{N,\pi}(x_0) = E\left\{ \sum_{k=0}^{N-1} \alpha^k g[x_k, \mu_k(x_k), w_k] \right\}.$$

This observation is significant in view of the fact that

$$\inf_{\pi \in \Pi} J_{N,\pi}(x) \leq \inf_{\pi \in \tilde{\Pi}} J_{N,\pi}(x) \qquad \forall x \in S,$$

where

$$\tilde{\Pi} = \{\pi \in \Pi \,|\, \pi = (\mu_0, \mu_1, \ldots),\ \mu_k \in M \text{ is } (\mathscr{S}, \mathscr{C})\text{-measurable},\ k = 0, 1, \ldots\}.$$

Thus, if an optimal (ε-optimal) policy π^* can be found for problem (F) and

$\pi^* \in \tilde{\Pi}$ (i.e., is measurable), then π^* is optimal (ε-optimal) for the problem

$$\text{minimize} \quad J_{N,\pi}(x)$$
$$\text{subject to} \quad \pi \in \tilde{\Pi},$$

which is a traditional stochastic optimal control problem.

These remarks illustrate how one can utilize the outer integration framework in an initial formulation of a particular problem and subsequently show via further (and hopefully simple) analysis that attention can be restricted to the class of measurable policies $\tilde{\Pi}$ for which the cost function admits a traditional interpretation. The main advantage that the outer integral formulation offers is simplicity. One does not need to introduce an elaborate topological and measure-theoretic structure such as the one of Part II in an initial formulation of the problem. In addition the policy iteration algorithm of Chapter 4 is applicable to the problem of this section but cannot be justified for the corresponding model of Part II. The outer integral formulation has, however, important limitations which become apparent in the treatment of problems with imperfect state information by means of sufficient statistics (Chapter 10).

2.3.4 Stochastic Optimal Control—Multiplicative Cost Functional

Consider the mapping $H : SCF \to R^*$ defined by

$$H(x, u, J) = E\{g(x, u, w)J[f(x, u, w)] | x, u\}. \tag{26}$$

We make the same assumptions on w, g, and f as in Section 2.3.2, i.e., w takes values in a *countable* set W with a given probability distribution depending on x and u. We assume further that

$$g(x, u, w) \geq 0 \qquad \forall x \in S, \quad u \in U(x), \quad w \in W. \tag{27}$$

In view of (27), the mapping H of (26) satisfies the monotonicity assumption.

We take

$$J_0(x) = 1 \quad \forall x \in S$$

and consider the problems (F) and (I). Problem (F) corresponds to the stochastic optimal control problem

$$\text{minimize} \quad J_{N,\pi}(x_0) = E\{g[x_0, \mu_0(x_0), w_0] \cdots g[x_{N-1}, \mu_{N-1}(x_{N-1}), w_{N-1}]\} \tag{28}$$

$$\text{subject to} \quad x_{k+1} = f[x_k, \mu_k(x_k), w_k], \qquad \mu_k \in M, \quad k = 0, 1, \ldots,$$

and problem (I) corresponds to the infinite horizon version of (28). The limit as $N \to \infty$ in (28) exists if $g(x, u, w) \geq 1$ for every x, u, w or $0 \leq g(x, u, w) \leq 1$

for every x, u, w. A special case of (28) is the exponential cost functional problem

$$\text{minimize} \quad E\left\{\exp\left[\sum_{k=0}^{N-1} g'[x_k, \mu_k(x_k), w_k]\right]\right\}$$

$$\text{subject to} \quad x_{k+1} = f[x_k, \mu_k(x_k), w_k], \qquad \mu_k \in M, \quad k = 0, 1, \ldots,$$

where g' is some function mapping SCW into $(-\infty, \infty]$.

2.3.5   Minimax Control

Consider the mapping $H : SCF \to R^*$ defined by

$$H(x, u, J) = \sup_{w \in W(x, u)} \{g(x, u, w) + \alpha J[f(x, u, w)]\} \qquad (29)$$

where the following are assumed:

(1)   The parameter w takes values in a set W and $W(x, u)$ is a nonempty subset of W for each $x \in S$, $u \in U(x)$.

(2)   The functions g and f map SCW into $[-\infty, \infty]$ and S respectively.

(3)   The scalar α is positive.

Clearly the monotonicity assumption is satisfied.

We take

$$J_0(x) = 0 \qquad \forall x \in S.$$

If $g(x, u, w) > -\infty$ for all x, u, w, the corresponding N-stage problem (F) can also be written as

$$\text{minimize} \quad J_{N, \pi}(x_0) = \sup_{w_k \in W[x_k, \mu_k(x_k)]} \left\{\sum_{k=0}^{N-1} \alpha^k g[x_k, \mu_k(x_k), w_k]\right\}$$

$$\text{subject to} \quad x_{k+1} = f[x_k, \mu_k(x_k), w_k], \qquad \mu_k \in M, \quad k = 0, 1, \ldots, \qquad (30)$$

and this is an N-stage minimax control problem. The infinite horizon version is

$$\text{minimize} \quad J_\pi(x_0) = \lim_{N \to \infty} \sup_{w_k \in W[x_k, \mu_k(x_k)]} \left\{\sum_{k=0}^{N-1} \alpha^k g[x_k, \mu_k(x_k), w_k]\right\}$$

$$\text{subject to} \quad x_{k+1} = f[x_k, \mu_k(x_k), w_k], \qquad \mu_k \in M, \quad k = 0, 1, \ldots. \qquad (31)$$

The limit in (31) exists under any one of the conditions (21), (22), or (23). This problem contains as a special case the problem of infinite time reachability examined in Bertsekas [B2]. Problems (30) and (31) arise also in the analysis of sequential zero-sum games.

Chapter 3

Finite Horizon Models

3.1   General Remarks and Assumptions

Consider the N-stage optimization problem

$$\text{minimize} \quad J_{N,\pi}(x) = (T_{\mu_0} \cdots T_{\mu_{N-1}})(J_0)(x)$$
$$\text{subject to} \quad \pi = (\mu_0, \mu_1, \ldots) \in \Pi,$$

where for every $\mu \in M$, $J \in F$, and $x \in S$ we have

$$T_\mu(J)(x) = H[x, \mu(x), J], \qquad T(J)(x) = \inf_{u \in U(x)} H(x, u, J).$$

Experience with a large variety of sequential optimization problems suggests that the N-stage optimal cost function J_N^* satisfies

$$J_N^* = \inf_{\pi \in \Pi} J_{N,\pi} = T^N(J_0),$$

and hence is obtained after N steps of the DP algorithm. In our more general setting, however, we shall need to place additional conditions on H in order to guarantee this equality. Consider the following two assumptions.

Assumption F.1   If $\{J_k\} \subset F$ is a sequence satisfying $J_{k+1} \leq J_k$ for all k and $H(x, u, J_1) < \infty$ for all $x \in S$, $u \in U(x)$, then

$$\lim_{k \to \infty} H(x, u, J_k) = H\left(x, u, \lim_{k \to \infty} J_k\right) \qquad \forall x \in S, \quad u \in U(x).$$

Assumption F.2   There exists a scalar $\alpha \in (0, \infty)$ such that for all scalars $r \in (0, \infty)$ and functions $J \in F$, we have

$$H(x, u, J) \leq H(x, u, J + r) \leq H(x, u, J) + \alpha r \qquad \forall x \in S, \quad u \in U(x).$$

We will also consider the following assumption, which is admittedly somewhat complicated. It will enable us to obtain a stronger result on the existence of nearly optimal policies (Proposition 3.2) than can be obtained under F.2. The assumption is satisfied for the stochastic optimal control problem of Section 2.3.3, as we show in the last section of this chapter.

Assumption F.3   There is a scalar $\beta \in (0, \infty)$ such that if $J \in F$, $\{J_n\} \subset F$, and $\{\varepsilon_n\} \subset R$ satisfy

$$\sum_{n=1}^{\infty} \varepsilon_n < \infty, \qquad\qquad \varepsilon_n > 0, \quad n = 1, 2, \dots,$$

$$J = \lim_{n \to \infty} J_n, \qquad\qquad J \leq J_n, \quad n = 1, 2, \dots,$$

$$J_n(x) \leq \begin{cases} J(x) + \varepsilon_n, & n = 1, 2, \dots \text{ and } x \in S \text{ with } J(x) > -\infty, \\ J_{n-1}(x) + \varepsilon_n, & n = 2, 3, \dots \text{ and } x \in S \text{ with } J(x) = -\infty, \end{cases}$$

$$H(x, u, J_1) < \infty, \qquad\qquad \forall x \in S, \quad u \in U(x),$$

then there exists a sequence $\{\mu_n\} \subset M$ such that

$$\lim_{n \to \infty} T_{\mu_n}(J_n) = T(J),$$

$$T_{\mu_n}(J_n)(x) \leq \begin{cases} T(J)(x) + \beta\varepsilon_n, & n = 1, 2, \dots, \ x \in S \text{ with } T(J)(x) > -\infty, \\ T_{\mu_{n-1}}(J_{n-1})(x) + \beta\varepsilon_n, & n = 2, 3, \dots, \ x \in S \text{ with } T(J)(x) = -\infty. \end{cases}$$

Each of our results will require *at most* one of the preceding assumptions. As we show in Section 3.3, at least one of these assumptions is satisfied by every specific model considered in Section 2.3.

3.2   Main Results

The central question regarding the finite horizon problem is whether $J_N^* = T^N(J_0)$, in which case the N-stage optimal cost function J_N^* can be obtained via the DP algorithm that successively computes $T(J_0), T^2(J_0), \dots$. A related question is whether optimal or nearly optimal policies exist. The results of this section provide conditions under which the answer to these questions is affirmative.

Proposition 3.1   (a)   Let F.1 hold and assume that $J_{k, \pi}(x) < \infty$ for all $x \in S$, $\pi \in \Pi$, and $k = 1, 2, \dots, N$. Then

$$J_N^* = T^N(J_0).$$

(b)   Let F.2 hold and assume that $J_k^*(x) > -\infty$ for all $x \in S$ and $k = 1, 2, \ldots, N$. Then

$$J_N^* = T^N(J_0),$$

and for every $\varepsilon > 0$, there exists an N-stage ε-optimal policy, i.e., a $\pi_\varepsilon \in \Pi$ such that

$$J_N^* \le J_{N, \pi_\varepsilon} \le J_N^* + \varepsilon.$$

Proof   (a)   For each $k = 0, 1, \ldots, N - 1$, consider a sequence $\{\mu_k^i\} \subset M$ such that

$$\lim_{i \to \infty} T_{\mu_k^i}[T^{N-k-1}(J_0)] = T^{N-k}(J_0), \qquad k = 0, \ldots, N-1,$$

$$T_{\mu_k^i}[T^{N-k-1}(J_0)] \ge T_{\mu_k^{i+1}}[T^{N-k-1}(J_0)], \quad k = 0, \ldots, N-1, \ i = 0, 1, \ldots.$$

By using F.1 and the assumption that $J_{k,\pi}(x) < \infty$, we have

$$\begin{aligned}
J_N^* &\le \inf_{i_0} \cdots \inf_{i_{N-1}} (T_{\mu_0^{i_0}} \cdots T_{\mu_{N-1}^{i_{N-1}}})(J_0) \\
&= \inf_{i_0} \cdots \inf_{i_{N-2}} (T_{\mu_0^{i_0}} \cdots T_{\mu_{N-2}^{i_{N-2}}}) \left[\inf_{i_{N-1}} T_{\mu_{N-1}^{i_{N-1}}}(J_0) \right] \\
&= \inf_{i_0} \cdots \inf_{i_{N-2}} (T_{\mu_0^{i_0}} \cdots T_{\mu_{N-2}^{i_{N-2}}})[T(J_0)] \\
&= T^N(J_0),
\end{aligned}$$

where the last equality is obtained by repeating the process used to obtain the previous equalities. On the other hand, it is clear from the definitions of Chapter 2 that $T^N(J_0) \le J_N^*$, and hence $J_N^* = T^N(J_0)$.

(b)   We use induction. The result clearly holds for $N = 1$. Assume that it holds for $N = k$, i.e., $J_k^* = T^k(J_0)$ and for a given $\varepsilon > 0$, there is a $\pi_\varepsilon \in \Pi$ with $J_{k, \pi_\varepsilon} \le J_k^* + \varepsilon$. Using F.2 we have for all $\mu \in M$,

$$J_{k+1}^* \le T_\mu(J_{k, \pi_\varepsilon}) \le T_\mu(J_k^*) + \alpha\varepsilon.$$

Hence $J_{k+1}^* \le T(J_k^*)$, and by using the induction hypothesis we obtain $J_{k+1}^* \le T^{k+1}(J_0)$. On the other hand, we have clearly $T^{k+1}(J_0) \le J_{k+1}^*$, and hence $T^{k+1}(J_0) = J_{k+1}^*$. For any $\bar\varepsilon > 0$, let $\bar\pi = (\bar\mu_0, \bar\mu_1, \ldots)$ be such that $J_{k, \bar\pi} \le J_k^* + (\bar\varepsilon/2\alpha)$, and let $\bar\mu \in M$ be such that $T_{\bar\mu}(J_k^*) \le T(J_k^*) + (\bar\varepsilon/2)$. Consider the policy $\bar\pi_{\bar\varepsilon} = (\bar\mu, \bar\mu_0, \bar\mu_1, \ldots)$. Then

$$J_{k+1, \bar\pi_{\bar\varepsilon}} = T_{\bar\mu}(J_{k, \bar\pi}) \le T_{\bar\mu}(J_k^*) + (\bar\varepsilon/2) \le T(J_k^*) + \bar\varepsilon = J_{k+1}^* + \bar\varepsilon.$$

The induction is complete.   Q.E.D.

Proposition 3.1(a) may be strengthened by using the following assumption in place of F.1.

Assumption F.1′ The function J_0 satisfies

$$J_0(x) \geq H(x, u, J_0) \qquad \forall x \in S, \quad u \in U(x),$$

and if $\{J_k\} \subset F$ is a sequence satisfying $J_{k+1} \leq J_k \leq J_0$ for all k, then

$$\lim_{k \to \infty} H(x, u, J_k) = H\left(x, u, \lim_{k \to \infty} J_k\right) \qquad \forall x \in S, \quad u \in U(x).$$

The following corollary is obtained by verbatim repetition of the proof of Proposition 3.1(a).

Corollary 3.1.1 Let F.1′ hold. Then

$$J_N^* = T^N(J_0).$$

Proposition 3.1 and Corollary 3.1.1 may fail to hold if their assumptions are slightly relaxed.

COUNTEREXAMPLE 1 Take $S = \{0\}$, $C = U(0) = (-1, 0]$, $J_0(0) = 0$, $H(0, u, J) = u$ if $-1 < J(0)$, $H(0, u, J) = J(0) + u$ if $J(0) \leq -1$. Then $(T_{\mu_0} \cdots T_{\mu_{N-1}})(J_0)(0) = \mu_0(0)$ and $J_N^*(0) = -1$, while $T^N(J_0)(0) = -N$ for every N. Here the assumptions $J_{k,\pi}(0) < \infty$ and $J_k^*(0) > -\infty$ are satisfied, but F.1, F.1′, and F.2 are violated.

COUNTEREXAMPLE 2 Take $S = \{0, 1\}$, $C = U(0) = U(1) = (-\infty, 0]$, $J_0(0) = J_0(1) = 0$, $H(0, u, J) = u$ if $J(1) = -\infty$, $H(0, u, J) = 0$ if $J(1) > -\infty$, and $H(1, u, J) = u$. Then $(T_{\mu_0} \cdots T_{\mu_{N-1}})(J_0)(0) = 0$, $(T_{\mu_0} \cdots T_{\mu_{N-1}})(J_0)(1) = \mu_0(1)$ for all $N \geq 1$. Hence, $J_N^*(0) = 0$, $J_N^*(1) = -\infty$. On the other hand, we have $T^N(J_0)(0) = T^N(J_0)(1) = -\infty$ for all $N \geq 2$. Here F.2 is satisfied, but F.1, F.1′, and the assumptions $J_{k,\pi}(x) < \infty$ and $J_k^*(x) > -\infty$ for $\forall x \in S$ are all violated.

The following counterexample is a stochastic optimal control problem with countable disturbance space as discussed in Section 2.3.2. We use the notation introduced there.

COUNTEREXAMPLE 3 Let $N = 2$, $S = \{0, 1\}$, $C = U(0) = U(1) = R$, $W = \{2, 3, \ldots\}$, $p(w = k | x, u) = k^{-2}(\sum_{n=2}^{\infty} n^{-2})^{-1}$ for $k = 2, 3, \ldots$, $x \in S$, $u \in C$, $f(0, u, w) = f(1, u, w) = 1$ for $\forall u \in C$, $w \in W$, $g(0, u, w) = w$, $g(1, u, w) = u$ for $\forall u \in C$, $w \in W$. Then a straightforward calculation shows that $J_2^*(0) = \infty$, $J_2^*(1) = -\infty$, while $T^2(J_0)(0) = -\infty$, $T^2(J_0)(1) = -\infty$. Here F.1 and F.2 are satisfied, but F.1′ and the assumptions $J_{k,\pi}(x) < \infty$ for all x, π, k, and $J_k^*(x) > -\infty$ for all x and k are all violated.

The next counterexample is a deterministic optimal control problem as discussed in Section 2.3.1. We use the notation introduced there.

COUNTEREXAMPLE 4 Let $N = 2$, $S = \{0, 1, \ldots\}$, $C = U(x) = (0, \infty)$ for $\forall x \in S$, $f(x, u) = 0$ for $\forall x \in S$, $u \in C$, $g(0, u) = -u$ for $\forall u \in U(0)$, $g(x, u) = x$ for $\forall u \in U(x)$ if $x \neq 0$. Then for $\pi \in \Pi$ and $x \neq 0$, we have $J_{2,\pi}(x) = x - \mu_1(0)$, so that $J_2^*(x) = -\infty$ for all $x \in S$. On the other hand clearly there is no two-stage ε-optimal policy for any $\varepsilon > 0$. Here F.1, F.2, and the assumption $J_{k,\pi}(x) < \infty$ for all x, π, k are satisfied, and indeed we have $J_2^*(x) = T^2(J_0)(x) = -\infty$ for $\forall x \in S$. However, the assumption $J_k^*(x) > -\infty$ for all x and k is violated.

As Counterexample 4 shows there may not exist an N-stage ε-optimal policy if we have $J_k^*(x) = -\infty$ for some k and $x \in S$. The following proposition establishes, under appropriate assumptions, the existence of a sequence of nearly optimal policies whose cost functions converge to the optimal cost function.

Proposition 3.2 Let F.3 hold and assume $J_{k,\pi}(x) < \infty$ for all $x \in S$, $\pi \in \Pi$, and $k = 1, 2, \ldots, N$. Then

$$J_N^* = T^N(J_0).$$

Furthermore, if $\{\varepsilon_n\}$ is a sequence of positive numbers with $\varepsilon_n \downarrow 0$, then there exists a sequence of policies $\{\pi_n\}$ exhibiting $\{\varepsilon_n\}$-dominated convergence to optimality. In particular, if in addition $J_N^*(x) > -\infty$ for all $x \in S$, then for every $\varepsilon > 0$ there exists an ε-optimal policy.

Proof We will prove by induction that for $K \leq N$ we have $J_K^* = T^K(J_0)$, and furthermore, given K and $\{\varepsilon_n\}$ with $\varepsilon_n \downarrow 0$, $\varepsilon_n > 0$ for $\forall n$, there exists a sequence $\{\pi_n\} \subset \Pi$ such that for all n,

$$\lim_{n \to \infty} J_{K,\pi_n} = J_K^*, \tag{1}$$

$$J_{K,\pi_n}(x) \leq \begin{cases} J_K^*(x) + \varepsilon_n & \forall x \in S \quad \text{with} \quad J_K^*(x) > -\infty, \quad (2) \\ J_{K,\pi_{n-1}}(x) + \varepsilon_n & \forall x \in S \quad \text{with} \quad J_K^*(x) = -\infty. \quad (3) \end{cases}$$

We show that this holds for $K = 1$. We have

$$J_1^*(x) = \inf_{\pi \in \Pi} J_{1,\pi}(x) = \inf_{\mu \in M} H[x, \mu(x), J_0] = T(J_0)(x) \qquad \forall x \in S.$$

It is also clear that, given $\{\varepsilon_n\}$, there exists a sequence $\{\pi_n\} \subset \Pi$ satisfying (1)–(3) for $K = 1$.

Assume that the result is true for $K = N - 1$. Let β be the scalar specified in F.3. Consider a sequence $\{\varepsilon_n\} \subset R$ with $\varepsilon_n > 0$ for $\forall n$ and $\lim_{n \to \infty} \varepsilon_n = 0$, and let $\{\hat{\pi}_n\} \subset \Pi$, $\hat{\pi}_n = (\mu_1^n, \mu_2^n, \ldots)$, be such that

$$\lim_{n \to \infty} J_{N-1,\hat{\pi}_n} = J_{N-1}^*, \tag{4}$$

$$J_{N-1,\hat{\pi}_n}(x) \leq \begin{cases} J_{N-1}^*(x) + \beta^{-1}\varepsilon_n & \forall x \in S \quad \text{with} \quad J_{N-1}^*(x) > -\infty, \quad (5) \\ J_{N-1,\hat{\pi}_{n-1}}(x) + \beta^{-1}\varepsilon_n & \forall x \in S \quad \text{with} \quad J_{N-1}^*(x) = -\infty. \quad (6) \end{cases}$$

The assumption $J_{k,\pi}(x) < \infty$ for all $x \in S$, $\pi \in \Pi$, $k = 1, 2, \ldots, N$, guarantees that we have

$$H(x, u, J_{N-1, \hat{\pi}_1}) < \infty \qquad \forall x \in S, \quad u \in U(x). \tag{7}$$

Without loss of generality we assume that $\sum_{n=1}^{\infty} \varepsilon_n < \infty$. Then Assumption F.3 together with (4) implies that there exists a sequence $\{\mu_0^n\} \subset M$ such that, for all n,

$$\lim_{n \to \infty} T_{\mu_0^n}(J_{N-1, \hat{\pi}_n}) = T(J_{N-1}^*), \tag{8}$$

$$T_{\mu_0^n}(J_{N-1, \hat{\pi}_n})(x) \le \begin{cases} T(J_{N-1}^*)(x) + \varepsilon_n & \text{if} \quad T(J_{N-1}^*)(x) > -\infty, \quad (9) \\ T_{\mu_0^{n-1}}(J_{N-1, \pi_{\hat{n}-1}})(x) + \varepsilon_n & \text{if} \quad T(J_{N-1}^*)(x) = -\infty. \quad (10) \end{cases}$$

We have by the induction hypothesis $J_{N-1}^* = T^{N-1}(J_0)$, and it is clear that $T^N(J_0) \le J_N^*$. Hence,

$$T(J_{N-1}^*) = T^N(J_0) \le J_N^*. \tag{11}$$

We also have

$$J_N^* \le \lim_{n \to \infty} T_{\mu_0^n}(J_{N-1, \hat{\pi}_n}) \tag{12}$$

Combining (8), (11), and (12), we obtain

$$J_N^* = T(J_{N-1}^*) = T^N(J_0). \tag{13}$$

Let $\pi_n = (\mu_0^n, \mu_1^n, \mu_2^n, \ldots)$. Then from (8)–(10) and (13), we obtain, for all n,

$$\lim_{n \to \infty} J_{N, \pi_n} = J_N^*,$$

$$J_{N, \pi_n}(x) \le \begin{cases} J_N^*(x) + \varepsilon_n & \forall x \in S \quad \text{with} \quad J_N^*(x) > -\infty, \\ J_{N, \pi_{n-1}}(x) + \varepsilon_n & \forall x \in S \quad \text{with} \quad J_N^*(x) = -\infty, \end{cases}$$

and the induction argument is complete.   Q.E.D.

Despite the need for various assumptions in order to guarantee $J_N^* = T^N(J_0)$, the following result, which establishes the validity of the DP algorithm as a means for constructing optimal policies, requires no assumption other than monotonicity of H.

Proposition 3.3   A policy $\pi^* = (\mu_0^*, \mu_1^*, \ldots)$ is uniformly N-stage optimal if and only if

$$(T_{\mu_k^*} T^{N-k-1})(J_0) = T^{N-k}(J_0), \qquad k = 0, \ldots, N-1. \tag{14}$$

Proof   Let (14) hold. Then we have, for $k = 0, 1, \ldots, N-1$,

$$(T_{\mu_k^*} \cdots T_{\mu_{N-1}^*})(J_0) = T^{N-k}(J_0).$$

On the other hand, we have $J_{N-k}^* \le (T_{\mu_k^*} \cdots T_{\mu_{N-1}^*})(J_0)$, while $T^{N-k}(J_0) \le J_{N-k}^*$. Hence, $J_{N-k}^* = (T_{\mu_k^*} \cdots T_{\mu_{N-1}^*})(J_0)$ and π^* is uniformly N-stage optimal. Conversely, let π^* be uniformly N-stage optimal. Then

$$T(J_0) = J_1^* = T_{\mu_{N-1}^*}(J_0)$$

by definition. We also have for every $\mu \in M$, $(T_\mu T)(J_0) = (T_\mu T_{\mu_{N-1}^*})(J_0)$, which implies that

$$T^2(J_0) = \inf_{\mu \in M} (T_\mu T)(J_0) = \inf_{\mu \in M} (T_\mu T_{\mu_{N-1}^*})(J_0)$$

$$\ge J_2^* = (T_{\mu_{N-2}^*} T_{\mu_{N-1}^*})(J_0) \ge T^2(J_0).$$

Therefore

$$T^2(J_0) = J_2^* = (T_{\mu_{N-2}^*} T_{\mu_{N-1}^*})(J_0) = (T_{\mu_{N-2}^*} T)(J_0).$$

Proceeding similarly, we show all the equations in (14).   Q.E.D.

As a corollary of Proposition 3.3, we have the following.

Corollary 3.3.1   (a)   There exists a uniformly N-stage optimal policy if and only if the infimum in the relation

$$T^{k+1}(J_0)(x) = \inf_{u \in U(x)} H[x, u, T^k(J_0)] \tag{15}$$

is attained for each $x \in S$ and $k = 0, 1, \ldots, N - 1$.

(b)   If there exists a uniformly N-stage optimal policy, then

$$J_N^* = T^N(J_0).$$

We now turn to establishing conditions for existence of a uniformly N-stage optimal policy. For this we need compactness assumptions. If C is a Hausdorff topological space, we say that a subset U of C is compact if every collection of open sets that covers U has a finite subcollection that covers U. The empty set in particular is considered to be compact. Any sequence $\{u_n\}$ belonging to a compact set $U \subset C$ has at least one accumulation point $\bar{u} \in U$, i.e., a point $\bar{u} \in U$ every (open) neighborhood of which contains an infinite number of elements of $\{u_n\}$. Furthermore, all accumulation points of $\{u_n\}$ belong to U. If $\{U_n\}$ is a sequence of nonempty compact subsets of C and $U_n \supset U_{n+1}$ for all n, then the intersection $\bigcap_{n=1}^\infty U_n$ is nonempty and compact. This yields the following lemma, which will be useful in what follows.

Lemma 3.1   Let C be a Hausdorff space, $f : C \to R^*$ a function, and U a subset of C. Assume that the set $U(\lambda)$ defined by

$$U(\lambda) = \{u \in U | f(u) \le \lambda\}$$

is compact for each $\lambda \in R$. Then f attains a minimum over U.

Proof   If $f(u) = \infty$ for all $u \in U$, then every $u \in U$ attains the minimum. If $f^* = \inf\{f(u)|u \in U\} < \infty$, let $\{\lambda_n\}$ be a scalar sequence such that $\lambda_n > \lambda_{n+1}$ for all n and $\lambda_n \to f^*$. Then the sets $U(\lambda_n)$ are nonempty, compact, and satisfy $U(\lambda_n) \supset U(\lambda_{n+1})$ for all n. Hence, the intersection $\bigcap_{n=1}^{\infty} U(\lambda_n)$ is nonempty and compact. Let u^* be any point in the intersection. Then $u^* \in U$ and $f(u^*) \le \lambda_n$ for all n, and it follows that $f(u^*) \le f^*$. Hence, f attains its minimum over U at u^*.   Q.E.D.

Direct application of Corollary 3.3.1 and Lemma 3.1 yields the following proposition.

Proposition 3.4   Let the control space C be a Hausdorff space and assume that for each $x \in S$, $\lambda \in R$, and $k = 0, 1, \ldots, N - 1$, the set

$$U_k(x, \lambda) = \{u \in U(x)|H[x, u, T^k(J_0)] \le \lambda\} \tag{16}$$

is compact. Then

$$J_N^* = T^N(J_0),$$

and there exists a uniformly N-stage optimal policy.

The compactness of the sets $U_k(x, \lambda)$ of (16) may be verified in a number of important special cases. As an illustration, we state two sets of assumptions which guarantee compactness of $U_k(x, \lambda)$ in the case of the mapping

$$H(x, u, J) = g(x, u) + \alpha(x, u)J[f(x, u)]$$

corresponding to a deterministic optimal control problem (Section 2.3.1). Assume that $0 \le \alpha(x, u)$, $b \le g(x, u) < \infty$ for some $b \in R$ and all $x \in S$, $u \in U(x)$, and take $J_0 \equiv 0$. Then compactness of $U_k(x, \lambda)$ is guaranteed if:

(a) $S = R^n$ (n-dimensional Euclidean space), $C = R^m$, $U(x) \equiv C$, f, g, and α are continuous in (x, u), and g satisfies $\lim_{k \to \infty} g(x_k, u_k) = \infty$ for every bounded sequence $\{x_k\}$ and every sequence $\{u_k\}$ for which $|u_k| \to \infty$ ($|\cdot|$ is a norm on R^m);

(b) $S = R^n$, $C = R^m$, f, g, and α are continuous, $U(x)$ is compact and nonempty for each $x \in R^n$, and $U(\cdot)$ is a continuous point-to-set mapping from R^n to the space of all nonempty compact subsets of R^m. The metric on this space is given by (3) of Appendix C.

The proof consists of verifying that the functions $T^k(J_0)$, $k = 0, 1, \ldots$, $N - 1$, are continuous, which in turn implies compactness of the sets $U_k(x, \lambda)$ of (16). Additional results along the lines of Proposition 3.4 will be given in Part II (cf. Corollary 8.5.2 and Proposition 8.6).

3.3 Application to Specific Models

We will now apply the results of the previous section to the models described in Section 2.3.

Stochastic Optimal Control—Outer Integral Formulation

Proposition 3.5 The mapping

$$H(x, u, J) = E^*\{g(x, u, w) + \alpha J[f(x, u, w)]|x, u\} \tag{17}$$

of Section 2.3.3 satisfies Assumptions F.2 and F.3.

Proof We have

$$H(x, u, J) = \int^* \{g(x, u, w) + \alpha J[f(x, u, w)]\}p(dw|x, u),$$

where $\int^*$ denotes the outer integral as in Appendix A. From Lemma A.3(b) we obtain for all $x \in S$, $u \in C$, $J \in F$, $r > 0$,

$$H(x, u, J) \leq H(x, u, J + r) \leq H(x, u, J) + 2\alpha r.$$

Hence, F.2 is satisfied.

We now show F.3. Let $J \in F$, $\{J_n\} \subset F$, $\{\varepsilon_n\} \subset R$ satisfy $\sum_{n=1}^{\infty} \varepsilon_n < \infty$, $\varepsilon_n > 0$, and for all n,

$$J = \lim_{n \to \infty} J_n, \qquad J \leq J_n, \tag{18}$$

$$J_n(x) \leq \begin{cases} J(x) + \varepsilon_n & \text{if } J(x) > -\infty, \tag{19} \\ J_{n-1}(x) + \varepsilon_n & \text{if } J(x) = -\infty, \tag{20} \end{cases}$$

$$H(x, u, J_1) < \infty, \qquad \forall x \in S, \quad u \in U(x). \tag{21}$$

Let $\{\bar{\mu}_n\} \subset M$ be such that for all n,

$$T_{\bar{\mu}_n}(J)(x) \leq \begin{cases} T(J)(x) + \varepsilon_n & \text{if } T(J)(x) > -\infty, \tag{22} \\ -1/\varepsilon_n & \text{if } T(J)(x) = -\infty, \tag{23} \end{cases}$$

$$T_{\bar{\mu}_n}(J) \leq T_{\bar{\mu}_{n-1}}(J). \tag{24}$$

Consider the set

$$A(J) = \{x \in S | \text{there exists } u \in U(x) \text{ with } p^*(\{w|J[f(x, u, w)] = -\infty\}|x, u) > 0\},$$

where p^* denotes p-outer measure (see Appendix A). Let $\bar{\mu} \in M$ be such that

$$p^*(\{w|J[f(x, \bar{\mu}(x), w)] = -\infty\}|x, \bar{\mu}(x)) > 0 \qquad \forall x \in A(J). \tag{25}$$

Define for all n

$$\mu_n(x) = \begin{cases} \bar{\mu}(x) & \text{if } x \in A(J), \\ \bar{\mu}_n(x) & \text{if } x \notin A(J). \end{cases} \tag{26}$$

We will show that $\{\mu_n\}$ thus defined satisfies the requirement of F.3 with $\beta = 1 + 2\alpha$.

For $x \in A(J)$, we have, from Corollary A.1.1 and (18)–(21),

$$\limsup_{n \to \infty} T_{\mu_n}(J_n)(x) = \limsup_{n \to \infty} T_{\bar{\mu}}(J_n)(x)$$

$$= \limsup_{n \to \infty} \int^* \{g[x, \bar{\mu}(x), w] + \alpha J_n[f(x, \bar{\mu}(x), w)]\}$$

$$\times p(dw|x, \bar{\mu}(x))$$

$$= \int^* \{g[x, \bar{\mu}(x), w] + \alpha J[f(x, \bar{\mu}(x), w)]\} p(dw|x, \bar{\mu}(x)).$$

It follows from Lemma A.3(g) and the fact that $T_{\bar{\mu}}(J)(x) < \infty$ [cf. (18) and (21)] that

$$\limsup_{n \to \infty} T_{\mu_n}(J_n)(x) = -\infty \leq T(J)(x). \tag{27}$$

For $x \notin A(J)$, we have, for all n,

$$p^*(\{w | J[f(x, \mu_n(x), w)] = -\infty\} | x, \mu_n(x)) = 0.$$

Take $B_n \in \mathcal{F}$ to contain $\{w | J[f(x, \mu_n(x), w)] = -\infty\}$ and satisfy

$$p(B_n | x, \mu_n(x)) = 0 \qquad \forall n.$$

Using Lemma A.3(e) and (b) and (19), we have

$$T_{\mu_n}(J_n)(x) = \int^* \chi_{W-B_n}(w)\{g[x, \mu_n(x), w] + \alpha J_n[f(x, \mu_n(x), w)]\} p(dw|x, \mu_n(x))$$

$$\leq \int^* \chi_{W-B_n}(w)\{g[x, \mu_n(x), w] + \alpha J[f(x, \mu_n(x), w)]\}$$

$$\times p(dw|x, \mu_n(x)) + 2\alpha\varepsilon_n$$

$$= T_{\mu_n}(J)(x) + 2\alpha\varepsilon_n. \tag{28}$$

Hence, for $x \notin A(J)$ we have from (28), (22), and (23) that

$$\limsup_{n \to \infty} T_{\mu_n}(J_n)(x) \leq \limsup_{n \to \infty} T_{\mu_n}(J)(x) = T(J)(x).$$

Combining (27) and this relation we obtain

$$\limsup_{n \to \infty} T_{\mu_n}(J_n)(x) \leq T(J)(x) \qquad \forall x \in S,$$

and since $T_{\mu_n}(J_n) \geq T(J)$ for all n, it follows that

$$\lim_{n \to \infty} T_{\mu_n}(J_n) = T(J). \tag{29}$$

If x is such that $T(J)(x) > -\infty$, it follows from (27) and (29) that we must have $x \notin A(J)$. Hence, from (28), (22), and Lemma A.3(b),

$$T_{\mu_n}(J_n)(x) \leq T_{\mu_n}(J)(x) + 2\alpha\varepsilon_n \leq T(J)(x) + (1 + 2\alpha)\varepsilon_n \quad \text{if} \quad T(J)(x) > -\infty. \tag{30}$$

If x is such that $T(J)(x) = -\infty$, there are two possibilities:

(a) $x \notin A(J)$ and
(b) $x \in A(J)$.

If $x \notin A(J)$, it follows from (28), (24), and (18) that

$$T_{\mu_n}(J_n)(x) \leq T_{\mu_n}(J)(x) + 2\alpha\varepsilon_n \leq T_{\mu_{n-1}}(J)(x) + 2\alpha\varepsilon_n$$
$$\leq T_{\mu_{n-1}}(J_{n-1})(x) + 2\alpha\varepsilon_n. \tag{31}$$

If $x \in A(J)$, then by (18)–(20) and Lemma A.3(b),

$$T_{\mu_n}(J_n)(x) = \int^* \{g[x, \bar{\mu}(x), w] + \alpha J_n[f(x, \bar{\mu}(x), w)]\} p(dw|x, \bar{\mu}(x))$$
$$\leq \int^* \{g[x, \bar{\mu}(x), w] + \alpha J_{n-1}[f(x, \bar{\mu}(x), w)]\} p(dw|x, \bar{\mu}(x)) + 2\alpha\varepsilon_n$$
$$= T_{\mu_{n-1}}(J_{n-1})(x) + 2\alpha\varepsilon_n. \tag{32}$$

It follows now from (29)–(32) that $\{\mu_n\}$ satisfies the requirement of F.3 with $\beta = 1 + 2\alpha$.   Q.E.D.

As mentioned earlier, mapping (17) contains as special cases the mappings of Sections 2.3.1 and 2.3.2. In fact, for those mappings F.1 is satisfied as well, as the reader may easily verify by using the monotone covergence theorem for ordinary integration.

Direct application of the results of the previous section and Proposition 3.5 yields the following.

Corollary 3.5.1 Let H be mapping (17) and let $J_0(x) = 0$ for $\forall x \in S$.

(a) If $J_{k,\pi}(x) < \infty$ for all $x \in S$, $\pi \in \Pi$, and $k = 1, 2, \ldots, N$, then $J_N^* = T^N(J_0)$ and for each sequence $\{\varepsilon_n\}$ with $\varepsilon_n \downarrow 0$, $\varepsilon_n > 0$ for $\forall n$, there exists a sequence of policies $\{\pi_n\}$ exhibiting $\{\varepsilon_n\}$-dominated convergence to optimality. In particular, if in addition $J_N^*(x) > -\infty$ for all $x \in S$, then for every $\varepsilon > 0$ there exists an ε-optimal policy.

(b) If $J_k^*(x) > -\infty$ for all $x \in S$, $k = 1, 2, \ldots, N$, then $J_N^* = T^N(J_0)$ and for each $\varepsilon > 0$ there exists an N-stage ε-optimal policy.

(c) Propositions 3.3 and 3.4 and Corollary 3.3.1 apply.

As Counterexample 3 in the previous section shows, it is possible to have $J_N^* \neq T^N(J_0)$ in the stochastic optimal control problem if the assumptions of parts (a) and (b) of Corollary 3.5.1 are not satisfied. Naturally for special classes of problems it may be possible to guarantee the equality $J_N^* = T^N(J_0)$ in other ways. For example, if the problem is such that existence of a uniformly N-stage optimal policy is assured, then we obtain $J_N^* = T^N(J_0)$ via Corollary 3.3.1(b). An important special case where we have $J_N^* = T^N(J_0)$ without any further assumptions is the deterministic optimal control problem of Section 2.3.1. This fact can be easily verified by the reader by using essentially the same argument as the one used to prove Proposition 3.1(a). However, if $J_N^*(x) = -\infty$ for some $x \in S$, even in the deterministic problem there may not exist an N-stage ε-optimal policy for a given ε (see Counterexample 4).

Stochastic Optimal Control—Multiplicative Cost Functional

Proposition 3.6   The mapping

$$H(x, u, J) = E\{g(x, u, w)J[f(x, u, w)]|x, u\} \tag{33}$$

of Section 2.3.4 satisfies F.1. If there exists a $b \in R$ such that $0 \leq g(x, u, w) \leq b$ for all $x \in S$, $u \in U(x)$, $w \in W$, then H satisfies F.2.

Proof   Assumption F.1 is satisfied by virtue of the monotone convergence theorem for ordinary integration (recall that W is countable). Also, if $0 \leq g(x, u, w) \leq b$, we have for every $J \in F$ and $r > 0$,

$$H(x, u, J + r) = E\{g(x, u, w)(J[f(x, u, w)] + r)|x, u\}$$
$$= E\{g(x, u, w)J[f(x, u, w)]|x, u\} + rE\{g(x, u, w)|x, u\}.$$

Thus F.2 is satisfied with $\alpha = b$.   Q.E.D.

By combining Propositions 3.6 and 3.1, we obtain the following.

Corollary 3.6.1   Let H be the mapping (33) and $J_0(x) = 1$ for $\forall x \in S$.

(a)   If $J_{k,\pi}(x) < \infty$ for all $x \in S$, $\pi \in \Pi$, $k = 1, 2, \ldots, N$, then $J_N^* = T^N(J_0)$.
(b)   If there exists a $b \in R$ such that $0 \leq g(x, u, w) \leq b$ for all $x \in S$, $u \in U(x)$, $w \in W$, then $J_N^* = T^N(J_0)$ and there exists an N-stage ε-optimal policy.
(c)   Propositions 3.3 and 3.4 and Corollary 3.3.1 apply.

We now provide two counterexamples showing that the conclusions of parts (a) and (b) of Corollary 3.6.1 may fail to hold if the corresponding assumptions are relaxed.

COUNTEREXAMPLE 5   Let everything be as in Counterexample 3 except that $C = (0, \infty)$ instead of $C = R$ (and, of course, $J_0(0) = J_0(1) = 1$ instead of $J_0(0) = J_0(1) = 0$). Then a straightforward calculation shows that $J_2^*(0) = \infty$, $J_2^*(1) = 0$, while $T^2(J_0)(0) = T^2(J_0)(1) = 0$. Here the assumption that $J_{k,\pi}(x) < \infty$ for all x, π, k is violated, and g is unbounded above.

COUNTEREXAMPLE 6 Let everything be as in Counterexample 4 except for the definition of g. Take $g(0, u) = u$ for $\forall u \in U(0)$ and $g(x, u) = x$ for $\forall u \in U(x)$ if $x \neq 0$. Then for every $\pi \in \Pi$ we have $J_{2,\pi}(x) = x\mu_1(0)$ for every $x \neq 0$, and $J_2^*(x) = 0$ for $\forall x \in S$. On the other hand, there is no two-stage ε-optimal policy for any $\varepsilon > 0$. Here the assumption $J_{k,\pi}(x) < \infty$ for all x, π, k is satisfied, and indeed we have $J_2^*(x) = T^2(J_0)(x) = 0$ for $\forall x \in S$. However, g is unbounded above.

Minimax Control

Proposition 3.7 The mapping

$$H(x, u, J) = \sup_{w \in W(x, u)} \{g(x, u, w) + \alpha J[f(x, u, w)]\} \tag{34}$$

of Section 2.3.5 satisfies F.2.

Proof We have for $r > 0$ and $J \in F$,

$$H(x, u, J + r) = \sup_{w \in W(x, u)} \{g(x, u, w) + \alpha J[f(x, u, w)] + \alpha r\}$$

$$= H(x, u, J) + \alpha r. \qquad \text{Q.E.D.}$$

Corollary 3.7.1 Let H be mapping (34) and $J_0(x) = 0$ for $\forall x \in S$.

(a) If $J_k^*(x) > -\infty$ for all $x \in S$, $k = 1, 2, \ldots, N$, then $J_N^* = T^N(J_0)$, and for each $\varepsilon > 0$ there exists an N-stage ε-optimal policy.

(b) Propositions 3.3 and 3.4 and Corollary 3.3.1 apply.

If we have $J_k^*(x) = -\infty$ for some $x \in S$, then it is clearly possible that there exists no N-stage ε-optimal policy for a given $\varepsilon > 0$, since this is true even for deterministic optimal control problems (Counterexample 4). It is also possible to construct examples very similar to Counterexample 3 which show that it is possible to have $J_N^* \neq T^N(J_0)$ if $J_k^*(x) = -\infty$ for some x and k.

Chapter 4

Infinite Horizon Models under a Contraction Assumption

4.1 General Remarks and Assumptions

Consider the infinite horizon problem

$$\text{minimize } J_\pi(x) = \lim_{N \to \infty} (T_{\mu_0} T_{\mu_1} \cdots T_{\mu_{N-1}})(J_0)(x)$$

subject to $\pi = (\mu_0, \mu_1, \ldots) \in \Pi$.

The following assumption is motivated by the contraction property of the mapping associated with discounted stochastic optimal control problems with bounded cost per stage (cf. DPSC, Chapter 6).

Assumption C (Contraction Assumption)   There is a closed subset $\bar{B}$ of the space B (Banach space of all bounded real-valued functions on S with the supremum norm) such that $J_0 \in \bar{B}$, and for all $J \in \bar{B}$, $\mu \in M$, the functions $T(J)$ and $T_\mu(J)$ belong to $\bar{B}$. Furthermore, for every $\pi = (\mu_0, \mu_1, \ldots) \in \Pi$, the limit

$$\lim_{N \to \infty} (T_{\mu_0} T_{\mu_1} \cdots T_{\mu_{N-1}})(J_0)(x) \tag{1}$$

exists and is a real number for each $x \in S$. In addition, there exists a positive integer m and scalars ρ, α, with $0 < \rho < 1$, $0 < \alpha$, such that

$$\|T_\mu(J) - T_\mu(J')\| \le \alpha\|J - J'\| \qquad \forall \mu \in M, \quad J, J' \in B, \tag{2}$$

$$\|(T_{\mu_0}T_{\mu_1}\cdots T_{\mu_{m-1}})(J) - (T_{\mu_0}T_{\mu_1}\cdots T_{\mu_{m-1}})(J')\| \le \rho\|J - J'\|$$
$$\forall \mu_0, \ldots, \mu_{m-1} \in M, \quad J, J' \in \bar{B}. \tag{3}$$

Condition (3) implies that the mapping $(T_{\mu_0}T_{\mu_1}\cdots T_{\mu_{m-1}})$ is a contraction mapping in $\bar{B}$ for all $\mu_k \in M$, $k = 0, 1, \ldots, m - 1$. When $m = 1$, the mapping T_μ is a contraction mapping for each $\mu \in M$. Note that (2) is required to hold on a possibly larger set of functions than (3). It is often convenient to take $\bar{B} = B$. This is the case for the problems of Sections 2.3.1, 2.3.2, and 2.3.5 assuming that $\alpha < 1$ and g is uniformly bounded above and below. We will demonstrate this fact in Section 4.4. In other problems such as, for example, the one of Section 2.3.3, the contraction property (3) can be verified only on a strict subset $\bar{B}$ of B.

4.2   Convergence and Existence Results

We first provide some preliminary results in the following proposition.

Proposition 4.1   Let Assumption C hold. Then:

(a)   For every $J \in \bar{B}$ and $\pi \in \Pi$, we have

$$J_\pi = \lim_{N \to \infty}(T_{\mu_0}\cdots T_{\mu_{N-1}})(J_0) = \lim_{N \to \infty}(T_{\mu_0}\cdots T_{\mu_{N-1}})(J).$$

(b)   For each positive integer N and each $J \in \bar{B}$, we have

$$\inf_{\pi \in \Pi}(T_{\mu_0}\cdots T_{\mu_{N-1}})(J) = T^N(J)$$

and, in particular,

$$J_N^* = \inf_{\pi \in \Pi}(T_{\mu_0}\cdots T_{\mu_{N-1}})(J_0) = T^N(J_0).$$

(c)   The mappings T^m and T_μ^m, $\mu \in M$, are contraction mappings in $\bar{B}$ with modulus ρ, i.e.,

$$\|T^m(J) - T^m(J')\| \le \rho\|J - J'\| \qquad \forall J, J' \in \bar{B},$$
$$\|T_\mu^m(J) - T_\mu^m(J')\| \le \rho\|J - J'\| \qquad \forall J, J' \in \bar{B}, \quad \mu \in M.$$

Proof   (a)   For any integer $k \ge 0$, write $k = nm + q$, where q, n are nonnegative integers and $0 \le q < m$. Then for any $J, J' \in \bar{B}$, using (2) and (3), we obtain

$$\|(T_{\mu_0}\cdots T_{\mu_{k-1}})(J) - (T_{\mu_0}\cdots T_{\mu_{k-1}})(J')\| \le \rho^n\alpha^q\|J - J'\|,$$

from which, by taking the limit as k (and hence also n) tends to infinity, we have

$$\lim_{k \to \infty}(T_{\mu_0} \cdots T_{\mu_{k-1}})(J_0) = \lim_{k \to \infty}(T_{\mu_0} \cdots T_{\mu_{k-1}})(J) \qquad \forall J \in \bar{B}.$$

(b)   Since $T^k(J) \in \bar{B}$ for all k by assumption, we have $T^k(J)(x) > -\infty$ for all $x \in S$ and k. For any $\varepsilon > 0$, let $\bar{\mu}_k \in M$, $k = 0, 1, \ldots, N - 1$, be such that

$$T_{\bar{\mu}_{N-1}}(J) \leq T(J) + \varepsilon,$$
$$(T_{\bar{\mu}_{N-2}}T)(J) \leq T^2(J) + \varepsilon,$$
$$\vdots$$
$$(T_{\bar{\mu}_0}T^{N-1})(J) \leq T^N(J) + \varepsilon.$$

Using (2) we obtain

$$T^N(J) \geq (T_{\bar{\mu}_0}T^{N-1})(J) - \varepsilon$$
$$\geq T_{\bar{\mu}_0}[(T_{\bar{\mu}_1}T^{N-2})(J) - \varepsilon] - \varepsilon$$
$$\geq (T_{\bar{\mu}_0}T_{\bar{\mu}_1}T^{N-2})(J) - \alpha\varepsilon - \varepsilon$$
$$\vdots$$
$$\geq (T_{\bar{\mu}_0}T_{\bar{\mu}_1} \cdots T_{\bar{\mu}_{N-1}})(J) - \left(\sum_{k=0}^{N-1} \alpha^k \varepsilon\right)$$
$$\geq \inf_{\pi \in \Pi}(T_{\mu_0} \cdots T_{\mu_{N-1}})(J) - \left(\sum_{k=0}^{N-1} \alpha^k \varepsilon\right).$$

Since $\varepsilon > 0$ is arbitrary, it follows that

$$T^N(J) \geq \inf_{\pi \in \Pi}(T_{\mu_0} \cdots T_{\mu_{N-1}})(J).$$

The reverse inequality clearly holds and the result follows.

(c)   The fact that T_μ^m is a contraction mapping is immediate from (3). We also have from (3) for all $\mu_k \in M$, $k = 0, \ldots, m - 1$, and $J, J' \in \bar{B}$,

$$(T_{\mu_0} \cdots T_{\mu_{m-1}})(J) \leq (T_{\mu_0} \cdots T_{\mu_{m-1}})(J') + \rho\|J - J'\|.$$

Taking the infimum of both sides over $\mu_k \in M$, $k = 0, 1, \ldots, m - 1$, and using part (b) we obtain

$$T^m(J) \leq T^m(J') + \rho\|J - J'\|.$$

A symmetric argument yields

$$T^m(J') \leq T^m(J) + \rho\|J - J'\|.$$

Combining the two inequalities, we obtain $\|T^m(J) - T^m(J')\| \leq \rho\|J - J'\|$.
$$\text{Q.E.D.}$$

In what follows we shall make use of the following fixed point theorem. (See [O5, p. 383]—the proof found there can be generalized to Banach spaces.)

Fixed Point Theorem If $\bar{B}$ is a closed subset of a Banach space with norm denoted by $\|\cdot\|$ and $L: \bar{B} \to \bar{B}$ is a mapping such that for some positive integer m and scalar $\rho \in (0, 1)$, $\|L^m(z) - L^m(z')\| \leq \rho \|z - z'\|$ for all $z, z' \in \bar{B}$, then L has a unique fixed point in $\bar{B}$, i.e., there exists a unique vector $z^* \in \bar{B}$ such that $L(z^*) = z^*$. Furthermore, for every $z \in \bar{B}$, we have

$$\lim_{N \to \infty} \|L^N(z) - z^*\| = 0.$$

The following proposition characterizes the optimal cost function J^* and the cost function J_μ corresponding to any stationary policy $(\mu, \mu, \ldots) \in \Pi$. It also shows that these functions can be obtained in the limit via successive application of T and T_μ on any $J \in \bar{B}$.

Proposition 4.2 Let Assumption C hold. Then:

(a) The optimal cost function J^* belongs to $\bar{B}$ and is the unique fixed point of T within $\bar{B}$, i.e., $J^* = T(J^*)$, and if $J' \in \bar{B}$ and $J' = T(J')$, then $J' = J^*$. Furthermore, if $J' \in \bar{B}$ is such that $T(J') \leq J'$, then $J^* \leq J'$, while if $J' \leq T(J')$, then $J' \leq J^*$.

(b) For every $\mu \in M$, the function J_μ belongs to $\bar{B}$ and is the unique fixed point of T_μ within $\bar{B}$.

(c) There holds

$$\lim_{N \to \infty} \|T^N(J) - J^*\| = 0 \qquad \forall J \in \bar{B},$$

$$\lim_{N \to \infty} \|T_\mu^N(J) - J_\mu\| = 0 \qquad \forall J \in \bar{B}, \quad \mu \in M.$$

Proof From part (c) of Proposition 4.1 and the fixed point theorem, we have that T and T_μ have unique fixed points in $\bar{B}$. The fixed point of T_μ is clearly J_μ, and hence part (b) is proved. Let $\tilde{J}^*$ be the fixed point of T. We have $\tilde{J}^* = T(\tilde{J}^*)$. For any $\bar{\varepsilon} > 0$, take $\bar{\mu} \in M$ such that

$$T_{\bar{\mu}}(\tilde{J}^*) \leq \tilde{J}^* + \bar{\varepsilon}.$$

From (2) it follows that $T_{\bar{\mu}}^2(\tilde{J}^*) \leq T_{\bar{\mu}}(\tilde{J}^*) + \alpha\bar{\varepsilon} \leq \tilde{J}^* + (1 + \alpha)\bar{\varepsilon}$. Continuing in the same manner, we obtain

$$T_{\bar{\mu}}^m(\tilde{J}^*) \leq \tilde{J}^* + (1 + \alpha + \cdots + \alpha^{m-1})\bar{\varepsilon}.$$

Using (3) we have

$$T_{\bar{\mu}}^{2m}(\tilde{J}^*) \leq T_{\bar{\mu}}^m(\tilde{J}^*) + \rho(1 + \alpha + \cdots + \alpha^{m-1})\bar{\varepsilon}$$
$$\leq \tilde{J}^* + (1 + \rho)(1 + \alpha + \cdots + \alpha^{m-1})\bar{\varepsilon}.$$

Proceeding similarly, we obtain, for all $k \geq 1$,

$$T_{\bar{\mu}}^{km}(\tilde{J}^*) \leq \tilde{J}^* + (1 + \rho + \cdots + \rho^{k-1})(1 + \alpha + \cdots + \alpha^{m-1})\bar{\varepsilon}.$$

Taking the limit as $k \to \infty$ and using the fact that $J_{\bar{\mu}} = \lim_{k \to \infty} T_{\bar{\mu}}^{km}(\tilde{J}^*)$, we have

$$J_{\bar{\mu}} \leq \tilde{J}^* + \frac{1}{1-\rho}(1 + \alpha + \cdots + \alpha^{m-1})\bar{\varepsilon}. \tag{4}$$

Taking $\bar{\varepsilon} = (1 - \rho)(1 + \alpha + \cdots + \alpha^{m-1})^{-1}\varepsilon$, we obtain

$$J_{\bar{\mu}} \leq \tilde{J}^* + \varepsilon.$$

Since $J^* \leq J_{\bar{\mu}}$ and $\varepsilon > 0$ is arbitrary, we see that $J^* \leq \tilde{J}^*$. We also have

$$J^* = \inf_{\pi \in \Pi} \lim_{N \to \infty} (T_{\mu_0} \cdots T_{\mu_{N-1}})(\tilde{J}^*) \geq \lim_{N \to \infty} T^N(\tilde{J}^*) = \tilde{J}^*.$$

Hence $J^* = \tilde{J}^*$ and J^* is the unique fixed point of T. Part (c) follows immediately from the fixed point theorem. The remaining part of (a) follows easily from part (c) and the monotonicity of the mapping T.   Q.E.D.

The next proposition relates to the existence and characterization of stationary optimal policies.

Proposition 4.3   Let Assumption C hold. Then:

(a)   A stationary policy $\pi^* = (\mu^*, \mu^*, \ldots) \in \Pi$ is optimal if and only if

$$T_{\mu^*}(J^*) = T(J^*).$$

Equivalently, π^* is optimal if and only if

$$T_{\mu^*}(J_{\mu^*}) = T(J_{\mu^*}).$$

(b)   If for each $x \in S$ there exists a policy which is optimal at x, then there exists a stationary optimal policy.

(c)   For any $\varepsilon > 0$, there exists a stationary ε-optimal policy, i.e., a $\pi_\varepsilon = (\mu_\varepsilon, \mu_\varepsilon, \ldots) \in \Pi$ such that

$$\|J^* - J_{\mu_\varepsilon}\| \leq \varepsilon.$$

Proof (a)   If π^* is optimal, then $J_{\mu^*} = J^*$ and the result follows from parts (a) and (b) of Proposition 4.2. If $T_{\mu^*}(J^*) = T(J^*)$, then $T_{\mu^*}(J^*) = J^*$, and hence $J_{\mu^*} = J^*$ by part (b) of Proposition 4.2. If $T_{\mu^*}(J_{\mu^*}) = T(J_{\mu^*})$, then $J_{\mu^*} = T(J_{\mu^*})$ and $J_{\mu^*} = J^*$ by part (a) of Proposition 4.2.

(b)   Let $\pi_x^* = (\mu_{0,x}^*, \mu_{1,x}^*, \ldots)$ be a policy which is optimal at $x \in S$. Then using part (a) of Proposition 4.1 and part (a) of Proposition 4.2, we have

$$J^*(x) = J_{\pi_x^*}(x) = \lim_{k \to \infty} (T_{\mu_{0,x}^*} \cdots T_{\mu_{k,x}^*})(J_0)(x)$$

$$= \lim_{k \to \infty} (T_{\mu_{0,x}^*} \cdots T_{\mu_{k,x}^*})(J^*)(x)$$

$$\geq \lim_{k \to \infty} (T_{\mu_{0,x}^*} T^k)(J^*)(x) = T_{\mu_{0,x}^*}(J^*)(x) \geq T(J^*)(x) = J^*(x).$$

Hence $T_{\mu_{0,x}^*}(J^*)(x) = T(J^*)(x)$ for each x. Define $\mu^* \in M$ by means of $\mu^*(x) = \mu_{0,x}^*(x)$. Then $T_{\mu^*}(J^*) = T(J^*)$ and the stationary policy $(\mu^*, \mu^*, \ldots)$ is optimal by part (a).

(c)   This part was proved earlier in the proof of part (a) of Proposition 4.2 [cf. (4)].   Q.E.D.

Part (a) of Proposition 4.3 shows that there exists a stationary optimal policy if and only if the infimum is attained for every $x \in S$ in the optimality equation

$$J^*(x) = T(J^*)(x) = \inf_{u \in U(x)} H(x, u, J^*).$$

Thus if the set $U(x)$ is a finite set for each $x \in S$, then there exists a stationary optimal policy. The following proposition strengthens this result and also shows that stationary optimal policies may be obtained in the limit from finite horizon optimal policies via the DP algorithm, which for any given $J \in \bar{B}$ successively computes $T(J), T^2(J), \ldots$.

Proposition 4.4   Let Assumption C hold and assume that the control space C is a Hausdorff space. Assume further that for some $J \in \bar{B}$ and some positive integer $\bar{k}$, the sets

$$U_k(x, \lambda) = \{u \in U(x) | H[x, u, T^k(J)] \leq \lambda\} \tag{5}$$

are compact for all $x \in S$, $\lambda \in R$, and $k \geq \bar{k}$. Then:

(a)   There exists a policy $\pi^* = (\mu_0^*, \mu_1^*, \ldots) \in \Pi$ attaining the infimum for all $x \in S$ and $k \geq \bar{k}$ in the DP algorithm with initial function J, i.e.,

$$(T_{\mu_k^*} T^k)(J) = T^{k+1}(J) \qquad \forall k \geq \bar{k}. \tag{6}$$

(b)   There exists a stationary optimal policy.
(c)   For every policy π^* satisfying (6), the sequence $\{\mu_k^*(x)\}$ has at least one accumulation point for each $x \in S$.
(d)   If $\mu^* : S \to C$ is such that $\mu^*(x)$ is an accumulation point of $\{\mu_k^*(x)\}$ for each $x \in S$, then the stationary policy $(\mu^*, \mu^*, \ldots)$ is optimal.

Proof   (a)   We have

$$T^{k+1}(J)(x) = \inf_{u \in U(x)} H[x, u, T^k(J)],$$

and the result follows from compactness of sets (5) and Lemma 3.1.
(b)   This part will follow immediately once we prove (c) and (d).
(c)   Let $\pi^* = (\mu_0^*, \mu_1^*, \ldots)$ satisfy (6) and define

$$\varepsilon_k = \sup\{\|T^i(J) - J^*\| \,|\, i \geq k\}, \qquad k = 0, 1, \ldots .$$

We have from (2), (6), and the fact that $T(J^*) = J^*$,

$$\|(T_{\mu_n^*}T^n)(J) - J^*\| = \|T^{n+1}(J) - T(J^*)\|$$
$$\leq \alpha\|T^n(J) - J^*\| \qquad \forall n \geq \bar{k},$$
$$\|(T_{\mu_n^*}T^n)(J) - (T_{\mu_n^*}T^k)(J)\| \leq \alpha\|T^n(J) - T^k(J)\|$$
$$\leq \alpha\|T^n(J) - J^*\| + \alpha\|T^k(J) - J^*\|$$
$$\forall n \geq \bar{k}, \quad k = 0, 1, \ldots.$$

From these two relations we obtain

$$H[x, \mu_n^*(x), T^k(J)] \leq H[x, \mu_n^*(x), T^n(J)] + 2\alpha\varepsilon_k$$
$$\leq J^*(x) + 3\alpha\varepsilon_k \qquad \forall n \geq k, \quad k \geq \bar{k}.$$

It follows that $\mu_n^*(x) \in U_k[x, J^*(x) + 3\alpha\varepsilon_k]$ for all $n \geq k$ and $k \geq \bar{k}$, and $\{\mu_n^*(x)\}$ has an accumulation point by the compactness of $U_k[x, J^*(x) + 3\alpha\varepsilon_k]$.

(d) If $\mu^*(x)$ is an accumulation point of $\{\mu_n^*(x)\}$, then $\mu^*(x) \in U_k[x, J^*(x) + 3\alpha\varepsilon_k]$ for all $k \geq \bar{k}$, or equivalently,

$$(T_{\mu^*}T^k)(J)(x) \leq J^*(x) + 3\alpha\varepsilon_k \qquad \forall x \in S, \quad k \geq \bar{k}.$$

By using (2), we have, for all k,

$$\|(T_{\mu^*}T^k)(J) - T_{\mu^*}(J^*)\| \leq \alpha\|T^k(J) - J^*\| \leq \alpha\varepsilon_k.$$

Combining the preceding two inequalities, we obtain

$$T_{\mu^*}(J^*)(x) \leq J^*(x) + 4\alpha\varepsilon_k \qquad \forall x \in S, \quad k \geq \bar{k}.$$

Since $\varepsilon_k \to 0$ [cf. Proposition 4.2(c)], we obtain $T_{\mu^*}(J^*) \leq J^*$. Using the fact that $J^* = T(J^*) \leq T_{\mu^*}(J^*)$, we obtain $T_{\mu^*}(J^*) = J^*$, which implies by Proposition 4.3 that the stationary policy $(\mu^*, \mu^*, \ldots)$ is optimal.   Q.E.D.

Examples where compactness of sets (5) can be verified were given at the end of Section 3.2. Another example is the lower semicontinuous stochastic optimal control model of Section 8.3.

4.3 Computational Methods

There are a number of computational methods which can be used to obtain the optimal cost function J^* and optimal or nearly optimal stationary policies. Naturally, these methods will be useful in practice only if they require a finite number of arithmetic operations. Thus, while "theoretical" algorithms which require an infinite number of arithmetic operations are of some interest, in practice we must modify these algorithms so that they become computationally implementable. In the algorithms we provide, we assume that for any $J \in \bar{B}$ and $\varepsilon > 0$ there is available a computational method which determines in a finite number of arithmetic operations functions $J_\varepsilon \in \bar{B}$ and $\mu_\varepsilon \in M$ such that

$$J_\varepsilon \le T(J) + \varepsilon, \qquad T_{\mu_\varepsilon}(J) \le T(J) + \varepsilon.$$

For many problems of interest, S is a compact subset of a Euclidean space, and such procedures may be based on discretization of the state space or the control space (or both) and piecewise constant approximations of various functions (see e.g., DPSC, Section 5.2). Based on this assumption (the limitations of which we fully realize), we shall provide computationally implementable versions of all "theoretical" algorithms we consider.

4.3.1   Successive Approximation

The successive approximation method consists of choosing a starting function $J \in \bar{B}$ and computing successively $T(J), T^2(J), \ldots, T^k(J), \ldots$. By part (c) of Proposition 4.2, we have $\lim_{k \to \infty} \|T^k(J) - J^*\| = 0$, and hence we obtain in the limit the optimal cost function J^*. Subsequently, stationary optimal policies (if any exist) may be obtained by minimization for each $x \in S$ in the optimality equation

$$J^*(x) = \inf_{x \in U(x)} H(x, u, J^*).$$

If this minimization cannot be carried out exactly or if only an approximation to J^* is available, then nearly optimal stationary policies can still be obtained, as the following proposition shows.

Proposition 4.5   Let Assumption C hold and assume that $\tilde{J}^* \in \bar{B}$ and $\mu \in M$ are such that

$$\|\tilde{J}^* - J^*\| \le \varepsilon_1, \qquad T_\mu(\tilde{J}^*) \le T(\tilde{J}^*) + \varepsilon_2,$$

where $\varepsilon_1 \ge 0, \varepsilon_2 \ge 0$ are scalars. Then

$$J^* \le J_\mu \le J^* + [(2\alpha\varepsilon_1 + \varepsilon_2)(1 + \alpha + \cdots + \alpha^{m-1})/(1 - \rho)].$$

Proof   Using (2) we obtain

$$T_\mu(J^*) - \alpha\varepsilon_1 \le T_\mu(\tilde{J}^*) \le T(\tilde{J}^*) + \varepsilon_2 \le T(J^*) + (\alpha\varepsilon_1 + \varepsilon_2),$$

and it follows that

$$T_\mu(J^*) \le J^* + (2\alpha\varepsilon_1 + \varepsilon_2).$$

Using this inequality and an argument identical to the one used to prove (4) in Proposition 4.2, we obtain our result.   Q.E.D.

An interesting corollary of this proposition is the following.

Corollary 4.5.1 Let Assumption C hold and assume that S is a finite set and $U(x)$ is a finite set for each $x \in S$. Then the successive approximation method yields an optimal stationary policy after a finite number of iterations in the sense that, for a given $J \in \bar{B}$, if $\pi^* = (\mu_0^*, \mu_1^*, \ldots) \in \Pi$ is such that

$$(T_{\mu_k^*} T^k)(J) = T^{k+1}(J), \qquad k = 0, 1, \ldots,$$

then there exists an integer $\bar{k}$ such that the stationary policy $(\mu_k^*, \mu_k^*, \ldots)$ is optimal for every $k \geq \bar{k}$.

Proof Under our finiteness assumptions, the set M is a finite set. Hence there exists a scalar $\varepsilon^* > 0$ such that $J_\mu \leq J^* + \varepsilon^*$ implies that $(\mu, \mu, \ldots)$ is optimal. Take $\bar{k}$ sufficiently large so that $\|T^k(J) - J^*\| \leq \bar{\varepsilon}$ for all $k \geq \bar{k}$, where $\bar{\varepsilon}$ satisfies $2\alpha\bar{\varepsilon}(1 + \alpha + \cdots + \alpha^{m-1})(1 - \rho)^{-1} \leq \varepsilon^*$, and use Proposition 4.5.   Q.E.D.

The successive approximation scheme can be sharpened considerably by making use of the monotonic error bounds of the following proposition.

Proposition 4.6 Let Assumption C hold and assume that for all scalars $r \neq 0$, $J \in B$, and $x \in S$, we have

$$\alpha_1 \leq [T^m(J + r)(x) - T^m(J)(x)]/r \leq \alpha_2, \tag{7}$$

where α_1, α_2 are two scalars satisfying $0 \leq \alpha_1 \leq \alpha_2 < 1$. Then for all $J \in \bar{B}$, $x \in S$, and $k = 1, 2, \ldots$, we have

$$T^{km}(J)(x) + b_k \leq T^{(k+1)m}(J)(x) + b_{k+1}$$
$$\leq J^*(x) \leq T^{(k+1)m}(J)(x) + \bar{b}_{k+1} \leq T^{km}(J)(x) + \bar{b}_k, \tag{8}$$

where

$$b_k = \min\left[\frac{\alpha_1}{1 - \alpha_1} d_k, \frac{\alpha_2}{1 - \alpha_2} d_k\right], \qquad \bar{b}_k = \max\left[\frac{\alpha_1}{1 - \alpha_1} \bar{d}_k, \frac{\alpha_2}{1 - \alpha_2} \bar{d}_k\right],$$

$$d_k = \inf_{x \in S}[T^{km}(J)(x) - T^{(k-1)m}(J)(x)], \qquad \bar{d}_k = \sup_{x \in S}[T^{km}(J)(x) - T^{(k-1)m}(J)(x)].$$

Note If $B = \bar{B}$ we can always take $\alpha_2 = \rho$, $\alpha_1 = 0$, but sharper bounds are obtained if scalars α_1 and α_2 with $0 < \alpha_1$ and/or $\alpha_2 < \rho$ are available.

Proof It is sufficient to prove (8) for $k = 1$, since the result for $k > 1$ then follows by replacing J by $T^{(k-1)m}(J)$. In order to simplify the notation, we assume $m = 1$. In order to prove the result for the general case simply replace T by T^m in the following arguments. We also use the notation

$$d_1 = d, \qquad \bar{d}_1 = \bar{d}, \qquad d_2 = d', \qquad \bar{d}_2 = \bar{d}'.$$

Relation (7) may also be written (for $m = 1$) as

$$T(J) + \min[\alpha_1 r, \alpha_2 r] \le T(J + r) \le T(J) + \max[\alpha_1 r, \alpha_2 r]. \tag{9}$$

We have for all $x \in S$,

$$J(x) + d \le T(J)(x). \tag{10}$$

Applying T on both sides of (10) and using (9) and (10), we obtain

$$J(x) + \min[d + \alpha_1 d, d + \alpha_2 d] \le T(J)(x) + \min[\alpha_1 d, \alpha_2 d]$$
$$\le T(J + d)(x) \le T^2(J)(x). \tag{11}$$

By adding $\min[\alpha_1^2 d, \alpha_2^2 d]$ to each side of these inequalities, using (9) (with J replaced by $T(J)$ and $r = \min[\alpha_1 d, \alpha_2 d]$), and then again (11), we obtain

$$J(x) + \min[d + \alpha_1 d + \alpha_1^2 d, d + \alpha_2 d + \alpha_2^2 d] \le T(J)(x) + \min[\alpha_1 d + \alpha_1^2 d, \alpha_2 d + \alpha_2^2 d]$$
$$\le T^2(J)(x) + \min[\alpha_1^2 d, \alpha_2^2 d]$$
$$\le T[T(J) + \min[\alpha_1 d, \alpha_2 d]](x)$$
$$\le T^3(J)(x).$$

Proceeding similarly, we have for every $k = 1, 2, \ldots,$

$$J(x) + \min\left[\sum_{i=0}^{k} \alpha_1^i d, \sum_{i=0}^{k} \alpha_2^i d\right] \le T(J)(x) + \min\left[\sum_{i=1}^{k} \alpha_1^i d, \sum_{i=1}^{k} \alpha_2^i d\right]$$
$$\le \cdots \le T^k(J)(x) + \min[\alpha_1^k d, \alpha_2^k d]$$
$$\le T^{k+1}(J)(x).$$

Taking the limit as $k \to \infty$, we have

$$J(x) + \min\left[\frac{1}{1-\alpha_1} d, \frac{1}{1-\alpha_2} d\right] \le T(J)(x) + \min\left[\frac{\alpha_1}{1-\alpha_1} d, \frac{\alpha_2}{1-\alpha_2} d\right]$$
$$\le T^2(J)(x) + \min\left[\frac{\alpha_1^2}{1-\alpha_1} d, \frac{\alpha_2^2}{1-\alpha_2} d\right]$$
$$\le J^*(x). \tag{12}$$

Also, we have from (11) that

$$\min[\alpha_1 d, \alpha_2 d] \le T^2(J)(x) - T(J)(x),$$

and by taking the infinum over $x \in S$, we see that

$$\min[\alpha_1 d, \alpha_2 d] \le d'.$$

It is easy to see that this relation implies

$$\min\left[\frac{\alpha_1^2}{1-\alpha_1}d, \frac{\alpha_2^2}{1-\alpha_2}d\right] \le \min\left[\frac{\alpha_1}{1-\alpha_1}d', \frac{\alpha_2}{1-\alpha_2}d'\right]. \tag{13}$$

Combining (12) and (13) and using the definition of b_1 and b_2, we obtain

$$T(J)(x) + b_1 \le T^2(J)(x) + b_2.$$

Also from (12) we have $T(J)(x) + b_1 \le J^*(x)$, and an identical argument shows that $T^2(J)(x) + b_2 \le J^*(x)$. Hence the left part of (8) is proved for $k=1, m=1$. The right part follows by an entirely similar argument.   Q.E.D.

Notice that the scalars b_k and $\bar{b}_k$ in (8) are readily available as a byproduct of the computation. Computational examples and further discussion of the error bounds of Proposition 4.6 may be found in DPSC, Section 6.2.

By using the error bounds of Proposition 4.6, we can obtain J^* to an arbitrary prespecified degree of accuracy in a finite number of iterations of the successive approximation method. However, we still do not have an implementable algorithm, since Proposition 4.6 requires the exact values of the functions $T^k(J)$. Approximations to $T^k(J)$ may, however, be obtained in a computationally implementable manner as shown in the following proposition, which also yields error bounds similar to those of Proposition 4.6.

Proposition 4.7   Let Assumption C hold. For a given $J \in \bar{B}$ and $\varepsilon > 0$, consider a sequence $\{J_k\} \subset \bar{B}$ satisfying

$$T(J) \le J_1 \le T(J) + \varepsilon,$$
$$T(J_k) \le J_{k+1} \le T(J_k) + \varepsilon, \qquad k = 1, 2, \ldots.$$

Then

$$\|T^{km}(J) - J_{km}\| \le \bar{\varepsilon}, \qquad k = 0, 1, \ldots, \tag{14}$$

where

$$\bar{\varepsilon} = \varepsilon(1 + \alpha + \cdots + \alpha^{m-1})/(1 - \rho)$$

Furthermore, if the assumptions of Proposition 4.6 hold, then for all $x \in S$ and $k = 1, 2, \ldots$

$$J_{km}(x) + \beta_k \le J^*(x) \le J_{km}(x) + \bar{\beta}_k,$$

where

$$\beta_k = \min\left[\frac{\alpha_1}{1-\alpha_1}\delta_k, \frac{\alpha_2}{1-\alpha_2}\delta_k\right] - \bar{\varepsilon}, \qquad \bar{\beta}_k = \max\left[\frac{\alpha_1}{1-\alpha_1}\bar{\delta}_k, \frac{\alpha_2}{1-\alpha_2}\bar{\delta}_k\right] + \bar{\varepsilon},$$

$$\delta_k = \inf_{x \in S}[J_{km}(x) - J_{(k-1)m}(x)] - 2\bar{\varepsilon}, \qquad \bar{\delta}_k = \sup_{x \in S}[J_{km}(x) - J_{(k-1)m}(x)] + 2\bar{\varepsilon}.$$

Proof We have

$$
\begin{aligned}
J_m &\leq T(J_{m-1}) + \varepsilon \leq T[T(J_{m-2}) + \varepsilon] + \varepsilon \\
&\leq T^2(J_{m-2}) + (1 + \alpha)\varepsilon \\
&\leq T^2[T(J_{m-3}) + \varepsilon] + (1 + \alpha)\varepsilon \\
&\leq T^3(J_{m-3}) + (1 + \alpha + \alpha^2)\varepsilon \\
&\vdots \\
&\leq T^{m-1}(J_1) + (1 + \alpha + \cdots + \alpha^{m-2})\varepsilon \\
&\leq T^m(J) + (1 + \alpha + \cdots + \alpha^{m-1})\varepsilon.
\end{aligned}
$$

An identical argument yields

$$ J_{2m} \leq T^m(J_m) + (1 + \alpha + \cdots + \alpha^{m-1})\varepsilon, $$

and we also have

$$ \|T^m(J_m) - T^{2m}(J)\| \leq \rho\|J_m - T^m(J)\|. $$

Using the preceding three inequalities we obtain

$$
\begin{aligned}
\|J_{2m} - T^{2m}(J)\| &\leq \|J_{2m} - T^m(J_m)\| + \|T^m(J_m) - T^{2m}(J)\| \\
&\leq (1 + \rho)\varepsilon(1 + \alpha + \cdots + \alpha^{m-1}).
\end{aligned}
$$

Proceeding similarly we obtain, for $k = 1, 2, \ldots,$

$$ \|J_{km} - T^{km}(J)\| \leq (1 + \rho + \cdots + \rho^{k-1})\varepsilon(1 + \alpha + \cdots + \alpha^{m-1}), $$

and (14) follows. The remaining part of the proposition follows by using (14) and the error bounds of Proposition 4.6.   Q.E.D.

Proposition 4.7 provides the basis for a computationally feasible algorithm to determine J^* to an arbitrary degree of accuracy, and nearly optimal stationary policies can be obtained using the result of Proposition 4.5.

4.3.2 Policy Iteration

The policy iteration algorithm in its theoretical form proceeds as follows. An initial function $\mu_0 \in M$ is chosen, the corresponding cost function J_{μ_0} is computed, and a new function $\mu_1 \in M$ satisfying $T_{\mu_1}(J_{\mu_0}) = T(J_{\mu_0})$ is obtained. More generally, given $\mu_k \in M$, one computes J_{μ_k} and a function $\mu_{k+1} \in M$ satisfying $T_{\mu_{k+1}}(J_{\mu_k}) = T(J_{\mu_k})$, and the process is repeated. When S is a finite set and $U(x)$ is a finite set for each $x \in S$, one can often compute J_{μ_k} in a finite number of arithmetic operations, and the algorithm can be carried out in a computationally implementable manner. Under these circumstances, one obtains an optimal stationary policy in a finite number of iterations, as the following proposition shows.

Proposition 4.8   Let Assumption C hold and assume that S is a finite set and $U(x)$ is a finite set for each $x \in S$. Then for any starting function $\mu_0 \in M$, the policy iteration algorithm yields a stationary optimal policy after a finite number of iterations, i.e., if $\{\mu_k\}$ is the generated sequence, there exists an integer $\overline{k}$ such that $(\mu_k, \mu_k, \ldots)$ is optimal for all $k \geq \overline{k}$.

Proof   We have, for all k,

$$T_{\mu_{k+1}}(J_{\mu_k}) = T(J_{\mu_k}) \leq T_{\mu_k}(J_{\mu_k}) = J_{\mu_k}.$$

Applying $T_{\mu_{k+1}}$ repeatedly on both sides, we obtain

$$T^N_{\mu_{k+1}}(J_{\mu_k}) \leq T^{N-1}_{\mu_{k+1}}(J_{\mu_k}) \leq \cdots \leq T_{\mu_{k+1}}(J_{\mu_k}) = T(J_{\mu_k})$$

$$\leq J_{\mu_k}, \qquad N = 1, 2, \ldots \tag{15}$$

By Proposition 4.2,

$$\lim_{N \to \infty} T^N_{\mu_{k+1}}(J_{\mu_k}) = J_{\mu_{k+1}}, \tag{16}$$

so $J_{\mu_{k+1}} \leq J_{\mu_k}$.

If $(\mu_k, \mu_k, \ldots)$ is an optimal policy, then $J_{\mu_{k+1}} = J_{\mu_k} = J^*$ and $(\mu_{k+1}, \mu_{k+1}, \ldots)$ is also optimal. Otherwise, we must have $J_{\mu_{k+1}}(x) < J_{\mu_k}(x)$ for some $x \in S$, for if $J_{\mu_{k+1}} = J_{\mu_k}$, then from (15) and (16) we have $T(J_{\mu_k}) = J_{\mu_k}$, which implies the optimality of $(\mu_k, \mu_k, \ldots)$. Hence, either $(\mu_k, \mu_k, \ldots)$ is optimal or else $(\mu_{k+1}, \mu_{k+1}, \ldots)$ is a strictly better policy. Since the set M is finite under our assumptions, the result follows.   Q.E.D.

When S and $U(x)$ are not finite sets, the policy iteration algorithm must be modified for a number of reasons. First, given μ_k, there may not exist a μ_{k+1} such that $T_{\mu_{k+1}}(J_{\mu_k}) = T(J_{\mu_k})$. Second, even if such a μ_{k+1} exists, it may not be possible to obtain $T_{\mu_{k+1}}(J_{\mu_k})$ and $J_{\mu_{k+1}}$ in a computationally implementable manner. For these reasons we consider the following *modified policy iteration algorithm.*

Step 1   Choose a function $\mu_0 \in M$ and positive scalars γ, δ, and ε.

Step 2   Given $\mu_k \in M$, find $\tilde{J}_{\mu_k} \in \overline{B}$ such that $\|\tilde{J}_{\mu_k} - J_{\mu_k}\| \leq \gamma \rho^k$.

Step 3   Find $\mu_{k+1} \in M$ such that $\|T_{\mu_{k+1}}(\tilde{J}_{\mu_k}) - T(\tilde{J}_{\mu_k})\| \leq \delta \rho^k$. If

$$\|T_{\mu_{k+1}}(\tilde{J}_{\mu_k}) - \tilde{J}_{\mu_k}\| \leq \varepsilon,$$

stop. Otherwise, replace μ_k by μ_{k+1} and return to Step 2.

Notice that Steps 2 and 3 of the algorithm are computationally implementable. The next proposition establishes the validity of the algorithm.

Proposition 4.9   Let Assumption C hold. Then the modified policy iteration algorithm terminates in a finite, say $\overline{k}$, number of iterations, and the final function $\mu_{\bar{k}}$ satisfies

$$\|J_{\mu_{\bar{k}}} - J^*\| \le \gamma \rho^{\bar{k}} + \frac{(\varepsilon + \delta\rho^{\bar{k}})(1 + \alpha + \cdots + \alpha^{m-1})}{1 - \rho}. \tag{17}$$

Proof   We first show that if the algorithm terminates at the $\bar{k}$th iteration, then (17) holds. Indeed we have

$$\|\tilde{J}_{\mu_{\bar{k}}} - J_{\mu_{\bar{k}}}\| \le \gamma \rho^{\bar{k}}, \tag{18}$$

$$\|T_{\mu_{\bar{k}+1}}(\tilde{J}_{\mu_{\bar{k}}}) - T(\tilde{J}_{\mu_{\bar{k}}})\| \le \delta\rho^{\bar{k}}, \tag{19}$$

$$\|T_{\mu_{\bar{k}+1}}(\tilde{J}_{\mu_{\bar{k}}}) - \tilde{J}_{\mu_{\bar{k}}}\| \le \varepsilon. \tag{20}$$

For any positive integer n, we have

$$\|\tilde{J}_{\mu_{\bar{k}}} - J^*\| \le \|\tilde{J}_{\mu_{\bar{k}}} - T^m(\tilde{J}_{\mu_{\bar{k}}})\| + \|T^m(\tilde{J}_{\mu_{\bar{k}}}) - T^{2m}(\tilde{J}_{\mu_{\bar{k}}})\| + \cdots$$
$$+ \|T^{(n-1)m}(\tilde{J}_{\mu_{\bar{k}}}) - T^{nm}(\tilde{J}_{\mu_{\bar{k}}})\| + \|T^{nm}(\tilde{J}_{\mu_{\bar{k}}}) - J^*\|.$$

From this relation we obtain, for all $n \ge 1$,

$$\|\tilde{J}_{\mu_{\bar{k}}} - J^*\| \le (1 + \rho + \cdots + \rho^{n-1})\|\tilde{J}_{\mu_{\bar{k}}} - T^m(\tilde{J}_{\mu_{\bar{k}}})\| + \|T^{nm}(\tilde{J}_{\mu_{\bar{k}}}) - J^*\|. \tag{21}$$

We also have

$$\|\tilde{J}_{\mu_{\bar{k}}} - T^m(\tilde{J}_{\mu_{\bar{k}}})\| \le \|\tilde{J}_{\mu_{\bar{k}}} - T(\tilde{J}_{\mu_{\bar{k}}})\| + \|T(\tilde{J}_{\mu_{\bar{k}}}) - T^2(\tilde{J}_{\mu_{\bar{k}}})\| + \cdots$$
$$+ \|T^{m-1}(\tilde{J}_{\mu_{\bar{k}}}) - T^m(\tilde{J}_{\mu_{\bar{k}}})\|,$$

from which we obtain, by using (2),

$$\|\tilde{J}_{\mu_{\bar{k}}} - T^m(\tilde{J}_{\mu_{\bar{k}}})\| \le (1 + \alpha + \cdots + \alpha^{m-1})\|\tilde{J}_{\mu_{\bar{k}}} - T(\tilde{J}_{\mu_{\bar{k}}})\|. \tag{22}$$

Combining (21) and (22), we obtain, for all $n \ge 1$,

$$\|\tilde{J}_{\mu_{\bar{k}}} - J^*\| \le (1 + \rho + \cdots + \rho^{n-1})(1 + \alpha + \cdots + \alpha^{m-1})\|\tilde{J}_{\mu_{\bar{k}}} - T(\tilde{J}_{\mu_{\bar{k}}})\|$$
$$+ \|T^{nm}(\tilde{J}_{\mu_{\bar{k}}}) - J^*\|.$$

Taking the limit as $n \to \infty$, we obtain

$$\|\tilde{J}_{\mu_{\bar{k}}} - J^*\| \le (1 + \alpha + \cdots + \alpha^{m-1})\|\tilde{J}_{\mu_{\bar{k}}} - T(\tilde{J}_{\mu_{\bar{k}}})\|/(1 - \rho). \tag{23}$$

Using (18), we also have

$$\|J_{\mu_{\bar{k}}} - J^*\| \le \|J_{\mu_{\bar{k}}} - \tilde{J}_{\mu_{\bar{k}}}\| + \|\tilde{J}_{\mu_{\bar{k}}} - J^*\| \le \gamma\rho^{\bar{k}} + \|\tilde{J}_{\mu_{\bar{k}}} - J^*\|. \tag{24}$$

From (19) and (20) we obtain

$$\|\tilde{J}_{\mu_{\bar{k}}} - T(\tilde{J}_{\mu_{\bar{k}}})\| \le \varepsilon + \delta\rho^{\bar{k}}. \tag{25}$$

By combining (23)–(25), we obtain (17).

To show that the algorithm will terminate in a finite number of iterations, assume the contrary, i.e., assume we have $\|T_{\mu_{k+1}}(\tilde{J}_{\mu_k}) - \tilde{J}_{\mu_k}\| > \varepsilon$ for all k, and the algorithm generates an infinite sequence $\{\mu_k\} \subset M$. We have, for all k,

$$\begin{aligned}
\left\| T_{\mu_{k+1}}(J_{\mu_k}) - T(J_{\mu_k}) \right\| &\leq \left\| T_{\mu_{k+1}}(J_{\mu_k}) - T_{\mu_{k+1}}(\tilde{J}_{\mu_k}) \right\| \\
&\quad + \left\| T_{\mu_{k+1}}(\tilde{J}_{\mu_k}) - T(\tilde{J}_{\mu_k}) \right\| + \left\| T(\tilde{J}_{\mu_k}) - T(J_{\mu_k}) \right\| \\
&\leq (\delta + 2\alpha\gamma)\rho^k.
\end{aligned}$$

This relation yields, for all k,

$$\begin{aligned}
T_{\mu_{k+1}}(J_{\mu_k}) \leq T(J_{\mu_k}) + (\delta + 2\alpha\gamma)\rho^k &\leq T_{\mu_k}(J_{\mu_k}) + (\delta + 2\alpha\gamma)\rho^k \\
&= J_{\mu_k} + (\delta + 2\alpha\gamma)\rho^k.
\end{aligned} \tag{26}$$

Applying $T_{\mu_{k+1}}$ to both sides of (26) and using (26) again, we obtain

$$\begin{aligned}
T^2_{\mu_{k+1}}(J_{\mu_k}) &\leq T_{\mu_{k+1}}(J_{\mu_k}) + \alpha(\delta + 2\alpha\gamma)\rho^k \leq T(J_{\mu_k}) + (1 + \alpha)(\delta + 2\alpha\gamma)\rho^k \\
&\leq J_{\mu_k} + (1 + \alpha)(\delta + 2\alpha\gamma)\rho^k.
\end{aligned}$$

Proceeding similarly, we obtain, for all k,

$$\begin{aligned}
T^m_{\mu_{k+1}}(J_{\mu_k}) &\leq T(J_{\mu_k}) + (1 + \alpha + \cdots + \alpha^{m-1})(\delta + 2\alpha\gamma)\rho^k \\
&\leq J_{\mu_k} + (1 + \alpha + \cdots + \alpha^{m-1})(\delta + 2\alpha\gamma)\rho^k.
\end{aligned}$$

Applying $T^m_{\mu_{k+1}}$ repeatedly to both sides, we obtain, for all n and k,

$$T^{nm}_{\mu_{k+1}}(J_{\mu_k}) \leq T(J_{\mu_k}) + (1 + \rho + \cdots + \rho^{n-1})(1 + \alpha + \cdots + \alpha^{m-1})(\delta + 2\alpha\gamma)\rho^k. \tag{27}$$

Denote

$$\lambda = (1 + \alpha + \cdots + \alpha^{m-1})(\delta + 2\alpha\gamma)/(1 - \rho).$$

Then by taking the limit in (27) as $n \to \infty$, we obtain

$$J_{\mu_{k+1}} \leq T(J_{\mu_k}) + \lambda\rho^k, \qquad k = 0, 1, \ldots.$$

By repeatedly applying T to both sides, we obtain

$$J_{\mu_{nm}} \leq T^m(J_{\mu_{(n-1)m}}) + \lambda(\alpha^{m-1} + \alpha^{m-2}\rho + \cdots + \rho^{m-1})\rho^{(n-1)m}. \tag{28}$$

Let $\bar{\lambda} = \lambda(\alpha^{m-1} + \alpha^{m-2}\rho + \cdots + \rho^{m-1})$. Then (28) can be written as

$$J_{\mu_{nm}} \leq T^m(J_{\mu_{(n-1)m}}) + \bar{\lambda}\rho^{(n-1)m}, \qquad n = 1, 2, \ldots. \tag{29}$$

Using (29) repeatedly, we have, for all n,

$$\begin{aligned}
J_{\mu_{nm}} &\leq T^m(J_{\mu_{(n-1)m}}) + \bar{\lambda}\rho^{(n-1)m} \\
&\leq T^m\big[T^m(J_{\mu_{(n-2)m}}) + \bar{\lambda}\rho^{(n-2)m}\big] + \bar{\lambda}\rho^{(n-1)m} \\
&\leq T^{2m}(J_{\mu_{(n-2)m}}) + \bar{\lambda}\big[\rho^{(n-1)m} + \rho\rho^{(n-2)m}\big] \\
&\;\;\vdots \\
&\leq T^{nm}(J_{\mu_0}) + \bar{\lambda}\big[\rho^{(n-1)m} + \rho\rho^{(n-2)m} + \rho^2\rho^{(n-3)m} + \cdots + \rho^{n-1}\big].
\end{aligned}$$

Since $\rho^k \rho^{(n-k-1)m} \leq \rho^{n-1}$ for all $k = 0, 1, \ldots, n-1$, this inequality yields

$$J^* \leq J_{\mu_{nm}} \leq T^{nm}(J_{\mu_0}) + n\rho^{n-1}\bar{\lambda}, \qquad n = 1, 2, \ldots.$$

Since $\lim_{n \to \infty}(n\rho^{n-1}) = 0$ and $\lim_{n \to \infty}\|T^{nm}(J_{\mu_0}) - J^*\| = 0$, the right side tends to J^* as $n \to \infty$, and it follows that

$$\lim_{n \to \infty} \|J_{\mu_{nm}} - J^*\| = 0.$$

Since by construction

$$
\begin{aligned}
\|T_{\mu_{nm+1}}(\tilde{J}_{\mu_{nm}}) - \tilde{J}_{\mu_{nm}}\| &\leq \|T_{\mu_{nm+1}}(\tilde{J}_{\mu_{nm}}) - T(\tilde{J}_{\mu_{nm}})\| \\
&\quad + \|T(\tilde{J}_{\mu_{nm}}) - T(J_{\mu_{nm}})\| + \|T(J_{\mu_{nm}}) - T(J^*)\| \\
&\quad + \|J^{*\prime} - J_{\mu_{nm}}\| + \|J_{\mu_{nm}} - \tilde{J}_{\mu_{nm}}\| \\
&\leq (\delta + \alpha\gamma + \gamma)\rho^{nm} + (1 + \alpha)\|J_{\mu_{nm}} - J^*\|,
\end{aligned}
$$

we conclude that

$$\lim_{n \to \infty} \|T_{\mu_{nm+1}}(\tilde{J}_{\mu_{nm}}) - \tilde{J}_{\mu_{nm}}\| = 0.$$

This contradicts our assumption that

$$\|T_{\mu_{k+1}}(\tilde{J}_{\mu_k}) - \tilde{J}_{\mu_k}\| > \varepsilon$$

for every k.   Q.E.D.

4.3.3   Mathematical Programming

Let the state space S be a finite set denoted by

$$S = \{x_1, x_2, \ldots, x_n\},$$

and assume $\bar{B} = B$. From part (a) of Proposition 4.2, we have that whenever $J \in B$ and $J \leq T(J)$, then $J \leq J^*$. Hence the values $J^*(x_1), \ldots, J^*(x_n)$ solve the mathematical programming problem

$$\text{maximize} \quad \sum_{i=1}^{n} \lambda_i$$

$$\text{subject to} \quad \lambda_i \leq H(x_i, u, J_\lambda), \qquad i = 1, \ldots, n, \quad u \in U(x_i),$$

where J_λ is the function taking values $J_\lambda(x_i) = \lambda_i, i = 1, \ldots, n$. If $U(x_i)$ is a finite set for each i, then this problem is a finite-dimensional (possibly nonlinear) programming problem having a finite number of inequality constraints. In fact, for the stochastic optimal control problem of Section 2.3.2, this problem is a linear programming problem, as the reader can easily verify (see also DPSC, Section 6.2). This linear program can be solved in a finite number of arithmetic operations.

4.4   Application to Specific Models

The results of the preceding sections apply in their entirety to the problems of Sections 2.3.3 and 2.3.5 if $\alpha < 1$ and g is a nonnegative bounded function. Under these circumstances Assumption C is satisfied, as we now show.

Stochastic Optimal Control—Outer Integral Formulation

Proposition 4.10   Consider the mapping

$$H(x, u, J) = E^*\{g(x, u, w) + \alpha J[f(x, u, w)]|x, u\} \tag{30}$$

of Section 2.3.3 and let $J_0(x) = 0$ for $\forall x \in S$. Assume that $\alpha < 1$ and for some $b \in R$ there holds

$$0 \le g(x, u, w) \le b \qquad \forall x \in S, \quad u \in U(x), \quad w \in W.$$

Then Assumption C is satisfied with $\bar{B}$ equal to the set of all nonnegative functions $J \in B$, the scalars in (2) and (3) equal to 2α and α, respectively, and $m = 1$.

Note   If the special cases of the mappings of Sections 2.3.1 and 2.3.2 are considered, then $\bar{B}$ can be taken equal to B, and the scalars in (2) and (3) can both be taken equal to α.

Proof   Clearly $J_0 \in \bar{B}$ and $T(J)$, $T_\mu(J) \in \bar{B}$ for all $J \in \bar{B}$ and $\mu \in M$. We also have, for any $\pi = (\mu_0, \mu_1, \ldots) \in \Pi$,

$$J_0 \le T_{\mu_0}(J_0) \le \cdots \le (T_{\mu_0} \cdots T_{\mu_k})(J_0) \le (T_{\mu_0} \cdots T_{\mu_{k+1}})(J_0) \le \cdots,$$

and hence $\lim_{N \to \infty}(T_{\mu_0} \cdots T_{\mu_{N-1}})(J_0)(x)$ exists for all $x \in S$. It is also easy to verify inductively using Lemma A.2 that

$$(T_{\mu_0} \cdots T_{\mu_{N-1}})(J_0)(x) \le \sum_{k=0}^{N-1} \alpha^k b \le b/(1 - \alpha) \qquad \forall x \in S, \quad N = 1, 2, \ldots.$$

Hence $\lim_{N \to \infty}(T_{\mu_0} \cdots T_{\mu_{N-1}})(J_0)(x)$ is a real number for every x.

We have for all $x \in S$, $J, J' \in B$, $\mu \in M$, and $w \in W$,

$$g[x, \mu(x), w] + \alpha J[f(x, \mu(x), w)] \le g[x, \mu(x), w]$$
$$+ \alpha J'[f(x, \mu(x), w)] + \alpha\|J - J'\|. \tag{31}$$

Hence, using Lemma A.3(b),

$$E^*\{g[x, \mu(x), w] + \alpha J[f(x, \mu(x), w)]|x, u\}$$
$$\le E^*\{g[x, \mu(x), w] + \alpha J'[f(x, \mu(x), w)]|x, u\} + 2\alpha\|J - J'\|. \tag{32}$$

Hence

$$T_\mu(J)(x) - T_\mu(J')(x) \le 2\alpha\|J - J'\|.$$

A symmetric argument yields $T_\mu(J')(x) - T_\mu(J)(x) \le 2\alpha\|J - J'\|$. Therefore,

$$|T_\mu(J)(x) - T_\mu(J')(x)| \le 2\alpha\|J - J'\| \qquad \forall x \in S, \qquad \mu \in M.$$

Taking the supremum of the left side over $x \in S$, we have

$$\|T_\mu(J) - T_\mu(J')\| \le 2\alpha\|J - J'\| \qquad \forall \mu \in M, \quad J, J' \in B, \tag{33}$$

which shows that (2) holds.

If $J, J' \in \bar{B}$, then from (31), Lemma A.2, and Lemma A.3(a), we obtain in place of (32)

$$E^*\{g[x, \mu(x), w] + \alpha J[f(x, \mu(x), w)]|x, u\}$$
$$\le E^*\{g[x, \mu(x), w] + \alpha J'[f(x, \mu(x), w)]|x, u\} + \alpha\|J - J'\|,$$

and proceeding as before, we obtain in place of (33)

$$\|T_\mu(J) - T_\mu(J')\| \le \alpha\|J - J'\| \qquad \forall \mu \in M, \quad J, J' \in \bar{B}.$$

This shows that (3) holds with $\rho = \alpha$.   Q.E.D.

Minimax Control

Proposition 4.11   Consider the mapping

$$H(x, u, J) = \sup_{w \in W(x, u)} \{g(x, u, w) + \alpha J[f(x, u, w)]\} \tag{34}$$

of Section 2.3.5 and let $J_0(x) = 0$ for $\forall x \in S$. Assume that $\alpha < 1$ and for some $b \in R$, there holds

$$0 \le g(x, u, w) \le b \qquad \forall x \in S, \quad u \in U(x), \quad w \in W.$$

Then Assumption C is satisfied with $\bar{B}$ equal to B, $m = 1$, and the scalars in (2) and (3) both equal to α.

Proof   The proof is entirely similar to the one of Proposition 4.10.
                                                                    Q.E.D.

For additional problems where the theory of this chapter is applicable, we refer the reader to DPSC. An example of an interesting problem where Assumption C is satisfied with $m > 1$ is the first passage problem described in Section 7.4 of DPSC.

Chapter 5

Infinite Horizon Models under Monotonicity Assumptions

5.1 General Remarks and Assumptions

Consider the infinite horizon problem

$$\text{minimize} \quad J_\pi(x) = \lim_{N \to \infty} (T_{\mu_0} T_{\mu_1} \cdots T_{\mu_{N-1}})(J_0)(x)$$

$$\text{subject to} \quad \pi = (\mu_0, \mu_1, \ldots) \in \Pi. \tag{1}$$

In this chapter we impose monotonicity assumptions on the function J_0 which guarantee that J_π is well defined for all $\pi \in \Pi$. For every result to be shown in this chapter, one of the following two assumptions will be in effect.

Assumption I (Uniform Increase Assumption) There holds

$$J_0(x) \le H(x, u, J_0) \qquad \forall x \in S, \quad u \in U(x). \tag{2}$$

Assumption D (Uniform Decrease Assumption) There holds

$$J_0(x) \ge H(x, u, J_0) \qquad \forall x \in S, \quad u \in U(x). \tag{3}$$

It is easy to see that under each of these assumptions the limit in (1) is well defined as a real number or $\pm \infty$. Indeed, in the case of Assumption I we have from (2) that

$$J_0 \leq T_{\mu_0}(J_0) \leq (T_{\mu_0} T_{\mu_1})(J_0) \leq \cdots \leq (T_{\mu_0} T_{\mu_1} \cdots T_{\mu_{N-1}})(J_0) \leq \cdots,$$

while in the case of Assumption D we have from (3) that

$$J_0 \geq T_{\mu_0}(J_0) \geq (T_{\mu_0} T_{\mu_1})(J_0) \geq \cdots \geq (T_{\mu_0} T_{\mu_1} \cdots T_{\mu_{N-1}})(J_0) \geq \cdots.$$

In both cases, the limit in (1) clearly exists in the extended real numbers for each $x \in S$.

In our analysis under Assumptions I or D we will occasionally need to assume one or more of the following continuity properties for the mapping H. Assumptions I.1 and I.2 will be used in conjunction with Assumption I, while Assumptions D.1 and D.2 will be used in conjunction with Assumption D.

Assumption I.1 If $\{J_k\} \subset F$ is a sequence satisfying $J_0 \leq J_k \leq J_{k+1}$ for all k, then

$$\lim_{k \to \infty} H(x, u, J_k) = H\left(x, u, \lim_{k \to \infty} J_k\right) \qquad \forall x \in S, \quad u \in U(x). \qquad (4)$$

Assumption I.2 There exists a scalar $\alpha > 0$ such that for all scalars $r > 0$ and functions $J \in F$ with $J_0 \leq J$, we have

$$H(x, u, J) \leq H(x, u, J + r) \leq H(x, u, J) + \alpha r \qquad \forall x \in S, \quad u \in U(x). \qquad (5)$$

Assumption D.1 If $\{J_k\} \subset F$ is a sequence satisfying $J_{k+1} \leq J_k \leq J_0$ for all k, then

$$\lim_{k \to \infty} H(x, u, J_k) = H\left(x, u, \lim_{k \to \infty} J_k\right) \qquad \forall x \in S, \quad u \in U(x). \qquad (6)$$

Assumption D.2 There exists a scalar $\alpha > 0$ such that for all scalars $r > 0$ and functions $J \in F$ with $J \leq J_0$, we have

$$H(x, u, J) - \alpha r \leq H(x, u, J - r) \leq H(x, u, J) \qquad \forall x \in S, \quad u \in U(x). \qquad (7)$$

5.2 The Optimality Equation

We first consider the question whether the optimality equation $J^* = T(J^*)$ holds. As a preliminary step we prove the following result, which is of independent interest.

Proposition 5.1 Let Assumptions I, I.1, and I.2 hold. Then given any $\varepsilon > 0$, there exists an ε-optimal policy, i.e., a $\pi_\varepsilon \in \Pi$, such that

$$J^* \leq J_{\pi_\varepsilon} \leq J^* + \varepsilon. \qquad (8)$$

Furthermore, if the scalar α in I.2 satisfies $\alpha < 1$, the policy π_ε can be taken to be stationary.

Proof Let $\{\varepsilon_k\}$ be a sequence such that $\varepsilon_k > 0$ for all k and

$$\sum_{k=0}^{\infty} \alpha^k \varepsilon_k = \varepsilon. \tag{9}$$

For each $x \in S$, consider a sequence of policies $\{\pi_k[x]\} \subset \Pi$ of the form

$$\pi_k[x] = (\mu_0^k[x], \mu_1^k[x], \ldots),$$

such that for $k = 0, 1, \ldots$

$$J_{\pi_k[x]}(x) \leq J^*(x) + \varepsilon_k \qquad \forall x \in S. \tag{10}$$

Such a sequence exists, since we have $J^*(x) > -\infty$ under our assumptions.

The (admittedly confusing) notation used here and later in the proof should be interpreted as follows. The policy $\pi_k[x] = (\mu_0^k[x], \mu_1^k[x], \ldots)$ is associated with x. Thus $\mu_i^k[x]$ denotes, for each $x \in S$ and k, a function in M, while $\mu_i^k[x](z)$ denotes the value of $\mu_i^k[x]$ at an element $z \in S$. In particular, $\mu_i^k[x](x)$ denotes the value of $\mu_i^k[x]$ at x.

Consider the functions $\bar{\mu}_k \in M$ defined by

$$\bar{\mu}_k(x) = \mu_0^k[x](x) \qquad \forall x \in S \tag{11}$$

and the functions $\bar{J}_k$ defined by

$$\bar{J}_k(x) = H\left[x, \bar{\mu}_k(x), \lim_{i \to \infty}(T_{\mu_1^k[x]} \cdots T_{\mu_i^k[x]})(J_0)\right] \qquad \forall x \in S, \quad k = 0, 1, \ldots. \tag{12}$$

By using (10), (11), I, and I.1, we obtain

$$\bar{J}_k(x) = \lim_{i \to \infty}(T_{\mu_0^k[x]} \cdots T_{\mu_i^k[x]})(J_0)(x)$$

$$= J_{\pi_k[x]}(x) \leq J^*(x) + \varepsilon_k, \qquad \forall x \in S, \quad k = 0, 1, \ldots. \tag{13}$$

We have from (12), (13), and I.2 for all $k = 1, 2, \ldots$ and $x \in S$

$$\begin{aligned}
T_{\bar{\mu}_{k-1}}(\bar{J}_k)(x) &= H[x, \bar{\mu}_{k-1}(x), \bar{J}_k] \\
&\leq H[x, \bar{\mu}_{k-1}(x), (J^* + \varepsilon_k)] \\
&\leq H[x, \bar{\mu}_{k-1}(x), J^*] + \alpha \varepsilon_k \\
&\leq H\left[x, \bar{\mu}_{k-1}(x), \lim_{i \to \infty}(T_{\mu_1^{k-1}[x]} \cdots T_{\mu_i^{k-1}[x]})(J_0)\right] + \alpha \varepsilon_k \\
&= \bar{J}_{k-1}(x) + \alpha \varepsilon_k,
\end{aligned}$$

and finally,

$$T_{\bar{\mu}_{k-1}}(\bar{J}_k) \leq \bar{J}_{k-1} + \alpha \varepsilon_k, \qquad k = 1, 2, \ldots. \tag{14}$$

Using this inequality and I.2, we obtain

$$T_{\bar{\mu}_{k-2}}[T_{\bar{\mu}_{k-1}}(\bar{J}_k)] \leq T_{\bar{\mu}_{k-2}}(\bar{J}_{k-1} + \alpha\varepsilon_k)$$
$$\leq T_{\bar{\mu}_{k-2}}(\bar{J}_{k-1}) + \alpha^2\varepsilon_k \leq \bar{J}_{k-2} + (\alpha\varepsilon_{k-1} + \alpha^2\varepsilon_k).$$

Continuing in the same manner, we obtain for $k = 1, 2, \ldots$

$$(T_{\bar{\mu}_0} \cdots T_{\bar{\mu}_{k-1}})(\bar{J}_k) \leq \bar{J}_0 + (\alpha\varepsilon_1 + \cdots + \alpha^k\varepsilon_k) \leq J^* + \left(\sum_{i=0}^{k} \alpha^i\varepsilon_i\right).$$

Since $J_0 \leq \bar{J}_k$, it follows that

$$(T_{\bar{\mu}_0} \cdots T_{\bar{\mu}_{k-1}})(J_0) \leq J^* + \left(\sum_{i=0}^{k} \alpha^i\varepsilon_i\right).$$

Denote $\pi_\varepsilon = (\bar{\mu}_0, \bar{\mu}_1, \ldots)$. Then by taking the limit in the preceding inequality and using (9), we obtain

$$J_{\pi_\varepsilon} \leq J^* + \varepsilon.$$

If $\alpha < 1$, we take $\varepsilon_k = \varepsilon(1 - \alpha)$ for all k and $\pi_k[x] = (\mu_0[x], \mu_1[x], \ldots)$ in (10). The stationary policy $\pi_\varepsilon = (\bar{\mu}, \bar{\mu}, \ldots)$, where $\bar{\mu}(x) = \mu_0[x](x)$ for all $x \in S$, satisfies $J_{\pi_\varepsilon} \leq J^* + \varepsilon$.   Q.E.D.

It is easy to see that the assumption $\alpha < 1$ is essential in order to be able to take π_ε stationary in the preceding proposition. As an example, take $S = \{0\}$, $U(0) = (0, \infty)$, $J_0(0) = 0$, $H(0, u, J) = u + J(0)$. Then $J^*(0) = 0$, but for any $\mu \in M$, we have $J_\mu(0) = \infty$.

By using Proposition 5.1 we can prove the optimality equation under I, I.1, and I.2.

Proposition 5.2 Let I, I.1, and I.2 hold. Then

$$J^* = T(J^*).$$

Furthermore, if $J' \in F$ is such that $J' \geq J_0$ and $J' \geq T(J')$, then $J' \geq J^*$.

Proof For every $\pi = (\mu_0, \mu_1, \ldots) \in \Pi$ and $x \in S$, we have, from I.1,

$$J_\pi(x) = \lim_{k \to \infty} (T_{\mu_0} T_{\mu_1} \cdots T_{\mu_k})(J_0)(x)$$

$$= T_{\mu_0}\left[\lim_{k \to \infty} (T_{\mu_1} \cdots T_{\mu_k})(J_0)\right](x) \geq T_{\mu_0}(J^*)(x) \geq T(J^*)(x).$$

By taking the infimum of the left-hand side over $\pi \in \Pi$, we obtain

$$J^* \geq T(J^*).$$

To prove the reverse inequality, let ε_1 and ε_2 be any positive scalars and let $\bar{\pi} = (\bar{\mu}_0, \bar{\mu}_1, \ldots)$ be such that

$$T_{\bar{\mu}_0}(J^*) \le T(J^*) + \varepsilon_1, \qquad J_{\bar{\pi}_1} \le J^* + \varepsilon_2,$$

where $\bar{\pi}_1 = (\bar{\mu}_1, \bar{\mu}_2, \ldots)$. Such a policy exists by Proposition 5.1. We have

$$J_{\bar{\pi}} = \lim_{k \to \infty} (T_{\bar{\mu}_0} T_{\bar{\mu}_1} \cdots T_{\bar{\mu}_k})(J_0)$$

$$= T_{\bar{\mu}_0} \left[\lim_{k \to \infty} (T_{\bar{\mu}_1} \cdots T_{\bar{\mu}_k})(J_0) \right]$$

$$= T_{\bar{\mu}_0}(J_{\bar{\pi}_1}) \le T_{\bar{\mu}_0}(J^*) + \alpha \varepsilon_2 \le T(J^*) + (\varepsilon_1 + \alpha \varepsilon_2).$$

Since $J^* \le J_{\bar{\pi}}$ and ε_1 and ε_2 can be taken arbitrarily small, it follows that

$$J^* \le T(J^*).$$

Hence $J^* = T(J^*)$.

Assume that $J' \in F$ satisfies $J' \ge J_0$ and $J' \ge T(J')$. Let $\{\varepsilon_k\}$ be any sequence with $\varepsilon_k > 0$ and consider a policy $\bar{\pi} = (\bar{\mu}_0, \bar{\mu}_1, \ldots) \in \Pi$ such that

$$T_{\bar{\mu}_k}(J') \le T(J') + \varepsilon_k, \qquad k = 0, 1, \ldots.$$

We have, from I.2,

$$J^* = \inf_{\pi \in \Pi} \lim_{k \to \infty} (T_{\mu_0} \cdots T_{\mu_k})(J_0)$$

$$\le \inf_{\pi \in \Pi} \liminf_{k \to \infty} (T_{\mu_0} \cdots T_{\mu_k})(J')$$

$$\le \liminf_{k \to \infty} (T_{\bar{\mu}_0} \cdots T_{\bar{\mu}_k})(J')$$

$$\le \liminf_{k \to \infty} (T_{\bar{\mu}_0} \cdots T_{\bar{\mu}_{k-1}})[T(J') + \varepsilon_k]$$

$$\le \liminf_{k \to \infty} (T_{\bar{\mu}_0} \cdots T_{\bar{\mu}_{k-2}} T_{\bar{\mu}_{k-1}})(J' + \varepsilon_k)$$

$$\le \liminf_{k \to \infty} [(T_{\bar{\mu}_0} \cdots T_{\bar{\mu}_{k-1}})(J') + \alpha^k \varepsilon_k]$$

$$\vdots$$

$$\le \lim_{k \to \infty} \left[T(J') + \left(\sum_{i=0}^{k} \alpha^i \varepsilon_i \right) \right] \le J' + \left(\sum_{i=0}^{\infty} \alpha^i \varepsilon_i \right).$$

Since we may choose $\sum_{i=0}^{\infty} \alpha^i \varepsilon_i$ as small as desired, it follows that $J^* \le J'$.

Q.E.D.

The following counterexamples show that I.1 and I.2 are essential in order for Proposition 5.2 to hold.

COUNTEREXAMPLE 1   Take $S = \{0, 1\}$, $C = U(0) = U(1) = (-1, 0]$, $J_0(0) = J_0(1) = -1$, $H(0, u, J) = u$ if $J(1) \leq -1$, $H(0, u, J) = 0$ if $J(1) > -1$, and $H(1, u, J) = u$. Then $(T_{\mu_0} \cdots T_{\mu_{N-1}})(J_0)(0) = 0$ and $(T_{\mu_0} \cdots T_{\mu_{N-1}})(J_0)(1) = \mu_0(1)$ for $N \geq 1$. Thus $J^*(0) = 0$, $J^*(1) = -1$, while $T(J^*)(0) = -1$, $T(J^*)(1) = -1$, and hence $J^* \neq T(J^*)$. Notice also that J_0 is a fixed point of T, while $J_0 \leq J^*$ and $J_0 \neq J^*$, so the second part of Proposition 5.2 fails when $J_0 = J'$. Here I and I.1 are satisfied, but I.2 is violated.

COUNTEREXAMPLE 2   Take $S = \{0, 1\}$, $C = U(0) = U(1) = \{0\}$, $J_0(0) = J_0(1) = 0$, $H(0, 0, J) = 0$ if $J(1) < \infty$, $H(0, 0, J) = \infty$ if $J(1) = \infty$, $H(1, 0, J) = J(1) + 1$. Then $(T_{\mu_0} \cdots T_{\mu_{N-1}})(J_0)(0) = 0$ and $(T_{\mu_0} \cdots T_{\mu_{N-1}})(J_0)(1) = N$. Thus $J^*(0) = 0$, $J^*(1) = \infty$. On the other hand, we have $T(J^*)(0) = T(J^*)(1) = \infty$ and $J^* \neq T(J^*)$. Here I and I.2 are satisfied, but I.1 is violated.

As a corollary to Proposition 5.1 we obtain the following.

Corollary 5.2.1   Let I, I.1, and I.2 hold. Then for every stationary policy $\pi = (\mu, \mu, \ldots)$, we have

$$J_\mu = T_\mu(J_\mu).$$

Furthermore, if $J' \in F$ is such that $J' \geq J_0$ and $J' \geq T_\mu(J')$, then $J' \geq J_\mu$.

Proof   Consider the variation of our problem where the control constraint set is $U_\mu(x) = \{\mu(x)\}$ rather than $U(x)$ for $\forall x \in S$. Application of Proposition 5.2 yields the result.   Q.E.D.

We now provide the counterpart of Proposition 5.2 under Assumption D.

Proposition 5.3   Let D and D.1 hold. Then

$$J^* = T(J^*).$$

Furthermore, if $J' \in F$ is such that $J' \leq J_0$ and $J' \leq T(J')$, then $J' \leq J^*$.

Proof   We first show the following lemma.

Lemma 5.1   Let D hold. Then

$$J^* = \lim_{N \to \infty} J_N^*, \tag{15}$$

where J_N^* is the optimal cost function for the N-stage problem.

Proof   Clearly we have $J^* \leq J_N^*$ for all N, and hence $J^* \leq \lim_{N \to \infty} J_N^*$. Also, for all $\pi = (\mu_0, \mu_1, \ldots) \in \Pi$, we have

$$(T_{\mu_0} \cdots T_{\mu_{N-1}})(J_0) \geq J_N^*,$$

and by taking the limit on both sides we obtain $J_\pi \geq \lim_{N \to \infty} J_N^*$, and hence $J^* \geq \lim_{N \to \infty} J_N^*$.   Q.E.D.

Proof (continued)   We return now to the proof of Proposition 5.3. An argument entirely similar to the one of the proof of Lemma 5.1 shows that under D we have for all $x \in S$

$$\lim_{N \to \infty} \inf_{u \in U(x)} H(x, u, J_N^*) = \inf_{u \in U(x)} \lim_{N \to \infty} H(x, u, J_N^*). \tag{16}$$

Using D.1, this equation yields

$$\lim_{N \to \infty} T(J_N^*) = T\left(\lim_{N \to \infty} J_N^* \right). \tag{17}$$

Since D and D.1 are equivalent to Assumption F.1' of Chapter 3, by Corollary 3.1.1 we have $J_N^* = T^N(J_0)$, from which we conclude that $T(J_N^*) = T^{N+1}(J_0) = J_{N+1}^*$. Combining this relation with (15) and (17), we obtain $J^* = T(J^*)$.

To complete the proof, let $J' \in F$ be such that $J' \leq J_0$ and $J' \leq T(J')$. Then we have

$$J^* = \inf_{\pi \in \Pi} \lim_{N \to \infty} (T_{\mu_0} \cdots T_{\mu_{N-1}})(J_0)$$

$$\geq \lim_{N \to \infty} \inf_{\pi \in \Pi} (T_{\mu_0} \cdots T_{\mu_{N-1}})(J_0)$$

$$\geq \lim_{N \to \infty} \inf_{\pi \in \Pi} (T_{\mu_0} \cdots T_{\mu_{N-1}})(J') \geq \lim_{N \to \infty} T^N(J') \geq J'.$$

Hence $J^* \geq J'$.   Q.E.D.

In Counterexamples 1 and 2 of Section 3.2, D is satisfied but D.1 is violated. In both cases we have $J^* \neq T(J^*)$, as the reader can easily verify.

A cursory examination of the proof of Proposition 5.3 reveals that the only point where we used D.1 was in establishing the relations

$$\lim_{N \to \infty} T(J_N^*) = T(\lim_{N \to \infty} J_N^*)$$

and $J_N^* = T^N(J_0)$. If these relations can be established independently, then the result of Proposition 5.3 follows. In this manner we obtain the following corollary.

Corollary 5.3.1   Let D hold and assume that D.2 holds, S is a finite set, and $J^*(x) > -\infty$ for all $x \in S$. Then $J^* = T(J^*)$. Furthermore, if $J' \in F$ is such that $J' \leq J_0$ and $J' \leq T(J')$, then $J' \leq J^*$.

Proof   A nearly verbatim repetition of the proof of Proposition 3.1(b) shows that, under D, D.2, and the assumption that $J^*(x) > -\infty$ for all $x \in S$, we have $J_N^* = T^N(J_0)$ for all $N = 1, 2, \ldots$. We will show that

$$\lim_{N \to \infty} H(x, u, J_N^*) = H\left(x, u, \lim_{N \to \infty} J_N^* \right) \qquad \forall x \in S, \quad u \in U(x).$$

Then using (16) we obtain (17), and the result follows as in the proof of Proposition 5.3. Assume the contrary, i.e., that for some $\tilde{x} \in S$, $\tilde{u} \in U(\tilde{x})$, and

$\varepsilon > 0$, there holds

$$H(\tilde{x}, \tilde{u}, J_k^*) - \varepsilon > H\left(\tilde{x}, \tilde{u}, \lim_{N \to \infty} J_N^*\right), \qquad k = 1, 2, \ldots.$$

From the finiteness of S and the fact that $J^*(x) = \lim_{N \to \infty} J_N^*(x) > -\infty$ for all x, we know that for some positive integer $\bar{k}$

$$J_k^* - (\varepsilon/\alpha) \leq \lim_{N \to \infty} J_N^* \qquad \forall k \geq \bar{k}.$$

By using D.2 we obtain for all $k \geq \bar{k}$

$$H(\tilde{x}, \tilde{u}, J_k^*) - \varepsilon \leq H(\tilde{x}, \tilde{u}, J_k^* - (\varepsilon/\alpha)) \leq H\left(\tilde{x}, \tilde{u}, \lim_{N \to \infty} J_N^*\right),$$

which contradicts the earlier inequality.   Q.E.D.

Similarly, as under Assumption I, we have the following corollary.

Corollary 5.3.2 Let D and D.1 hold. Then for every stationary policy $\pi = (\mu, \mu, \ldots)$, we have

$$J_\mu = T_\mu(J_\mu).$$

Furthermore, if $J' \in F$ is such that $J' \leq J_0$ and $J' \leq T_\mu(J')$, then $J' \leq J_\mu$.

It is worth noting that Propositions 5.2 and 5.3 can form the basis for computation of J^* when the state space S is a finite set with n elements denoted by $x_1, x_2, \ldots, x_n$. It follows from Proposition 5.2 that, under I, I.1, and I.2, $J^*(x_1), \ldots, J^*(x_n)$ solve the problem

$$\text{minimize} \quad \sum_{i=1}^n \lambda_i$$

$$\text{subject to} \quad \lambda_i \geq \inf_{u \in U(x_i)} H(x_i, u, J_\lambda), \qquad i = 1, \ldots, n,$$

$$\lambda_i \geq J_0(x_i), \qquad\qquad\qquad i = 1, \ldots, n,$$

where J_λ is the function taking values $J_\lambda(x_i) = \lambda_i$, $i = 1, \ldots, n$. Under D and D.1, or D, D.2, and the assumption that $J^*(x) > -\infty$ for $\forall x \in S$, the corresponding problem is

$$\text{maximize} \quad \sum_{i=1}^n \lambda_i$$

$$\text{subject to} \quad \lambda_i \leq H(x_i, u, J_\lambda), \qquad i = 1, \ldots, n, \quad u \in U(x_i)$$

$$\lambda_i \leq J_0(x_i), \qquad\qquad\quad i = 1, \ldots, n.$$

When $U(x_i)$ is also a finite set for all i, then the preceding problem becomes a finite-dimensional (possibly nonlinear) programming problem.

5.3   Characterization of Optimal Policies

We have the following necessary and sufficient conditions for optimality of a stationary policy.

Proposition 5.4   Let I, I.1, and I.2 hold. Then a stationary policy $\pi^* = (\mu^*, \mu^*, \ldots)$ is optimal if and only if

$$T_{\mu^*}(J^*) = T(J^*). \tag{18}$$

Furthermore, if for each $x \in S$, there exists a policy which is optimal at x, then there exists a stationary optimal policy.

Proof   If π^* is optimal, then $J_{\mu^*} = J^*$ and the result follows from Proposition 5.2 and Corollary 5.2.1. Conversely, if $T_{\mu^*}(J^*) = T(J^*)$, then since $J^* = T(J^*)$ (Proposition 5.2), it follows that $T_{\mu^*}(J^*) = J^*$. Hence by Corollary 5.2.1, $J_{\mu^*} \leq J^*$ and it follows that π^* is optimal.

If $\pi_x^* = (\mu_{0.x}^*, \mu_{1.x}^*, \ldots)$ is optimal at $x \in S$, we have, from I.1,

$$J^*(x) = J_{\pi_x^*}(x) = \lim_{k \to \infty} (T_{\mu_{0.x}^*} \cdots T_{\mu_{k.x}^*})(J_0)(x)$$

$$= T_{\mu_{0.x}^*}\left[\lim_{k \to \infty} (T_{\mu_{1.x}^*} \cdots T_{\mu_{k.x}^*})(J_0)\right](x)$$

$$\geq T_{\mu_{0.x}^*}(J^*)(x) \geq T(J^*)(x) = J^*(x).$$

Hence $T_{\mu_{0.x}^*}(J^*)(x) = T(J^*)(x)$ for all $x \in S$. Define $\mu^* \in M$ by $\mu^*(x) = \mu_{0.x}^*(x)$. Then $T_{\mu^*}(J^*) = T(J^*)$ and, by the result just proved, the stationary policy $(\mu^*, \mu^*, \ldots)$ is optimal.   Q.E.D.

Proposition 5.5   Let D and D.1 hold. Then a stationary policy $\pi^* = (\mu^*, \mu^*, \ldots)$ is optimal if and only if

$$T_{\mu^*}(J_{\mu^*}) = T(J_{\mu^*}). \tag{19}$$

Proof   If π^* is optimal, then $J_{\mu^*} = J^*$, and the result follows from Proposition 5.3 and Corollary 5.3.2. Conversely, if $T_{\mu^*}(J_{\mu^*}) = T(J_{\mu^*})$, then we obtain, from Corollary 5.3.2, that $J_{\mu^*} = T(J_{\mu^*})$, and Proposition 5.3 yields $J_{\mu^*} \leq J^*$. Hence π^* is optimal.   Q.E.D.

Examples where π^* satisfies (18) or (19) but is not optimal under D or I, respectively, are given in DPSC, Section 6.4.

Proposition 5.4 shows that there exists a stationary optimal policy if and only if the infimum in the optimality equation

$$J^*(x) = \inf_{u \in U(x)} H(x, u, J^*)$$

is attained for every $x \in S$. When the infimum is not attained for some $x \in S$, the optimality equation can still be used to yield an ε-optimal policy, which can be taken to be stationary whenever the scalar α in I.2 is strictly less than one. This is shown in the following proposition.

Proposition 5.6   Let I, I.1, and I.2 hold. Then:

(a)   If $\varepsilon > 0$, $\{\varepsilon_i\}$ satisfies $\sum_{k=0}^{\infty} \alpha^k \varepsilon_k = \varepsilon$, $\varepsilon_i > 0$, $i = 0, 1, \ldots$, and $\pi^* = (\mu_0^*, \mu_1^*, \ldots) \in \Pi$ is such that

$$T_{\mu_k^*}(J^*) \le T(J^*) + \varepsilon_k, \qquad k = 0, 1, \ldots,$$

then

$$J^* \le J_{\pi^*} \le J^* + \varepsilon.$$

(b)   If $\varepsilon > 0$, the scalar α in I.2 is strictly less than one, and $\mu^* \in M$ is such that

$$T_{\mu^*}(J^*) \le T(J^*) + \varepsilon(1 - \alpha),$$

then

$$J^* \le J_{\mu^*} \le J^* + \varepsilon.$$

Proof   (a)   Since $T(J^*) = J^*$, we have $T_{\mu_k^*}(J^*) \le J^* + \varepsilon_k$, and applying $T_{\mu_{k-1}^*}$ to both sides we obtain

$$(T_{\mu_{k-1}^*} T_{\mu_k^*})(J^*) \le T_{\mu_{k-1}^*}(J^*) + \alpha\varepsilon_k \le J^* + (\varepsilon_{k-1} + \alpha\varepsilon_k).$$

Applying $T_{\mu_{k-2}^*}$ throughout and repeating the process, we obtain, for every $k = 1, 2, \ldots$,

$$(T_{\mu_0^*} \cdots T_{\mu_k^*})(J^*) \le J^* + \left(\sum_{i=0}^{k} \alpha^i \varepsilon_i\right).$$

Since $J_0 \le J^*$, it follows that

$$(T_{\mu_0^*} \cdots T_{\mu_k^*})(J_0) \le J^* + \left(\sum_{i=0}^{k} \alpha^i \varepsilon_i\right), \qquad k = 1, 2, \ldots.$$

By taking the limit as $k \to \infty$, we obtain $J_{\pi^*} \le J^* + \varepsilon$.

(b)   This part is proved by taking $\varepsilon_k = \varepsilon(1 - \alpha)$ and $\mu_k^* = \mu^*$ for all k in the preceding proof.   Q.E.D.

A weak counterpart of part (a) of Proposition 5.6 under D is given in Corollary 5.7.1. We have been unable to obtain a counterpart of part (b) or conditions for existence of a stationary optimal policy under D.

5.4   Convergence of the Dynamic Programming Algorithm—
Existence of Stationary Optimal Policies

The DP algorithm consists of successive generation of the functions $T(J_0), T^2(J_0), \ldots$ Under Assumption I we have $T^k(J_0) \leq T^{k+1}(J_0)$ for all k, while under Assumption D we have $T^{k+1}(J_0) \leq T^k(J_0)$ for all k. Thus we can define a function $J_\infty \in F$ by

$$J_\infty(x) = \lim_{N \to \infty} T^N(J_0)(x) \qquad \forall x \in S. \tag{20}$$

We would like to investigate the question whether $J_\infty = J^*$. When Assumption D holds, the following proposition shows that we have $J_\infty = J^*$ under mild assumptions.

Proposition 5.7   Let D hold and assume that either D.1 holds or else $J_N^* = T^N(J_0)$ for all N, where J_N^* is the optimal cost function of the N-stage problem. Then

$$J_\infty = J^*.$$

Proof   By Lemma 5.1 we have that D implies $J^* = \lim_{N \to \infty} J_N^*$. Corollary 3.1.1 shows also that under our assumptions $J_N^* = T^N(J_0)$. Hence $J^* = \lim_{N \to \infty} T^N(J_0) = J_\infty$.   Q.E.D.

The following corollary is a weak counterpart of Proposition 5.1 and part (a) of Proposition 5.6 under D.

Corollary 5.7.1   Let D hold and assume that D.2 holds, S is a finite set, and $J^*(x) > -\infty$ for all $x \in S$. Then for any $\varepsilon > 0$, there exists an ε-optimal policy, i.e., a $\pi_\varepsilon \in \Pi$ such that

$$J^* \leq J_{\pi_\varepsilon} \leq J^* + \varepsilon$$

Proof   For each N, denote $\varepsilon_N = \varepsilon/2(1 + \alpha + \cdots + \alpha^{N-1})$ and let $\pi_N = \{\mu_0^N, \mu_1^N, \ldots, \mu_{N-1}^N, \mu, \mu, \ldots\}$ be such that $\mu \in M$, and for $k = 0, 1, \ldots, N - 1$, $\mu_k^N \in M$ and

$$(T_{\mu_k^N} T^{N-k-1})(J_0) \leq T^{N-k}(J_0) + \varepsilon_N.$$

We have $T_{\mu_{N-1}^N}(J_0) \leq T(J_0) + \varepsilon_N$, and applying $T_{\mu_{N-2}^N}$ to both sides, we obtain

$$(T_{\mu_{N-2}^N} T_{\mu_{N-1}^N})(J_0) \leq (T_{\mu_{N-2}^N} T)(J_0) + \alpha \varepsilon_N \leq T^2(J_0) + (1 + \alpha)\varepsilon_N.$$

Continuing in the same manner, we have

$$(T_{\mu_0^N} \cdots T_{\mu_{N-1}^N})(J_0) \leq T^N(J_0) + (1 + \alpha + \cdots + \alpha^{N-1})\varepsilon_N,$$

from which we obtain, for $N = 0, 1, \ldots,$

$$J_{\pi_N} \leq T^N(J_0) + (\varepsilon/2).$$

As in the proof of Corollary 5.3.1 our assumptions imply that $J_N^* = T^N(J_0)$ for all N, so by Proposition 5.7, $\lim_{N \to \infty} T^N(J_0) = J^*$. Let $\bar{N}$ be such that $T^{\bar{N}}(J_0) \le J^* + (\varepsilon/2)$. Such an $\bar{N}$ exists by the finiteness of S and the fact that $J^*(x) > -\infty$ for all $x \in S$. Then we obtain $J_{\pi_{\bar{N}}} \le J^* + \varepsilon$, and $\pi_{\bar{N}}$ is the desired policy.   Q.E.D.

Under Assumptions I, I.1, and I.2, the equality $J_\infty = J^*$ may fail to hold even in simple deterministic optimal control problems, as shown in the following counterexample.

COUNTEREXAMPLE 3   Let $S = [0, \infty), C = U(x) = (0, \infty)$ for $\forall x \in S, J_0(x) = 0$ for $\forall x \in S$, and

$$H(x, u, J) = x + J(2x + u)   \forall x \in S, \quad u \in U(x).$$

Then it is easy to verify that

$$J^*(x) = \inf_{\pi \in \Pi} J_\pi(x) = \infty   \forall x \in S,$$

while

$$T^N(J_0)(0) = 0, \qquad N = 1, 2, \ldots.$$

Hence $J_\infty(0) = 0$ and $J_\infty(0) < J^*(0)$.

In this example, we have $J^*(x) = \infty$ for all $x \in S$. Other examples exist where $J^* \ne J_\infty$ and $J^*(x) < \infty$ for all $x \in S$ (see [S14, p. 880]). The following preliminary result shows that in order to have $J_\infty = J^*$, it is necessary and sufficient to have $J_\infty = T(J_\infty)$.

Proposition 5.8 Let I, I.1, and I.2 hold. Then

$$J_\infty \le T(J_\infty) \le T(J^*) = J^*. \tag{21}$$

Furthermore, the equalities

$$J_\infty = T(J_\infty) = T(J^*) = J^* \tag{22}$$

hold if and only if

$$J_\infty = T(J_\infty). \tag{23}$$

Proof Clearly we have $J_\infty \le J_\pi$ for all $\pi \in \Pi$, and it follows that $J_\infty \le J^*$. Furthermore, by Proposition 5.2 we have $T(J^*) = J^*$. Also, we have, for all $k \ge 0$,

$$T(J_\infty) = \inf_{u \in U(x)} H(x, u, J_\infty) \ge \inf_{u \in U(x)} H[x, u, T^k(J_0)] = T^{k+1}(J_0).$$

Taking the limit of the right side, we obtain $T(J_\infty) \ge J_\infty$, which, combined with $J_\infty \le J^*$ and $T(J^*) = J^*$, proves (21). If (22) holds, then (23) also holds.

Conversely, let (23) hold. Then since we have $J_\infty \geq J_0$, we see from Proposition 5.2 that $J_\infty \geq J^*$, which combined with (21) proves (22).   Q.E.D.

In what follows we provide a necessary and sufficient condition for $J_\infty = T(J_\infty)$ [and hence also (22)] to hold under Assumptions I, I.1, and I.2. We subsequently obtain a useful sufficient condition for $J_\infty = T(J_\infty)$ to hold, which at the same time guarantees the existence of a stationary optimal policy.

For any $J \in F$, we denote by $E(J)$ the *epigraph* of J, i.e., the subset of SR given by

$$E(J) = \{(x, \lambda) | J(x) \leq \lambda\}. \tag{24}$$

Under I we have $T^k(J_0) \leq T^{k+1}(J_0)$ for all k and also $J_\infty = \lim_{k \to \infty} T^k(J_0)$, so it follows easily that

$$E(J_\infty) = \bigcap_{k=1}^{\infty} E[T^k(J_0)]. \tag{25}$$

Consider for each $k \geq 1$ the subset C_k of SCR given by

$$C_k = \{(x, u, \lambda) | H[x, u, T^{k-1}(J_0)] \leq \lambda, x \in S, u \in U(x)\}. \tag{26}$$

Denote by $P(C_k)$ the projection of C_k on SR, i.e.,

$$P(C_k) = \{(x, \lambda) | \exists u \in U(x) \text{ s.t. } (x, u, \lambda) \in C_k\}.^\dagger \tag{27}$$

Consider also the set

$$\overline{P(C_k)} = \{(x, \lambda) | \exists\{\lambda_n\} \text{ s.t. } \lambda_n \to \lambda, (x, \lambda_n) \in P(C_k), n = 0, 1, \ldots\}. \tag{28}$$

The set $\overline{P(C_k)}$ is obtained from $P(C_k)$ by adding for each x the point $[x, \overline{\lambda}(x)]$ where $\overline{\lambda}(x)$ is the perhaps missing end point of the half-line $\{\lambda | (x, \lambda) \in P(C_k)\}$. We have the following lemma.

Lemma 5.2   Let I hold. Then for all $k \geq 1$

$$P(C_k) \subset \overline{P(C_k)} = E[T^k(J_0)]. \tag{29}$$

Furthermore, we have

$$P(C_k) = \overline{P(C_k)} = E[T^k(J_0)] \tag{30}$$

if and only if the infimum is attained for each $x \in S$ in the equation

$$T^k(J_0)(x) = \inf_{u \in U(x)} H[x, u, T^{k-1}(J_0)]. \tag{31}$$

---

† The symbol $\exists$ means "there exists" and the initials "s.t." stand for "such that."

Proof If $(x, \lambda) \in E[T^k(J_0)]$, we have

$$T^k(J_0)(x) = \inf_{u \in U(x)} H[x, u, T^{k-1}(J_0)] \leq \lambda.$$

Let $\{\varepsilon_n\}$ be a sequence such that $\varepsilon_n > 0$, $\varepsilon_n \to 0$, and let $\{u_n\} \subset U(x)$ be a sequence such that

$$H[x, u_n, T^{k-1}(J_0)] \leq T^k(J_0)(x) + \varepsilon_n \leq \lambda + \varepsilon_n.$$

Then $(x, u_n, \lambda + \varepsilon_n) \in C_k$ and $(x, \lambda + \varepsilon_n) \in P(C_k)$ for all n. Since $\lambda + \varepsilon_n \to \lambda$, by (28) we obtain $(x, \lambda) \in \overline{P(C_k)}$. Hence

$$E[T^k(J_0)] \subset \overline{P(C_k)}. \tag{32}$$

Conversely, let $(x, \lambda) \in \overline{P(C_k)}$. Then by (26)–(28) there exists a sequence $\{\lambda_n\}$ with $\lambda_n \to \lambda$ and a corresponding sequence $\{u_n\} \subset U(x)$ such that

$$T^k(J_0)(x) \leq H[x, u_n, T^{k-1}(J_0)] \leq \lambda_n.$$

Taking the limit as $n \to \infty$, we obtain $T^k(J_0)(x) \leq \lambda$ and $(x, \lambda) \in E[T^k(J_0)]$. Hence

$$\overline{P(C_k)} \subset E[T^k(J_0)],$$

and using (32) we obtain (29).

To prove that (30) is equivalent to the attainment of the infimum in (31), assume first that the infimum is attained by $\mu^*_{k-1}(x)$ for each $x \in S$. Then for each $(x, \lambda) \in E[T^k(J_0)]$,

$$H[x, \mu^*_{k-1}(x), T^{k-1}(J_0)] \leq \lambda,$$

which implies by (27) that $(x, \lambda) \in P(C_k)$. Hence $E[T^k(J_0)] \subset P(C_k)$ and, in view of (29), we obtain (30). Conversely, if (30) holds, we have $[x, T^k(J_0)(x)] \in P(C_k)$ for every x for which $T^k(J_0)(x) < \infty$. Then by (26) and (27), there exists a $\mu^*_{k-1}(x) \in U(x)$ such that

$$H[x, \mu^*_{k-1}(x), T^{k-1}(J_0)] \leq T^k(J_0)(x) = \inf_{u \in U(x)} H[x, u, T^{k-1}(J_0)].$$

Hence the infimum in (31) is attained for all x for which $T^k(J_0)(x) < \infty$. It is also trivially attained by all $u \in U(x)$ whenever $T^k(J_0)(x) = \infty$, and the proof is complete.   Q.E.D.

Consider now the set $\bigcap_{k=1}^{\infty} C_k$ and define similarly as in (27) and (28) the sets

$$P\left(\bigcap_{k=1}^{\infty} C_k\right) = \left\{ (x, \lambda) \,\middle|\, \exists u \in U(x) \text{ s.t. } (x, u, \lambda) \in \bigcap_{k=1}^{\infty} C_k \right\}, \tag{33}$$

$$\overline{P\left(\bigcap_{k=1}^{\infty} C_k\right)} = \left\{ (x, \lambda) \,\middle|\, \exists \{\lambda_n\} \text{ s.t. } \lambda_n \to \lambda, (x, \lambda_n) \in P\left(\bigcap_{k=1}^{\infty} C_k\right) \right\}. \tag{34}$$

Using (25) and Lemma 5.2, it is easy to see that

$$P\left(\bigcap_{k=1}^{\infty} C_k\right) \subset \bigcap_{k=1}^{\infty} P(C_k) \subset \bigcap_{k=1}^{\infty} \overline{P(C_k)} = \bigcap_{k=1}^{\infty} E[T^k(J_0)] = E(J_\infty), \quad (35)$$

$$\overline{P\left(\bigcap_{k=1}^{\infty} C_k\right)} \subset \bigcap_{k=1}^{\infty} \overline{P(C_k)} = \bigcap_{k=1}^{\infty} E[T^k(J_0)] = E(J_\infty). \quad (36)$$

We have the following proposition.

Proposition 5.9   Let I, I.1, and I.2 hold. Then:

(a)   We have $J_\infty = T(J_\infty)$ (equivalently $J_\infty = J^*$) if and only if

$$\overline{P\left(\bigcap_{k=1}^{\infty} C_k\right)} = \bigcap_{k=1}^{\infty} \overline{P(C_k)}. \quad (37)$$

(b)   We have $J_\infty = T(J_\infty)$ (equivalently $J_\infty = J^*$), and the infimum in

$$J_\infty(x) = \inf_{u \in U(x)} H(x, u, J_\infty) \quad (38)$$

is attained for each $x \in S$ (equivalently there exists a stationary optimal policy) if and only if

$$P\left(\bigcap_{k=1}^{\infty} C_k\right) = \bigcap_{k=1}^{\infty} \overline{P(C_k)}. \quad (39)$$

Proof   (a)   Assume $J_\infty = T(J_\infty)$ and let (x, λ) be in $E(J_\infty)$, i.e.,

$$\inf_{u \in U(x)} H(x, u, J_\infty) = J_\infty(x) \le \lambda.$$

Let $\{\varepsilon_n\}$ be any sequence with $\varepsilon_n > 0$, $\varepsilon_n \to 0$. Then there exists a sequence $\{u_n\}$ such that

$$H(x, u_n, J_\infty) \le \lambda + \varepsilon_n, \quad n = 1, 2, \ldots,$$

and so

$$H[x, u_n, T^{k-1}(J_0)] \le \lambda + \varepsilon_n, \quad k, n = 1, 2, \ldots.$$

It follows that $(x, u_n, \lambda + \varepsilon_n) \in C_k$ for all k, n and $(x, u_n, \lambda + \varepsilon_n) \in \bigcap_{k=1}^{\infty} C_k$ for all n. Hence $(x, \lambda + \varepsilon_n) \in P(\bigcap_{k=1}^{\infty} C_k)$ for all n, and since $\lambda + \varepsilon_n \to \lambda$, we obtain $(x, \lambda) \in \overline{P(\bigcap_{k=1}^{\infty} C_k)}$. Therefore

$$E(J_\infty) \subset \overline{P\left(\bigcap_{k=1}^{\infty} C_k\right)},$$

and using (36) we obtain (37).

Conversely, let (37) hold. Then by (36) we have $\overline{P(\bigcap_{k=1}^{\infty} C_k)} = E(J_\infty)$. Let $x \in S$ be such that $J_\infty(x) < \infty$. Then $[x, J_\infty(x)] \in \overline{P(\bigcap_{k=1}^{\infty} C_k)}$, and there exists a sequence $\{\lambda_n\}$ with $\lambda_n \to J_\infty(x)$ and a sequence $\{u_n\} \subset U(x)$ such that

$$H[x, u_n, T^{k-1}(J_0)] \leq \lambda_n, \qquad k, n = 1, 2, \ldots.$$

Taking the limit with respect to k and using I.1, we obtain

$$H(x, u_n, J_\infty) \leq \lambda_n,$$

and since $T(J_\infty)(x) \leq H(x, u_n, J_\infty)$, it follows that

$$T(J_\infty)(x) \leq \lambda_n.$$

Taking the limit as $n \to \infty$, we obtain

$$T(J_\infty)(x) \leq J_\infty(x)$$

for all $x \in S$ such that $J_\infty(x) < \infty$. Since the preceding inequality holds also for all $x \in S$ with $J_\infty(x) = \infty$, we have

$$T(J_\infty) \leq J_\infty.$$

On the other hand, by Proposition 5.8, we have

$$J_\infty \leq T(J_\infty).$$

Combining the two inequalities, we obtain $J_\infty = T(J_\infty)$.

(b)   Assume $J_\infty = T(J_\infty)$ and that the infimum in (38) is attained for each $x \in S$. Then there exists a function $\mu^* \in M$ such that for each $(x, \lambda) \in E(J_\infty)$

$$H[x, \mu^*(x), J_\infty] \leq \lambda.$$

Hence $H[x, \mu^*(x), T^{k-1}(J_0)] \leq \lambda$ for $k = 1, 2, \ldots$, and we have $[x, \mu^*(x), \lambda] \in \bigcap_{k=1}^\infty C_k$. As a result, $(x, \lambda) \in P(\bigcap_{k=1}^\infty C_k)$. Hence

$$E(J_\infty) \subset P\left(\bigcap_{k=1}^\infty C_k\right),$$

and (39) follows from (35).

Conversely, let (39) hold. We have for all $x \in S$ with $J_\infty(x) < \infty$ that

$$[x, J_\infty(x)] \in E(J_\infty) = P\left(\bigcap_{k=1}^\infty C_k\right).$$

Thus there exists a $\mu^*(x) \in U(x)$ such that

$$[x, \mu^*(x), J_\infty(x)] \in \bigcap_{k=1}^\infty C_k,$$

from which we conclude that

$$H[x, \mu^*(x), T^{k-1}(J_0)] \leq J_\infty(x), \quad k = 0, 1, \ldots.$$

Taking the limit and using I.1, we see that

$$T(J_\infty)(x) \le H[x, \mu^*(x), J_\infty] \le J_\infty(x).$$

It follows that $T(J_\infty) \le J_\infty$, and since $J_\infty \le T(J_\infty)$ by Proposition 5.8, we finally obtain $J_\infty = T(J_\infty)$. Furthermore, the last inequality shows that $\mu^*(x)$ attains the infimum in (38) when $J_\infty(x) < \infty$. When $J_\infty(x) = \infty$, every $u \in U(x)$ attains the infimum, and the proof is complete.   Q.E.D.

In view of Proposition 5.8, the equality $J_\infty = T(J_\infty)$ is equivalent to the validity of interchanging infimum and limit as follows

$$J_\infty = \lim_{k \to \infty} \inf_{\pi \in \Pi} (T_{\mu_0} \cdots T_{\mu_k})(J_0) = \inf_{\pi \in \Pi} \lim_{k \to \infty} (T_{\mu_0} \cdots T_{\mu_k})(J_0) = J^*.$$

Thus Proposition 5.9 states that interchanging infimum and limit is in fact equivalent to the validity of interchanging projection and intersection in the manner of (37) or (39).

The following proposition provides a compactness assumption which guarantees that (39) holds.

Proposition 5.10   Let I, I.1, and I.2 hold and let the control space C be a Hausdorff space. Assume that there exists a nonnegative integer $\bar{k}$ such that for each $x \in S$, $\lambda \in R$, and $k \ge \bar{k}$, the set

$$U_k(x, \lambda) = \{u \in U(x) | H[x, u, T^k(J_0)] \le \lambda\} \tag{40}$$

is compact. Then

$$P\left(\bigcap_{k=1}^\infty C_k \right) = \bigcap_{k=1}^\infty \overline{P(C_k)} \tag{41}$$

and (by Propositions 5.8 and 5.9)

$$J_\infty = T(J_\infty) = T(J^*) = J^*.$$

Furthermore, there exists a stationary optimal policy.

Proof   By (35) it will be sufficient to show that

$$P\left(\bigcap_{k=1}^\infty C_k \right) \supset \bigcap_{k=1}^\infty P(C_k), \qquad \bigcap_{k=1}^\infty P(C_k) = \bigcap_{k=1}^\infty \overline{P(C_k)}. \tag{42}$$

Let (x, λ) be in $\bigcap_{k=1}^\infty P(C_k)$. Then there exists a sequence $\{u_n\} \subset U(x)$ such that

$$H[x, u_n, T^k(J_0)] \le H[x, u_n, T^n(J_0)] \le \lambda \qquad \forall n \ge k,$$

or equivalently

$$u_n \in U_k(x, \lambda) \qquad \forall n \ge k, \quad k = 0, 1, \ldots.$$

Since $U_k(x, \lambda)$ is compact for $k \geq \bar{k}$, it follows that the sequence $\{u_n\}$ has an accumulation point $\bar{u}$ and

$$\bar{u} \in U_k(x, \lambda) \qquad \forall k \geq \bar{k}.$$

But $U_0(x, \lambda) \supset U_1(x, \lambda) \supset \ldots$, so $\bar{u} \in U_k(x, \lambda)$ for $k = 0, 1, \ldots$. Hence

$$H[x, \bar{u}, T^k(J_0)] \leq \lambda, \qquad k = 0, 1, \ldots,$$

and $(x, \bar{u}, \lambda) \in \bigcap_{k=1}^{\infty} C_k$. It follows that $(x, \lambda) \in P(\bigcap_{k=1}^{\infty} C_k)$ and

$$P\left(\bigcap_{k=1}^{\infty} C_k\right) \supset \bigcap_{k=1}^{\infty} P(C_k).$$

Also, from the compactness of $U_k(x, \lambda)$ and the result of Lemma 3.1, it follows that the infimum in (31) is attained for every $x \in S$ and $k > \bar{k}$. By Lemma 5.2, $P(C_k) = \overline{P(C_k)}$ for $k > \bar{k}$, and since $P(C_1) \supset P(C_2) \supset \cdots$ and $\overline{P(C_1)} \supset \overline{P(C_2)} \supset \cdots$, we have

$$\bigcap_{k=1}^{\infty} P(C_k) = \bigcap_{k=1}^{\infty} \overline{P(C_k)}.$$

Thus (42) is proved.   Q.E.D.

The following proposition shows also that a stationary optimal policy may be obtained in the limit by means of the DP algorithm.

Proposition 5.11   Let the assumptions of Proposition 5.10 hold. Then:

(a)   There exists a policy $\pi^* = (\mu_0^*, \mu_1^*, \ldots) \in \Pi$ attaining the infimum in the DP algorithm for all $k \geq \bar{k}$, i.e.,

$$(T_{\mu_k^*} T^k)(J_0) = T^{k+1}(J_0) \qquad \forall k \geq \bar{k}. \tag{43}$$

(b)   For every policy π^* satisfying (43), the sequence $\{\mu_k^*(x)\}$ has at least one accumulation point for each $x \in S$ with $J^*(x) < \infty$.

(c)   If $\mu^* : S \to C$ is such that $\mu^*(x)$ is an accumulation point of $\{\mu_k^*(x)\}$ for all $x \in S$ with $J^*(x) < \infty$ and $\mu^*(x) \in U(x)$ for all $x \in S$ with $J^*(x) = \infty$, then the stationary policy $(\mu^*, \mu^*, \ldots)$ is optimal.

Proof   (a)   This follows from Lemma 3.1.

(b)   For any $\pi^* = (\mu_0^*, \mu_1^*, \ldots)$ satisfying (43) and $x \in S$ such that $J^*(x) < \infty$, we have

$$H[x, \mu_n^*(x), T^k(J_0)] \leq H[x, \mu_n^*(x), T^n(J_0)] \leq J^*(x) \qquad \forall k \geq \bar{k}, \ n \geq k.$$

Hence,

$$\mu_n^*(x) \in U_k[x, J^*(x)] \qquad \forall k \geq \bar{k}, \ n \geq k.$$

Since $U_k[x, J^*(x)]$ is compact, $\{\mu_n^*(x)\}$ has at least one accumulation point. Furthermore, each accumulation point $\mu^*(x)$ of $\{\mu_n^*(x)\}$ belongs to $U(x)$ and satisfies

$$H[x, \mu^*(x), T^k(J_0)] \leq J^*(x) \qquad \forall k \geq \bar{k}. \tag{44}$$

By taking the limit in (44) and using I.1, we obtain

$$H[x, \mu^*(x), J_\infty] = H[x, \mu^*(x), J^*] \leq J^*(x)$$

for all $x \in S$ with $J^*(x) < \infty$. This relation holds trivially for all $x \in S$ with $J^*(x) = \infty$. Hence $T_{\mu^*}(J^*) \leq J^* = T(J^*)$, which implies that $T_{\mu^*}(J^*) = T(J^*)$. It follows from Proposition 5.4 that $(\mu^*, \mu^*, \ldots)$ is optimal.   Q.E.D.

The compactness of the sets $U_k(x, \lambda)$ of (40) may be verified in a number of special cases, some examples of which are given at the end of Section 3.2. Another example is the lower semicontinuous model of Section 8.3, whose infinite horizon version is treated in Corollary 9.17.2.

5.5   Application to Specific Models

We now show that all the results of this chapter apply to the stochastic optimal control problems of Section 2.3.3 and 2.3.4. However, only a portion of the results apply to the minimax control problem of Section 2.3.5, since D.1 is not satisfied in the absence of additional assumptions.

Stochastic Optimal Control—Outer Integral Formulation

Proposition 5.12   Consider the mapping

$$H(x, u, J) = E^*\{g(x, u, w) + \alpha J[f(x, u, w)]|x, u\} \tag{45}$$

of Section 2.3.3 and let $J_0(x) = 0$ for $\forall x \in S$. If

$$0 \leq g(x, u, w) \qquad \forall x \in S, \quad u \in U(x), \quad w \in W, \tag{46}$$

then Assumptions I, I.1, and I.2 are satisfied with the scalar in I.2 equal to α. If

$$g(x, u, w) \leq 0 \qquad \forall x \in S, \quad u \in U(x), \quad w \in W, \tag{47}$$

then Assumptions D, D.1, and D.2 are satisfied with the scalar in D.2 equal to α.

Proof   Assumptions I and D are trivially satisfied in view of (46) or (47), respectively, and the fact that $J_0(x) = 0$ for $\forall x \in S$. Assumptions I.1 and D.1 are satisfied in view of the monotone convergence theorem for outer integration (Proposition A.1). From Lemma A.2 we have under (46) for all

$r > 0$ and $J \in F$ with $J_0 \le J$

$$H(x, u, J + r) = E^*\{g(x, u, w) + \alpha J[f(x, u, w)] + \alpha r | x, u\}$$
$$= E^*\{g(x, u, w) + \alpha J[f(x, u, w)] | x, u\} + \alpha r$$
$$= H(x, u, J) + \alpha r.$$

Hence I.2 is satisfied as stated in the proposition. Under (47), we have from Lemmas A.2 and A.3(c) that for all $r > 0$ and $J \in F$ with $J \le J_0$

$$H(x, u, J - r) = H(x, u, J) - \alpha r,$$

and D.2 is satisfied.   Q.E.D.

Thus all the results of the previous sections apply to stochastic optimal control problems with additive cost functionals. In fact, under additional countability assumptions it is possible to exploit the additive structure of these problems and obtain results relating to the existence of optimal or nearly optimal stationary policies under Assumption D. These results are stated in the following proposition. A proof of part (a) may be found in Blackwell [B10]. Proofs of parts (b) and (c) may be found in Ornstein [O4] and Frid [F2].

Proposition 5.13   Consider the mapping

$$H(x, u, J) = E\{g(x, u, w) + J[f(x, u, w)] | x, u\}$$

of Section 2.3.2 (W is countable), and let $J_0(x) = 0$ for all $x \in S$. Assume that S is countable, $J^*(x) > -\infty$ for all $x \in S$, and g satisfies

$$b \le g(x, u, w) \le 0 \qquad \forall x \in S, \quad u \in U(x), \quad w \in W,$$

where $b \in (-\infty, 0)$ is some scalar. Then:

(a)   If for each $x \in S$ there exists a policy which is optimal at x, then there exists a stationary optimal policy.

(b)   For every $\varepsilon > 0$ there exists a $\mu_\varepsilon \in M$ such that

$$J_{\mu_\varepsilon}(x) \le (1 - \varepsilon) J^*(x) \qquad \forall x \in S.$$

(c)   If there exists a scalar $\beta \in (-\infty, 0)$ such that $\beta \le J^*(x)$ for all $x \in S$, then for every $\varepsilon > 0$, there exists a stationary ε-optimal policy, i.e., a $\mu_\varepsilon \in M$ such that

$$J^* \le J_{\mu_\varepsilon} \le J^* + \varepsilon.$$

We note that the conclusion of part (a) may fail to hold if we have $J^*(x) = -\infty$ for some $x \in S$, even if S is finite, as shown by a counterexample found in Blackwell [B10]. The conclusions of parts (b) and (c) may fail to hold if S is uncountable, as shown by a counterexample due to Ornstein [O4]. The conclusion of part (c) may fail to hold if J^* is unbounded below, as shown by a counterexample due to Blackwell [B8]. We also note that the results of Proposition 5.13 can be strengthened considerably in the special case of a deterministic optimal control problem (cf. the mapping of Section 2.3.1). These results are given in Bertsekas and Shreve [B6].

Stochastic Optimal Control—Multiplicative Cost Functional

Proposition 5.14   Consider the mapping

$$H(x, u, J) = E\{g(x, u, w)J[f(x, u, w)]|x, u\}$$

of Section 2.3.4 and let $J_0(x) = 1$ for $\forall x \in S$.

(a)   If there exists a $b \in R$ such that

$$1 \leq g(x, u, w) \leq b \qquad \forall x \in S, \quad u \in U(x), \quad w \in W,$$

then Assumptions I, I.1, and I.2 are satisfied with the scalar in I.2 equal to b.

(b)   If

$$0 \leq g(x, u, w) \leq 1 \qquad \forall x \in S, \quad u \in U(x), \quad w \in W,$$

then Assumptions D, D.1, and D.2 are satisfied with the scalar in D.2 equal to unity.

Proof   This follows easily from the assumptions and the monotone convergence theorem for ordinary integration.   Q.E.D.

Minimax Control

Proposition 5.15   Consider the mapping

$$H(x, u, J) = \sup_{w \in W(x, u)} \{g(x, u, w) + \alpha J[f(x, u, w)]\}$$

of Section 2.3.5 and let $J_0(x) = 0$ for $\forall x \in S$.

(a)   If

$$0 \leq g(x, u, w) \qquad \forall x \in S, \quad u \in U(x), \quad w \in W,$$

then Assumptions I, I.1, and I.2 are satisfied with the scalar in I.2 equal to α.

(b)   If

$$g(x, u, w) \leq 0 \qquad \forall x \in S, \quad u \in U(x), \quad w \in W,$$

then Assumptions D and D.2 are satisfied with the scalar in D.2 equal to α.

Proof   The proof is entirely similar to the one of Proposition 5.12.

Q.E.D.

Chapter 6

A Generalized Abstract Dynamic Programming Model

As we discussed in Section 2.3.2, there are certain difficulties associated with the treatment of stochastic control problems in which the space W of the stochastic parameter is uncountable. For this reason we resorted to outer integration in the model of Section 2.3.3. The alternative explored in this chapter is to modify the entire framework so that policies $\pi = (\mu_0, \mu_1, \ldots)$ consist of functions μ_k from a strict subset of M—for example, those functions which are appropriately measurable. This approach is related to the one we employ in Part II. Unfortunately, however, many of our earlier results and particularly those of Chapter 5 cannot be proved within the generalized framework to be introduced. The results we provide are sufficient, however, for a satisfactory treatment of finite horizon models and infinite horizon models under contraction assumptions. Some of the results of Chapter 5 proved under Assumption D also have counterparts within the generalized framework. The reader, aided by our subsequent discussion, should be able to easily recognize these results.

Certain aspects of the framework of this chapter may seem somewhat artificial to the reader at this point. The motivation for our line of analysis stems primarily from ideas that are developed in Part II, and the reader may wish to return to this chapter after gaining some familiarity with Part II.

The results provided in the following sections are applied to a stochastic optimal control problem with multiplicative cost functional in Section 11.3. The analysis given there illustrates clearly the ideas underlying our development in this chapter.

6.1   General Remarks and Assumptions

Consider the sets S, C, $U(x)$, M, Π, and F introduced in Section 2.1. We consider in addition two subsets F^* and $\tilde{F}$ of the set F of extended real-valued functions on S satisfying

$$F^* \subset \tilde{F} \subset F$$

and a subset $\tilde{M}$ of the set M of functions $\mu : S \to C$ satisfying $\mu(x) \in U(x)$ for $\forall x \in S$. The subset of policies in Π corresponding to $\tilde{M}$ is denoted by $\tilde{\Pi}$, i.e.,

$$\tilde{\Pi} = \{(\mu_0, \mu_1, \ldots) \in \Pi \,|\, \mu_k \in \tilde{M}, \ k = 0, 1, \ldots\}.$$

In place of the mapping H of Section 2.1, we consider a mapping $H : SC\tilde{F} \to R^*$ satisfying for all $x \in S$, $u \in U(x)$, J, $J' \in \tilde{F}$, the *monotonicity assumption*

$$H(x, u, J) \le H(x, u, J') \qquad \text{if} \quad J \le J'.$$

Thus the mapping H in this chapter is of the same nature as the one of Chapters 2–5, the only difference being that H is defined on $SC\tilde{F}$ rather than on SCF. Thus if $\tilde{F}$ consists of appropriately measurable functions and H corresponds to a stochastic optimal control problem such as the one of Section 2.3.3 (with g measurable), then H can be defined in terms of ordinary integration rather than outer integration.

For $\mu \in \tilde{M}$ we consider the mapping $T_\mu : \tilde{F} \to F$ defined by

$$T_\mu(J)(x) = H[x, \mu(x), J] \qquad \forall x \in S.$$

Consider also the mapping $T : \tilde{F} \to F$ defined by

$$T(J)(x) = \inf_{u \in U(x)} H(x, u, J) \qquad \forall x \in S.$$

We are given a function $J_0 : S \to R^*$ satisfying

$$J_0 \in F^*, \qquad J_0(x) > -\infty \qquad \forall x \in S$$

and we are interested in the N-stage problem

$$\begin{aligned} \text{minimize} \quad & J_{N,\pi}(x) = (T_{\mu_0} \cdots T_{\mu_{N-1}})(J_0)(x) \\ \text{subject to} \quad & \pi \in \tilde{\Pi} \end{aligned} \tag{1}$$

and its infinite horizon counterpart

$$\text{minimize} \quad J_\pi(x) = \lim_{N \to \infty} (T_{\mu_0} \cdots T_{\mu_{N-1}})(J_0)(x)$$
$$\text{subject to} \quad \pi \in \tilde{\Pi}. \tag{2}$$

We use the notation,

$$J_N^* = \inf_{\pi \in \tilde{\Pi}} J_{N,\pi}, \qquad J^* = \inf_{\pi \in \tilde{\Pi}} J_\pi,$$

and employ the terminology of Chapter 2 regarding optimal, ε-optimal, and stationary policies, as well as sequences of polices exhibiting $\{\varepsilon_n\}$-dominated convergence to optimality.

The following conditions regarding the sets F^*, $\tilde{F}$, and $\tilde{M}$ will be assumed in every result of this chapter.

A.1   For each $x \in S$ and $u \in U(x)$, there exists a $\mu \in \tilde{M}$ such that $\mu(x) = u$. (This implies, in particular, that for every $J \in \tilde{F}$ and $x \in S$

$$T(J)(x) = \inf_{u \in U(x)} H(x, u, J) = \inf_{\mu \in \tilde{M}} H[x, \mu(x), J]).$$

A.2   For all $J \in F^*$ and $r \in R$, we have

$$T(J) \in F^*, \qquad (J + r) \in F^*.$$

A.3   For all $J \in \tilde{F}$, $\mu \in \tilde{M}$, and $r \in R$, we have

$$T_\mu(J) \in \tilde{F}, \qquad (J + r) \in \tilde{F}.$$

A.4   For each $J \in F^*$ and $\varepsilon > 0$, there exists a $\mu_\varepsilon \in \tilde{M}$ such that for all $x \in S$

$$T_{\mu_\varepsilon}(J)(x) \le \begin{cases} T(J)(x) + \varepsilon & \text{if} \quad T(J)(x) > -\infty, \\ -1/\varepsilon & \text{if} \quad T(J)(x) = -\infty. \end{cases}$$

In Section 6.3 the following assumption will also be used.

A.5   For every sequence $\{J_k\} \subset \tilde{F}$ that converges pointwise, we have $\lim_{k \to \infty} J_k \in \tilde{F}$. If, in addition, $\{J_k\} \subset F^*$, then $\lim_{k \to \infty} J_k \in F^*$.

Note that in the special case where $F^* = \tilde{F} = F$ and $\tilde{M} = M$, we obtain the framework of Chapters 2–5, and all the preceding assumptions are satisfied. Thus this chapter deals with an extension of the framework of Chapters 2–5.

We now provide some examples of sets F^*, $\tilde{F}$, and $\tilde{M}$ which are useful in connection with the mapping

$$H(x, u, J) = \int^* \{g(x, u, w) + \alpha J[f(x, u, w)]\} p(dw | x, u)$$

associated with the stochastic optimal control problem of Section 2.3.3. We take $J_0(x) = 0$ for $\forall x \in S$. The terminology employed is explained in Chapter 7.

EXAMPLE 1   Let S, C, and W be Borel spaces, $\mathscr{F}$ the Borel σ-algebra on W, f a Borel-measurable function mapping SCW into S, g a lower semi-analytic function mapping SCW into R^*, $p(dw|x, u)$ a Borel-measurable stochastic kernel on W given SC, and α a positive scalar. Let the set

$$\Gamma = \{(x, u) \in SC \,|\, x \in S, u \in U(x)\}$$

be analytic. Take F^* to be the set of extended real-valued, lower semi-analytic functions on S, $\tilde{F}$ the set of extended real-valued, universally measurable functions on S, and $\tilde{M}$ the set of universally measurable mappings from S to C with graph in Γ (i.e., $\mu \in \tilde{M}$ if μ is universally measurable and $(x, \mu(x)) \in \Gamma$ for $\forall x \in S$). This example is the subject of Chapters 8 and 9.

EXAMPLE 2   Same as Example 1 except that $\tilde{M}$ is the set of all analytically measurable mappings from S to C with graph in Γ. This example is treated in Section 11.2.

EXAMPLE 3   Same as Example 1 except for the following: $p(dw|x, u)$ and f are continuous, g real-valued, upper semicontinuous, and bounded above, Γ an open subset of SC, $\tilde{F}$ the set of extended, real-valued, Borel-measurable functions on S which are bounded above, F^* the set of extended real-valued, upper semicontinuous functions on S which are bounded above, and $\tilde{M}$ the set of Borel measurable mappings from S to C with graph in Γ. This is the upper semicontinuous model of Definition 8.8.

EXAMPLE 4   Same as Example 3 except for the following: C is in addition compact, g real-valued, lower semicontinuous, and bounded below, Γ a closed subset of SC, $\tilde{F}$ the set of extended real-valued, Borel-measurable functions on S which are bounded below, and F^* the set of extended real-valued, lower semicontinuous functions on S which are bounded below. This is a special case of the lower semicontinuous model of Definition 8.7.

All these examples satisfy Assumptions A.1–A.4 stated earlier (see also Sections 7.5 and 7.7). The first two satisfy Assumption A.5 as well.

6.2 Analysis of Finite Horizon Models

Simple modifications of some of the assumptions and proofs in Chapter 3 provide a satisfactory analysis of the finite horizon problem (1). We first modify appropriately some of the assumptions of Section 3.1.

Assumption $\tilde{F}$.2   Same in statement as Assumption F.2 of Section 3.1 except that F is replaced by $\tilde{F}$.

Assumption $\tilde{F}.3$ Same in statement as Assumption F.3 of Section 3.1 except we require that $J \in F^*$, $\{J_n\} \subset \tilde{F}$, and $\{\mu_n\} \subset \tilde{M}$, instead of $J \in F$, $\{J_n\} \subset F$, and $\{\mu_n\} \subset M$.

It can be easily seen that $\tilde{F}.2$ is satisfied in Examples 1–4 of the previous section. It is also possible to show (see the proof of Proposition 8.4) that $\tilde{F}.3$ is satisfied in Example 1, where universally measurable policies are employed.

By nearly verbatim repetition of the proofs of Proposition 3.1(b) and Proposition 3.2 we obtain the following.

Proposition 6.1   (a)   Let Assumptions A.1–A.4 and $\tilde{F}.2$ hold and assume that $J_k^*(x) > -\infty$ for all $x \in S$ and $k = 1, 2, \ldots, N$. Then

$$J_N^* = T^N(J_0),$$

and for every $\varepsilon > 0$ there exists an N-stage ε-optimal policy, i.e., a $\pi_\varepsilon \in \tilde{\Pi}$ such that

$$J_N^* \le J_{N, \pi_\varepsilon} \le J_N^* + \varepsilon.$$

(b)   Let Assumptions A.1–A.4 and $\tilde{F}.3$ hold and assume that $J_{k, \pi}(x) < \infty$ for all $x \in S$, $\pi \in \tilde{\Pi}$, and $k = 1, 2, \ldots, N$. Then

$$J_N^* = T^N(J_0).$$

Furthermore, given any sequence $\{\varepsilon_n\}$ with $\varepsilon_n \downarrow 0$, $\varepsilon_n > 0$ for $\forall n$, there exists a sequence of policies exhibiting $\{\varepsilon_n\}$-dominated covergence to optimality. In particular, if in addition $J_N^*(x) > -\infty$ for all $x \in S$, then for every $\varepsilon > 0$ there exists an ε-optimal policy.

Similarly, by modifying the proofs of Proposition 3.3 and Corollary 3.3.1(b), we obtain the following.

Proposition 6.2   Let Assumptions A.1–A.4 hold.

(a)   A policy $\pi^* = (\mu_0^*, \mu_1^*, \ldots) \in \tilde{\Pi}$ is uniformly N-stage optimal if and only if $(T_{\mu_k^*} T^{N-k-1})(J_0) = T^{N-k}(J_0)$, $k = 0, 1, \ldots, N - 1$.

(b)   If there exists a uniformly N-stage optimal policy, then

$$J_N^* = T^N(J_0).$$

Analogs of Corollary 3.3.1(a) and Proposition 3.4 can be proved if $\tilde{M}$ is rich enough so that the following assumption holds.

Exact Selection Assumption   For every $J \in F^*$, if the infimum in

$$T(J) = \inf_{u \in U(x)} H(x, u, J)$$

is attained for every $x \in S$, then there exists a $\mu^* \in \tilde{M}$ such that $T_{\mu^*}(J) = T(J)$.

In Examples 1 and 4 of the previous section the exact selection assumption is satisfied (see Propositions 7.50 and 7.33). The following proposition is proved similarly to Corollary 3.3.1(a) and Proposition 3.4.

Proposition 6.3 Let Assumptions A.1–A.4 and the exact selection assumption hold.

(a) There exists a uniformly N-stage optimal policy if and only if the infimum in the relation

$$T^{k+1}(J_0)(x) = \inf_{u \in U(x)} H[x, u, T^k(J_0)]$$

is attained for each $x \in S$ and $k = 0, 1, \ldots, N - 1$.

(b) Let the control space C be a Hausdorff space and assume that for each $x \in S$, $\lambda \in R$, and $k = 0, 1, \ldots, N - 1$, the set

$$U_k(x, \lambda) = \{u \in U(x) | H[x, u, T^k(J_0)] \le \lambda\}$$

is compact. Then

$$J_N^* = T^N(J_0),$$

and there exists a uniformly N-stage optimal policy.

6.3 Analysis of Infinite Horizon Models under a Contraction Assumption

We consider the following modified version of Assumption C of Section 4.1.

Assumption $\tilde{C}$ There is a closed subset $\bar{B}$ of the space B such that:

(a) $J_0 \in \bar{B} \cap F^*$,
(b) For all $J \in \bar{B} \cap F^*$, the function $T(J)$ belongs to $\bar{B} \cap F^*$,
(c) For all $J \in \bar{B} \cap \tilde{F}$ and $\mu \in \tilde{M}$, the function $T_\mu(J)$ belongs to $\bar{B} \cap \tilde{F}$.

Furthermore, for every $\pi = (\mu_0, \mu_1, \ldots) \in \tilde{\Pi}$, the limit

$$\lim_{N \to \infty} (T_{\mu_0} T_{\mu_1} \cdots T_{\mu_{N-1}})(J_0)(x)$$

exists and is a real number for each $x \in S$. In addition, there exists a positive integer m and scalars ρ and α with $0 < \rho < 1$, $0 < \alpha$ such that

$$\|T_\mu(J) - T_\mu(J')\| \le \alpha\|J - J'\| \qquad \forall \mu \in \tilde{M}, \quad J, J' \in B \cap \tilde{F},$$

$$\|(T_{\mu_0} T_{\mu_1} \cdots T_{\mu_{m-1}})(J) - (T_{\mu_0} T_{\mu_1} \cdots T_{\mu_{m-1}})(J')\| \le \rho\|J - J'\|$$
$$\forall \mu_0, \ldots, \mu_{m-1} \in \tilde{M}, \quad J, J' \in \bar{B} \cap \tilde{F}.$$

If Assumptions A.1–A.5 and $\tilde{C}$ are made, then almost all the results of Chapter 3 have counterparts within our extended framework. The key fact is that, since $\tilde{F}$ and F^* are closed under pointwise limits (Assumption A.5), it follows that $B \cap \tilde{F}$, $\bar{B} \cap \tilde{F}$, $B \cap F^*$, and $\bar{B} \cap F^*$ are closed subsets of B. This is true in view of the fact that convergence of a sequence in B (i.e., in sup norm) implies pointwise convergence. As a result the contraction mapping fixed point theorem can be used in exactly the same manner as in Chapter 3 to establish that, for each $\mu \in \tilde{M}$, J_μ is the unique fixed point of T_μ in $\bar{B} \cap \tilde{F}$ and J^* is the unique fixed point of T in $\bar{B} \cap F^*$. Only the modified policy iteration algorithm and the associated Proposition 4.9 have no counterparts in this extended framework. The reason is that our assumptions do not guarantee that Step 3 of the policy iteration algorithm can be carried out. Rather than provide a complete list of the analogs of all propositions in Chapter 4 we state selectively and without proof some of the main results that can be obtained within the extended framework.

Proposition 6.4 Let Assumptions A.1–A.5 and $\tilde{C}$ hold.

(a) The function J^* belongs to $\bar{B} \cap F^*$ and is the unique fixed point of T within $\bar{B} \cap F^*$. Furthermore, if $J' \in \bar{B} \cap F^*$ is such that $T(J') \le J'$, then $J^* \le J'$, while if $J' \le T(J')$, then $J' \le J^*$.

(b) For every $\mu \in \tilde{M}$, the function J_μ belongs to $\bar{B} \cap \tilde{F}$ and is the unique fixed point of T_μ within $\bar{B} \cap \tilde{F}$.

(c) There holds

$$\lim_{N \to \infty} \left\| T^N(J) - J^* \right\| = 0 \qquad \forall J \in \bar{B} \cap F^*,$$

$$\lim_{N \to \infty} \left\| T_\mu^N(J) - J_\mu \right\| = 0 \qquad \forall J \in \bar{B} \cap \tilde{F}, \quad \mu \in \tilde{M}.$$

(d) A stationary policy $\pi^* = (\mu^*, \mu^*, \ldots) \in \tilde{\Pi}$ is optimal if and only if

$$T_{\mu^*}(J^*) = T(J^*).$$

Equivalently, π^* is optimal if and only if $J_{\mu^*} \in \bar{B} \cap F^*$ and

$$T_{\mu^*}(J_{\mu^*}) = T(J_{\mu^*}).$$

(e) For any $\varepsilon > 0$, there exists a stationary ε-optimal policy, i.e., a $\pi_\varepsilon = (\mu_\varepsilon, \mu_\varepsilon, \ldots) \in \tilde{\Pi}$ such that

$$\left\| J^* - J_{\mu_\varepsilon} \right\| \le \varepsilon.$$

Proposition 6.5 Let Assumptions A.1–A.5 and $\tilde{C}$ hold. Assume further that the exact selection assumption of the previous section holds.

(a) If for each $x \in S$ there exists a policy which is optimal at x, then there exists an optimal stationary policy.

(b)   Let C be a Hausdorff space. If for some $J \in \bar{B} \cap F^*$ and for some positive integer $\bar{k}$ the set

$$U_k(x, \lambda) = \{u \in U(x) | H[x, u, T^k(J)] \leq \lambda\}$$

is compact for all $x \in S$, $\lambda \in R$, and $k \geq \bar{k}$, then there exists an optimal stationary policy.

Part II

Stochastic Optimal Control Theory

Chapter 7

Borel Spaces and Their Probability Measures

This chapter provides the mathematical background required for analysis of the dynamic programming models of the subsequent chapters. The key concept, which is developed in Section 7.3 with the aid of the topological concepts discussed in Section 7.2, is that of a Borel space. In Section 7.4 the set of probability measures on a Borel space is shown to be itself a Borel space, and the relationships between these two spaces are explored. Our general framework for dynamic programming hinges on the properties of analytic sets collected in Section 7.6 and used in Section 7.7 to define and characterize lower semianalytic functions. These functions result from executing the dynamic programming algorithm, so we will want to measurably select at or near their infima to construct optimal or nearly optimal policies. The possibilities for this are also discussed in Section 7.7. A similar analysis in a more specialized case is contained in Section 7.5, which is presented first for pedagogical reasons.

Our presentation is aimed at the reader who is acquainted with the basic notions of topology and measure theory, but is unfamiliar with some of the specialized results relating to separable metric spaces and probability measures on their Borel σ-algebras.

7.1 Notation

We collect here for easy reference many of the symbols used in Part II.

Operations on Sets

Let A and B be subsets of a space X. The *complement* of A in X is denoted by A^c. The *set-theoretic difference* $A - B$ is $A \cap B^c$. We will sometimes write $X - A$ is place of A^c. The *symmetric difference* $A \triangle B$ is $(A - B) \cup (B - A)$. If X is a topological space, $\bar{A}$ will denote the *closure* of A. If $A_1, A_2, \ldots$ is a sequence of sets such that $A_1 \subset A_2 \subset \cdots$ and $A = \bigcup_{n=1}^{\infty} A_n$, we write $A_n \uparrow A$. If $A_1 \supset A_2 \supset \cdots$ and $A = \bigcap_{n=1}^{\infty} A_n$, we write $A_n \downarrow A$. If $X_1, X_2, \ldots$ is a sequence of spaces, the *Cartesian products* of $X_1, X_2, \ldots, X_n$ and of $X_1, X_2, \ldots$ are denoted by $X_1 X_2 \cdots X_n$ and $X_1 X_2 \cdots$, respectively. If the given spaces have topologies, the product space will have the product topology. Under this topology, convergence in the product space is componentwise convergence in the factor spaces. If the given spaces have σ-algebras $\mathscr{F}_{X_1}, \mathscr{F}_{X_2}, \ldots$, the product σ-algebras are denoted by $\mathscr{F}_{X_1} \mathscr{F}_{X_2} \cdots \mathscr{F}_{X_n}$ and $\mathscr{F}_{X_1} \mathscr{F}_{X_2} \cdots$, respectively.

If X and Y are arbitrary spaces and $E \subset XY$, then for each $x \in X$, the *x-section* of E is

$$E_x = \{y \in Y | (x, y) \in E\}. \tag{1}$$

If $\mathscr{P}$ is a class of subsets of a space X, we denote by $\sigma(\mathscr{P})$ the smallest σ-algebra containing $\mathscr{P}$. We denote by $\mathscr{P}_\sigma$ or $\mathscr{P}_\delta$ the class of all subsets which can be obtained by union or intersection, respectively, of countably many sets in $\mathscr{P}$. If $\mathscr{F}$ is the collection of closed subsets of a topological space X, then $\mathscr{F}_\delta = \mathscr{F}$, and the members of $\mathscr{F}_\sigma$ are called the F_σ-subsets of X. If $\mathscr{G}$ is the collection of open subsets of X, the members of $\mathscr{G}_\delta$ are called the G_δ-subsets of X.

If $(X, \mathscr{P})$ is a paved space, i.e., $\mathscr{P}$ is a nonempty collection of subsets of X, and S is a Suslin scheme for $\mathscr{P}$ (Definition 7.15), then $N(S)$ is the nucleus of S. The collection of all nuclei of Suslin schemes for $\mathscr{P}$ is denoted $\mathscr{S}(\mathscr{P})$.

Special Sets

The symbol R represents the *real numbers* with the usual topology. We use R^* to denote the *extended real numbers* $[-\infty, +\infty]$ with the topology discussed following Definition 7.7 in Section 7.3. Similarly, Q is the set of *rational numbers* and Q^* is the set of extended rational numbers $Q \cup \{\pm\infty\}$. If X and Y are sets and $f : X \to Y$, the *graph* of f is

$$\mathrm{Gr}(f) = \{(x, f(x)) | x \in X\}. \tag{2}$$

If $A \subset X$ and $\mathscr{C}$ is a collection of subsets of X, we define $f(A) = \{f(x)|x \in A\}$ and

$$f(\mathscr{C}) = \{f(C)|C \in \mathscr{C}\}. \tag{3}$$

If $B \subset Y$ and $\mathscr{C}$ is a collection of subsets of Y, we define $f^{-1}(B) = \{x \in X|f(x) \in B\}$ and

$$f^{-1}(\mathscr{C}) = \{f^{-1}(C)|C \in \mathscr{C}\}. \tag{4}$$

If X is a topological space, $\mathscr{F}_X$ is the collection of closed subsets of X and $\mathscr{B}_X$ the *Borel σ-algebra* on X (Definition 7.6). The *space of probability measures* on $(X, \mathscr{B}_X)$ is denoted by $P(X)$; $C(X)$ is the *Banach space of bounded, real-valued, continuous functions* on X with the supremum norm

$$\|f\| = \sup_{x \in X}|f(x)|;$$

for any metric d on X which is consistent with its topology, $U_d(X)$ is the *space of bounded real-valued functions on X which are uniformly continuous with respect to d*. If X is a Borel space (Definition 7.7), $\mathscr{A}_X$ is its *analytic σ-algebra* (Definition 7.19) and $\mathscr{U}_X$ its *universal σ-algebra* (Definition 7.18).

We let N denote the *set of positive integers* with the discrete topology. The *Baire null space* $\mathscr{N}$ is the product of countably many copies of N. The *Hilbert cube* $\mathscr{H}$ is the product of countably many copies of $[0, 1]$. We will denote by Σ the collection of *finite sequences of positive integers*. We impose no topology on Σ. If $s \in \Sigma$ and $z = (\zeta_1, \zeta_2, \ldots) \in \mathscr{N}$, we write $s < z$ to mean $s = (\zeta_1, \zeta_2, \ldots, \zeta_k)$ for some k.

Mappings

If X and Y are spaces, proj_X is the *projection mapping* from XY onto X. If E is a subset of X, the *indicator function* of E is given by

$$\chi_E(x) = \begin{cases} 1 & \text{if } x \in E, \\ 0 & \text{if } x \notin E. \end{cases} \tag{5}$$

If $f: X \to [-\infty, +\infty]$, the *positive and negative parts* of f are the functions

$$f^+(x) = \max\{0, f(x)\}, \tag{6}$$
$$f^-(x) = \max\{0, -f(x)\}. \tag{7}$$

If $f_n: X \to Y$ is a sequence of functions, Y is a topological space, and $\lim_{n \to \infty} f_n(x) = f(x)$ for all $x \in X$, then we write $f_n \to f$. If, in addition, $Y = [-\infty, +\infty]$ and $f_1(x) \le f_2(x) \le \cdots$ for all $x \in X$, we write $f_n \uparrow f$, while if $f_1(x) \ge f_2(x) \ge \cdots$ for all $x \in X$, we write $f_n \downarrow f$. In general, when the arguments of extended real-valued functions are omitted, the statements are to be interpreted pointwise. For example, $(\sup_n f_n)(x) = \sup_n f_n(x)$ for all $x \in X$, $f_1 \leq f_2$ if and only if $f_1(x) \leq f_2(x)$ for all $x \in X$, and $f + \varepsilon$ is the function $(f + \varepsilon)(x) = f(x) + \varepsilon$ for all $x \in X$.

Miscellaneous

If (X, d) is a nonempty metric space, $x \in X$, and Y is a nonempty subset of X, we define the *distance from x to Y* by

$$d(x, Y) = \inf_{y \in Y} d(x, y). \tag{8}$$

We define the diameter of Y by

$$\text{diam}(Y) = \sup_{x, y \in Y} d(x, y). \tag{9}$$

If $(X, \mathscr{F})$ is a measurable space and $\mathscr{F}$ contains all singleton sets, then for $x \in X$ we denote by p_x the probability measure on $(X, \mathscr{F})$ which assigns mass one to the set $\{x\}$.

7.2   Metrizable Spaces

Definition 7.1   Let $(X, \mathscr{T})$ be a topological space. A metric d on X is *consistent* with $\mathscr{T}$ if every set of the form $\{y \in X \mid d(x, y) < c\}$, $x \in X$, $c > 0$, is in $\mathscr{T}$, and every nonempty set in $\mathscr{T}$ is the union of sets of this form. The space $(X, \mathscr{T})$ is *metrizable* if such a metric exists.

The distinction between metric and metrizable spaces is a fine one: In a metric space we have settled on a metric, while in a metrizable space the choice is still open. If one metric consistent with the given topology exists, then a multitude of them can be found. For example, if d is a metric on X consistent with $\mathscr{T}$, the metric ρ defined by

$$\rho(x, y) = d(x, y)/[1 + d(x, y)] \qquad \forall x, y \in X$$

is also consistent with $\mathscr{T}$. In what follows, we abbreviate the notation for metrizable spaces, writing X in place of $(X, \mathscr{T})$.

If $(X, \mathscr{T})$ is a topological space and $Y \subset X$, unless otherwise specified, we will understand Y to be a topological space with open sets $G \cap Y$, where G ranges over $\mathscr{T}$. This is called the *relative topology*. If $(Z, \mathscr{S})$ is another topological space, $\varphi : Z \to X$ is one-to-one and continuous, and φ^{-1} is continuous on $\varphi(Z)$ with the relative topology, we say that φ is a *homeomorphism* and Z is *homeomorphic* to $\varphi(Z)$. When there exists a homeomorphism from Z into X, we also say that Z can be *homeomorphically embedded* in X. Given a metric d on X consistent with its topology and a homeomorphism $\varphi : Z \to X$

as just described, we may define a metric d_1 on Z by

$$d_1(z_1, z_2) = d(\varphi(z_1), \varphi(z_2)). \tag{10}$$

It can be easily verified that the metric d_1 is consistent with the topology $\mathscr{S}$. This implies that every topological space homeomorphic to a metrizable space (or subset of a metrizable space) is itself metrizable.

Our attention will be focused on metrizable spaces and their Borel σ-algebras. The presence of a metric in such spaces permits simple proofs of facts whose proofs are quite complicated or even impossible in more general topological spaces. We give two of these as lemmas for later reference.

Lemma 7.1 (Urysohn's lemma) Let X be a metrizable space and A and B disjoint, nonempty, closed subsets of X. Then there exists a continuous function $f : X \rightarrow [0, 1]$ such that $f(a) = 0$ for every $a \in A$, $f(b) = 1$ for every $b \in B$, and $0 < f(x) < 1$ for every $x \notin A \cup B$. If d is a metric consistent with the topology on X and $\inf_{a \in A, b \in B} d(a, b) > 0$, then f can be chosen to be uniformly continuous with respect to the metric d.

Proof Let d be a metric on X consistent with its topology and define

$$f(x) = d(x, A)/[d(x, A) + d(x, B)],$$

where the distance from a point to a nonempty closed set is defined by (8). This distance is zero if and only if the point is in the set, and the mapping of (8) is Lipschitz-continuous by (6) of Appendix C. This f has the required properties. If $\inf_{a \in A, b \in B} d(a, b) > 0$, then $d(x, A) + d(x, B)$ is bounded away from zero, and the uniform continuity of f follows. Q.E.D.

Lemma 7.2 Let X be a metrizable space. Every closed subset of X is a G_δ and every open subset is an F_σ.

Proof We prove the first statement; the second follows by complementation. Let F be closed. We may assume without loss of generality that F is nonempty. Let d be a metric on X consistent with its topology. The continuity of the function $x \rightarrow d(x, F)$ implies that

$$G_n = \{x \in X \mid d(x, F) < 1/n\}$$

is open. But $F = \bigcap_{n=1}^{\infty} G_n$. Q.E.D.

Definition 7.2 Let X be a metrizable topological space. The space X is *separable* if it contains a countable dense set.

It is easily verified that any subspace of a separable metrizable space is separable and metrizable. A collection of subsets of a topological space $(X, \mathscr{T})$ is a *base* for the topology if every open set can be written as a union of sets from the collection. It is a *subbase* if a base can be obtained by taking finite intersections of sets from the collection. If $\mathcal{T}$ has a countable base, $(X, \mathcal{T})$ is said to be *second countable*. A topological space is *Lindelöf* if every collection of open sets which covers the space contains a countable subcollection which also covers the space. It is a standard result that in metrizable spaces, separability, second countability, and the Lindelöf property are equivalent. The following proposition is a direct consequence of this fact.

Proposition 7.1 Let $(X, \mathcal{T})$ be a separable, metrizable, topological space and $\mathcal{B}$ a base for the topology $\mathcal{T}$. Then $\mathcal{B}$ contains a countable subcollection $\mathcal{B}_0$ which is also a base for $\mathcal{T}$.

Proof Let $\mathcal{C}$ be a countable base for the topology $\mathcal{T}$. Every set $C \in \mathcal{C}$ has the form $C = \bigcup_{\alpha \in I(C)} B_\alpha$, where $I(C)$ is an index set and $B_\alpha \in \mathcal{B}$ for every $\alpha \in I(C)$. Since C is Lindelöf, we may assume $I(C)$ is countable. Let $\mathcal{B}_0 = \bigcup_{C \in \mathcal{C}} \{B_\alpha | \alpha \in I(C)\}$. Q.E.D.

The *Hilbert cube* $\mathcal{H}$ is the product of countably many copies of the unit interval (with the product topology). The unit interval is separable and metrizable, and, as we will show later (Proposition 7.4), these properties carry over to the Hilbert cube. In a sense, $\mathcal{H}$ is the canonical separable metrizable space, as the following proposition shows.

Proposition 7.2 (Urysohn's theorem) Every separable metrizable space is homeomorphic to a subset of the Hilbert cube $\mathcal{H}$.

Proof Let (X, d) be a separable metric space with a countable dense set $\{x_k\}$. Define functions

$$\varphi_k(x) = \min\{1, d(x, x_k)\}, \qquad k = 1, 2, \ldots,$$

and $\varphi: X \to \mathcal{H}$ by

$$\varphi(x) = (\varphi_1(x), \varphi_2(x), \ldots).$$

Each φ_k is continuous, so φ is continuous. (Convergence in $\mathcal{H}$ is component-wise.) If $\varphi(x) = \varphi(y)$, then letting $x_{k_j} \to x$, we see that $\lim_{j \to \infty} d(y, x_{k_j}) = 0$, so $x = y$ and φ is one-to-one. It remains to show that φ^{-1} is continuous, i.e., $\varphi(y_n) \to \varphi(y)$ implies $y_n \to y$. But if $\varphi(y_n) \to \varphi(y)$, choose $\varepsilon > 0$ and x_k such that $d(y, x_k) < \varepsilon$. Since $d(y_n, x_k) \to d(y, x_k)$ as $n \to \infty$, for n sufficiently large $d(y_n, x_k) < \varepsilon$. Then $d(y, y_n) < 2\varepsilon$. Q.E.D.

If X is a separable metrizable space and $\varphi: X \to \mathcal{H}$ is the homeomorphism whose existence is guaranteed by Proposition 7.2, then by identifying $x \in X$ with $\varphi(x) \in \mathcal{H}$, we can regard X as a subset of $\mathcal{H}$. Indeed, we can regard X as a topological subspace of $\mathcal{H}$, since the images of open sets in X under the mapping φ are just the relatively open subsets of $\varphi(X)$ considered as a subspace of $\mathcal{H}$. Note, however, that although X is both open and closed in itself,

$\varphi(X)$ may be neither open nor closed in $\mathscr{H}$. In fact, it may have no topological characterization at all. Likewise, a set with special structure in X, say a G_δ, may not have this structure when considered as a subset of $\mathscr{H}$. The next definition and proposition shed some light on this issue.

Definition 7.3 Let X be a topological space. The space X is *topologically complete* if there is a metric d on X consistent with its topology such that the metric space (X, d) is complete, i.e., if $\{x_k\} \subset X$ is a d-Cauchy sequence $[d(x_n, x_m) \to 0$ as $n, m \to \infty]$, then $\{x_k\}$ converges to an element of X.

Proposition 7.3 (Alexandroff's theorem) Let X be a topologically complete space, Z a metrizable space, and $\varphi : X \to Z$ a homeomorphism. Then $\varphi(X)$ is a G_δ-subset of Z. Conversely, if Y is a G_δ-subset of Z and Z is topologically complete, then Y is topologically complete.

Proof For the proof of the first part of the proposition, we treat X as a subset of Z. There are two metrics to consider, a metric d on Z consistent with its topology and a metric d_1 on X which makes it complete. Define

$$U_n = \{z \in Z | d(z, \bar{X}) < 1/n \text{ and } \exists \text{ an open neighborhood } V(z) \text{ of } z \text{ such that}$$
$$\sup_{x, y \in V(z) \cap X} d_1(x, y) < 1/n\}.$$

For $n = 1, 2, \ldots$, given $z \in U_n$ and $V(z)$ as just defined, we have

$$V(z) \cap \{y \in Z | d(y, \bar{X}) < 1/n\} \subset U_n,$$

so U_n is open. We show $X = \bigcap_{n=1}^{\infty} U_n$.

For $z \in X$, define

$$W(z) = \{y \in X | d_1(y, z) < 1/3n\}.$$

Then $W(z)$ is relatively open in X, thus of the form $W(z) = V(z) \cap X$, where $V(z)$ is an open neighborhood in Z of z. Also,

$$\sup_{x, y \in V(z) \cap X} d_1(x, y) < 1/n,$$

so $z \in U_n$. Therefore $X \subset \bigcap_{n=1}^{\infty} U_n$. Now suppose $z \in \bigcap_{n=1}^{\infty} U_n$. Then $d(z, \bar{X}) = 0$, and since $\bar{X}$ is closed, we have $z \in \bar{X}$. There is a sequence $\{x_k\} \subset X$ such that $x_k \to z$. Let $V_n(z)$ be an open neighborhood in Z of z for which

$$\sup_{x, y \in V_n(z) \cap X} d_1(x, y) < 1/n. \tag{11}$$

For each n, there is an index k_n such that $x_k \in V_n(z)$ for $k \geq k_n$. From (11) we see that $d_1(x_i, x_j) < 1/n$ for $i, j \geq k_n$, so $\{x_k\}$ is Cauchy in the complete space (X, d_1) and hence has a limit in X. But the limit is z by assumption, so $X = \bigcap_{n=1}^{\infty} U_n$.

For the converse part of the theorem, suppose (Z, d) is a complete metric space and $Y = \bigcap_{n=1}^{\infty} U_n$, where each U_n is open in Z. Define a metric d_1 on Y by

$$d_1(y, z) = d(y, z) + \sum_{n=1}^{\infty} \min\{1/2^n, |[1/d(y, Z - U_n)] - [1/d(z, Z - U_n)]|\}.$$

If $\{y_k\}$ is Cauchy in (Y, d_1), then it is also Cauchy in (Z, d), and thus has a limit $y \in Z$. For each n,

$$|[1/d(y_i, Z - U_n)] - [1/d(y_j, Z - U_n)]| \to 0$$

as $i, j \to \infty$, so $[1/d(y_k, Z - U_n)]$ remains bounded as $k \to \infty$. It follows that $y \in U_n$ for every n, hence $y \in Y$.   Q.E.D.

As we remarked earlier without proof, the Hilbert cube inherits metrizability and separability from the unit interval. It also inherits topological completeness. This is a special case of the fact, which we now prove, that completeness and separability of metrizable spaces are preserved when taking countable products.

Proposition 7.4 Let $X_1, X_2, \ldots$ be a sequence of metrizable spaces and $Y_n = X_1 X_2 \cdots X_n$, $Y = X_1 X_2 \cdots$. Then Y and each Y_n is metrizable. If each X_k is separable or topologically complete, then Y and each Y_n is separable or topologically complete, respectively.

Proof If d_k is a metric on X_k consistent with its topology, then

$$d(y, \hat{y}) = \sum_{k=1}^{\infty} \min\{1/2^k, d_k(\eta_k, \hat{\eta}_k)\},$$

where $y = (\eta_1, \eta_2, \ldots)$, $\hat{y} = (\hat{\eta}_1, \hat{\eta}_2, \ldots)$, is a metric on Y consistent with the product topology. If each (X_k, d_k) is complete, clearly (Y, d) is complete. If $\mathscr{G}_k$ is a countable base for the topology on X_k, the collection of sets of the form $G_1 G_2 \cdots G_n X_{n+1} X_{n+2} \cdots$, where G_k ranges over $\mathscr{G}_k$ and n ranges over the positive integers, is a countable base for the product topology on Y. The arguments for the product spaces Y_n are similar.   Q.E.D.

Combining Propositions 7.2–7.4, we see that every separable, topologically complete space is homeomorphic to a G_δ-subset of the Hilbert cube, and conversely, every G_δ-subset of the Hilbert cube is separable and topologically complete. We state a second consequence of these propositions as a corollary.

Corollary 7.4.1 Every separable, topologically complete space can be homeomorphically embedded as a dense G_δ-set in a compact metric space.

Proof Let X be separable and topologically complete and let φ be a homeomorphism from X into $\mathscr{H}$. Since $\mathscr{H}$ is metrizable, $\varphi(X)$ is a G_δ-subset of $\mathscr{H}$ (Proposition 7.3) and thus a dense G_δ-subset of $\overline{\varphi(X)}$. Tychonoff's theorem implies that $\mathscr{H}$ is compact, so $\overline{\varphi(X)}$ is compact.   Q.E.D.

If X and Z are topological spaces, φ a homeomorphism from Z *onto* X, and d a metric on X consistent with its topology such that (X, d) is complete, then d_1 defined by (10) is a metric on Z consistent with its topology, and (Z, d_1) is also complete. Thus topological completeness is preserved under homeomorphisms. The same is true for separability, as is well known. Topological completeness is somewhat different from separability, however, in that one must produce a metric to verify it. It is quite possible that a space has two metrics consistent with its topology, is a complete metric space with one, but is not a complete metric space with the other. For example, let $X = \{1, \frac{1}{2}, \frac{1}{3}, \ldots\}$ have the discrete topology,

$$d_1(x, y) = \begin{cases} 0 & \text{if } x = y, \\ 1 & \text{if } x \neq y, \end{cases}$$

and

$$d_2(x, y) = |x - y|.$$

Then (X, d_1) is complete, but (X, d_2) is not. A more surprising example is that the set $\mathscr{N}_0$ of irrational numbers between 0 and 1 with the usual topology is topologically complete. To see this, write $\mathscr{N}_0 = \bigcap_{r \in Q}([0, 1] - \{r\})$, where Q is the set of rational numbers. It follows that $\mathscr{N}_0$ is a G_δ-subset of $[0, 1]$ and is thus topologically complete by Proposition 7.3. Another proof is obtained as follows. Let N be the set of positive integers with the discrete topology and $\mathscr{N}$ the product of countably many copies of N. The space $\mathscr{N}$ is called the *Baire null space* and is topologically complete (Proposition 7.4). The topological completeness of $\mathscr{N}_0$ follows from the fact that $\mathscr{N}$ and $\mathscr{N}_0$ are homeomorphic. We give the rather lengthy proof of this because it is not readily available elsewhere. This homeomorphism will be used only to construct a counterexample (Example 1 in Chapter 8), so it may be skipped by the reader without loss of continuity.

Proposition 7.5 The topological spaces $\mathscr{N}_0$ and $\mathscr{N}$ are homeomorphic.

Proof Let Σ be the set of finite sequences of positive integers. If $z \in \Sigma \cup \mathscr{N}$, we will represent its components by ζ_k. Similarly, $\hat{\zeta}_k$ will represent the components of an element $\hat{z}$ of $\Sigma \cup \mathscr{N}$. The *length* of $z \in \Sigma \cup \mathscr{N}$ is defined to be the number of its components. If z has length greater than or equal to k, we define $z_k = (\zeta_k, \zeta_{k+1}, \ldots)$ or $z_k = (\zeta_k, \ldots, \zeta_m)$, depending on whether z has infinite length or length $m < \infty$.

For $z \in \Sigma \cup \mathcal{N}$, define a sequence whose initial terms are

$$x_1(z) = \zeta_1^{-1},$$
$$x_2(z) = (\zeta_1 + \zeta_2^{-1})^{-1},$$
$$x_3(z) = (\zeta_1 + (\zeta_2 + \zeta_3^{-1})^{-1})^{-1}.$$

If z has length $k < \infty$, we define $x_1(z), x_2(z), \ldots, x_k(z)$ as shown, and $x_{k+j}(z) = x_k(z), j = 1, 2, \ldots$.

Claim 1 The sequence $\{x_k(z)\}$ converges to an element of $(0,1]$ for $\forall z \in \Sigma \cup \mathcal{N}$.

If z has finite length, the claim is trivial. If z has infinite length, then

$$0 < x_{2n}(z) < x_{2n+2}(z) < x_{2n+1}(z) < x_{2n-1}(z) \le 1, \qquad n = 1, 2, \ldots, \quad (12)$$

so for every n

$$x_{2n}(z) \le \liminf_{k \to \infty} x_k(z) \le \limsup_{k \to \infty} x_k(z) \le x_{2n-1}(z). \tag{13}$$

Now

$$0 < x_{2n-1}(z) - x_{2n}(z)$$
$$= [\zeta_1 + x_{2n-2}(z_2)]^{-1} - [\zeta_1 + x_{2n-1}(z_2)]^{-1}$$
$$= [\zeta_1 + x_{2n-2}(z_2)]^{-1}[\zeta_1 + x_{2n-1}(z_2)]^{-1}[x_{2n-1}(z_2) - x_{2n-2}(z_2)]$$
$$\le [\zeta_1 + (\zeta_2 + 1)^{-1}]^{-2}[x_{2n-1}(z_2) - x_{2n-2}(z_2)]$$
$$\le [\zeta_1 + (\zeta_2 + 1)^{-1}]^{-2}[\zeta_2 + (\zeta_3 + 1)^{-1}]^{-2}[x_{2n-3}(z_3) - x_{2n-2}(z_3)]$$
$$\vdots$$
$$\le [\zeta_1 + (\zeta_2 + 1)^{-1}]^{-2}[\zeta_2 + (\zeta_3 + 1)^{-1}]^{-2} \cdots [\zeta_{2n-2} + (\zeta_{2n-1} + 1)^{-1}]^{-2}.$$

Since

$$[\zeta_{2k-1} + (\zeta_{2k} + 1)^{-1}]^{-2}[\zeta_{2k} + (\zeta_{2k+1} + 1)^{-1}]^{-2} \le \tfrac{4}{9}, \qquad k = 1, \ldots, n-1,$$

we have

$$0 < x_{2n-1}(z) - x_{2n}(z) \le (\tfrac{4}{9})^{n-1},$$

and Claim 1 follows.

Define $\varphi : \Sigma \cup \mathcal{N} \to (0,1]$ by

$$\varphi(z) = \lim_{k \to \infty} x_k(z).$$

Note that if $z \in \mathcal{N}$, then $0 < \varphi(z) < 1$. Also, if z has length at least k, then

$$\varphi(z) = \varphi[(\zeta_1, \zeta_2, \ldots, \zeta_{k-1}, 1/\varphi(z_k))]. \tag{14}$$

Claim 2 If $z \in \mathcal{N}$ and $\varphi(z) = \varphi(\hat{z})$, then $z = \hat{z}$.

Suppose $\varphi(z) = \varphi(\hat{z})$ and $z \neq \hat{z}$. We can use (14) to assume without loss of generality that $\zeta_1 \neq \hat{\zeta}_1$ or else $\hat{z}$ has length one and $\zeta_1 = \hat{\zeta}_1$. In the latter case, (12) implies

$$\varphi(\hat{z}) = 1/\hat{\zeta}_1 = 1/\zeta_1 = x_1(z) > x_3(z) \geq \varphi(z),$$

and a contradiction is reached. In the former case, if $\hat{z}$ has length one, then from (14)

$$1/\hat{\zeta}_1 = \varphi(\hat{z}) = \varphi(z) = 1/[\zeta_1 + \varphi(z_2)],$$

so

$$\hat{\zeta}_1 = \zeta_1 + \varphi(z_2),$$

which is impossible, since $0 < \varphi(z_2) < 1$. If $\hat{z}$ has length greater than one, then

$$1/[\hat{\zeta}_1 + \varphi(\hat{z}_2)] = \varphi(\hat{z}) = \varphi(z) = 1/[\zeta_1 + \varphi(z_2)],$$

and

$$\hat{\zeta}_1 + \varphi(\hat{z}_2) = \zeta_1 + \varphi(z_2).$$

This is also impossible, since $0 < \varphi(\hat{z}_2) \leq 1$ and $0 < \varphi(z_2) < 1$.

Claim 3 Every rational number in $(0, 1]$ has the form $\varphi(\hat{z})$, where $\hat{z} \in \Sigma$.

Let r_1/q be a rational number in $(0, 1]$ reduced to lowest terms, r_1 and q positive integers. Then

$$r_1/q = (q/r_1)^{-1} = [q_1 + (r_2/r_1)]^{-1},$$

where q_1 and r_2 are positive integers and $r_2 < r_1$. Likewise,

$$r_2/r_1 = (r_1/r_2)^{-1} = [q_2 + (r_3/r_2)]^{-1},$$

where q_2 and r_3 are positive integers and $r_3 < r_2$. Continuing, we eventually obtain $r_n = 1$ and have

$$r_1/q = \varphi[(q_1, q_2, \ldots, q_{n-1}, r_{n-1})].$$

Claims 2 and 3 imply that if $z \in \mathcal{N}$, then $\varphi(z)$ is irrational. Put another way, φ maps $\mathcal{N}$ into $\mathcal{N}_0$. But given $y \in \mathcal{N}_0$, it is possible to choose positive integers $\zeta_1, \zeta_2, \ldots$, such that

$$(\zeta_1 + 1)^{-1} < y < \zeta_1^{-1},$$
$$(\zeta_1 + \zeta_2^{-1})^{-1} < y < (\zeta_1 + (\zeta_2 + 1)^{-1})^{-1},$$
$$(\zeta_1 + (\zeta_2 + (\zeta_3 + 1)^{-1})^{-1})^{-1} < y < (\zeta_1 + (\zeta_2 + \zeta_3^{-1})^{-1})^{-1},$$

etc., so that defining $z = (\zeta_1, \zeta_2, \ldots)$, we have

$$x_{2k}(z) < y < x_{2k-1}(z), \qquad k = 1, 2, \ldots.$$

It follows that $\varphi(z) = y$, so φ maps $\mathcal{N}$ onto $\mathcal{N}_0$ and, by Claim 2, is one-to-one on $\mathcal{N}$.

We show that φ restricted to $\mathcal{N}$ is open and continuous. Let $V \subset \mathcal{N}$ be open. We may assume without loss of generality that

$$V = \{z \in \mathcal{N} \mid (\zeta_1, \ldots, \zeta_n) = (\hat{\zeta}_1, \ldots, \hat{\zeta}_n)\}.$$

Then

$$\varphi(V) = \{(\hat{\zeta}_1 + (\hat{\zeta}_2 + \cdots + (\hat{\zeta}_n + \varphi(z))^{-1} \cdots)^{-1})^{-1} \mid z \in \mathcal{N}\},$$

and since $\{\varphi(z) \mid z \in \mathcal{N}\} = \mathcal{N}_0$, $\varphi(V)$ is open. Since convergence in $\mathcal{N}$ is componentwise and $x_n(z)$ depends only on the first n components of $z \in \mathcal{N}$, continuity of φ on $\mathcal{N}$ follows from (13).   Q.E.D.

We now examine properties of metrizable spaces related to the notion of total boundedness.

Definition 7.4 A metric space (X, d) is *totally bounded* if, given $\varepsilon > 0$, there exists a finite subset F_ε of X for which

$$X = \bigcup_{x \in F_\varepsilon} \{y \in X \mid d(x, y) < \varepsilon\}.$$

A totally bounded metric space is necessarily separable, since $\bigcup_{n=1}^\infty F_{1/n}$ is a countable dense subset. Total boundedness depends on the metric, however, and a space which is totally bounded (and separable) with one metric may not be totally bounded with another. Like separability, total boundedness is preserved under passage to subspaces, i.e., if (X, d) is totally bounded and $Y \subset X$, then (Y, d) is totally bounded. To see this, take $\varepsilon > 0$ and let $F_{\varepsilon/2}$ be a finite subset of X such that

$$X = \bigcup_{x \in F_{\varepsilon/2}} \{y \in X \mid d(x, y) < \varepsilon/2\}.$$

Choose a point, if possible, in each of the sets

$$Y \cap \{y \in X \mid d(x, y) < \varepsilon/2\}, \qquad x \in F_{\varepsilon/2},$$

and call the collection of these points G_ε. Then

$$Y = \bigcup_{y \in G_\varepsilon} \{z \in Y \mid d(y, z) < \varepsilon\}.$$

We use this fact to prove the following classical result relating completeness, compactness, and total boundedness.

Proposition 7.6 A metric space is compact if and only if it is complete and totally bounded.

Proof If (X, d) is a compact metric space, then every Cauchy sequence has an accumulation point. The Cauchy property implies that the sequence converges to this point, and completeness follows. Also, for $\varepsilon > 0$, the collection of sets

$$\{y \in X \,|\, d(x, y) < \varepsilon\}, \qquad x \in X,$$

contains a finite cover of X. Hence, (X, d) is totally bounded.

If (X, d) is complete and totally bounded and $S = \{s_j\}$ is a sequence in (X, d), then an infinite subsequence $S_1 \subset S$ must lie in some set $B_1 = \{y \in X \,|\, d(x_1, y) < 1\}$. Since B_1 is totally bounded, an infinite subsequence $S_2 \subset S_1$ must lie in some set $B_2 = \{y \in B_1 \,|\, d(x_2, y) < \frac{1}{2}\}$. Continuing in this manner, we have for each n an infinite sequence $S_{n+1} \subset S_n$ lying in $B_{n+1} = \{y \in B_n \,|\, d(x_{n+1}, y) < 1/(n+1)\}$. Let $j_1 < j_2 < \cdots$ be such that $s_{j_n} \in S_n$. Then $\{s_{j_n}\}$ is Cauchy and thus convergent. Therefore S has an accumulation point, and the compactness of (X, d) follows.   Q.E.D.

Corollary 7.6.1 The Hilbert cube is totally bounded under any metric consistent with its topology, and every separable metrizable space has a totally bounded metrization.

Proof The Hilbert cube is compact by Tychonoff's theorem. Urysohn's theorem (Proposition 7.2) can be used to homeomorphically embed a given separable metrizable space into the Hilbert cube.   Q.E.D.

As mentioned previously, total boundedness implies separability. By combining this fact with Proposition 7.6, we obtain the following corollary.

Corollary 7.6.2 A compact metric space is complete and separable.

If X is a metrizable space, the set of all bounded, continuous, real-valued functions on X is denoted $C(X)$. As is well known, $C(X)$ is a Banach space under the norm

$$\|f\| = \sup_{x \in X} |f(x)|,$$

and we will always take $C(X)$ to have the metric and topology corresponding to this norm. If d is a metric on X consistent with its topology, we denote by $U_d(X)$ the collection of functions in $C(X)$ which are uniformly continuous with respect to d. We take $U_d(X)$ to have the relative topology of $C(X)$. We conclude this section with a discussion of the properties $C(X)$ and $U_d(X)$ inherit from X.

Proposition 7.7 If X is a compact metrizable space, then $C(X)$ is separable.

Proof The space X is separable (Corollary 7.6.2). Let $\{x_k\}$ be a countable dense subset of X and let $F_1, F_2, \ldots$ be an enumeration of the collection of sets of the form $\{y \in X \,|\, d(x_k, y) \leq 1/n\}$, where k and n range over the positive integers. For any disjoint pair F_i and F_j, let f_{ij} be a continuous function taking values in $[0,1]$ such that $f_{ij}(x) = 0$ for $x \in F_i$ and $f_{ij}(x) = 1$ for $x \in F_j$. If F_i and F_j are not disjoint, let f_{ij} be identically one. Let $\mathscr{C}$ consist of the functions f_{ij} as i and j range over the positive integers. The collection $\mathscr{C}$ clearly separates points in X, i.e., given $x \neq y$, there exists $f \in \mathscr{C}$ for which $f(x) \neq f(y)$. Let $\mathscr{P}$ be the collection of finite-degree polynomials over $\mathscr{C}$, i.e., a typical element in $\mathscr{P}$ has the form

$$\sum_{(i_1,\ldots,i_n),(j_1,\ldots,j_n)} \alpha(i_1,\ldots,i_n;j_1,\ldots,j_n)f_{i_1}^{j_1}\cdots f_{i_n}^{j_n},$$

where $\alpha(i_1,\ldots,i_n;j_1,\ldots,j_n) \in R$, $f_{i_1},\ldots,f_{i_n} \in \mathscr{C}$, and the summation is finite. Then $\mathscr{P}$ is a vector space under addition and the product of two elements in $\mathscr{P}$ is again in $\mathscr{P}$. With these operations $\mathscr{P}$ is an *algebra*, and by the Stone–Weierstrass theorem, $\mathscr{P}$ is dense in $C(X)$. Let $\mathscr{P}_0$ be the collection of finite-degree polynomials over $\mathscr{C}$ with rational coefficients. An easy approximation argument shows that $\mathscr{P}_0$ is dense in $\mathscr{P}$, and thus dense in $C(X)$ as well. Since $\mathscr{P}_0$ is countable, $C(X)$ is separable.　　　Q.E.D.

Definition 7.5　　Let (X, d_1) and (Y, d_2) be metric spaces. A mapping $\varphi : X \to Y$ is an *isometry* if

$$d_1(x_1, x_2) = d_2(\varphi(x_1), \varphi(x_2)) \qquad \forall x_1, x_2 \in X.$$

In this case we say that (X, d_1) and $(\varphi(X), d_2)$ are *isometric* spaces.

If (X, d_1) and (Y, d_2) are as in Definition 7.5, we may regard the former as a subspace of the later, and the distances between points in X are unaffected by this embedding. Thus an isometry is a metric-preserving homeomorphism.

Proposition 7.8　　Let (X, d) be a metric space. There exists a complete metric space (X_d, d_1), called the *completion of* (X, d), and an isometry $\varphi : X \to X_d$ such that $\varphi(X)$ is dense in X_d.

Proof　　The construction of the completion of a metric space is standard, so we content ourselves with a sketch of it. Given the metric space (X, d), define an equivalence relation $\sim$ on the set of Cauchy sequences in (X, d) by

$$\{x_n\} \sim \{x_n'\} \Leftrightarrow \lim_{n \to \infty} d(x_n, x_n') = 0.$$

Let X_d be the set of equivalence classes of Cauchy sequences in (X, d) under this relation and let d_1 be defined on $X_d X_d$ by

$$d_1(x, y) = \lim_{n \to \infty} d(x_n, y_n), \tag{15}$$

where $\{x_n\}$ and $\{y_n\}$ are chosen to represent the equivalence classes x and y. It is straightforward to verify that the limit in (15) exists for every pair of Cauchy sequences $\{x_n\}$ and $\{y_n\}$, and it is independent of the particular sequences chosen to represent the equivalence classes x and y. Furthermore, (X_d, d_1) can be shown to be a complete metric space, and the mapping φ which takes $x \in X$ into the equivalence class in X_d containing the Cauchy sequence $(x, x, \ldots)$ is an isometry. The image of X under φ is dense in X_d.

Q.E.D.

We can regard X_d as consisting of X together with limits of all Cauchy sequences in X. We are really interested in the case in which (X, d) is totally bounded, for which we have the following result.

Corollary 7.8.1 Let (X, d) be a totally bounded metric space. There exists a compact metric space (X_d, d_1) and an isometry $\varphi : X \to X_d$ such that $\varphi(X)$ is dense in X_d.

Proof In light of Propositions 7.6 and 7.8, it suffices to prove that the completion (X_d, d_1) of (X, d) is totally bounded. Choose $\varepsilon > 0$. Regarding (X, d) as a subspace of (X_d, d_1), choose a finite set F_ε of X for which

$$X = \bigcup_{x \in F_\varepsilon} \{y \in X \mid d(x, y) < \varepsilon/2\}.$$

Since X is dense in X_d, we have

$$X_d = \bigcup_{x \in F_\varepsilon} \{y \in X_d \mid d_1(x, y) < \varepsilon\}. \qquad \text{Q.E.D.}$$

If X is a separable metrizable space, it is not necessarily true that $C(X)$ is separable (unless X is compact, in which case we have Proposition 7.7). For example, let $f : R \to [0, 1]$ be defined as

$$f(x) = \begin{cases} 0 & \text{if } |x| \geq \tfrac{1}{2}, \\ 1 + 2x & \text{if } -\tfrac{1}{2} \leq x \leq 0, \\ 1 - 2x & \text{if } 0 \leq x \leq \tfrac{1}{2}, \end{cases}$$

and given an infinite sequence $b = (\beta_1, \beta_2, \ldots)$ of zeroes and ones, define

$$f_b(x) = \sum_{\{n \mid \beta_n = 1\}} f(x - n).$$

We have constructed an uncountable collection of functions f_b in $C(R)$ such that if $b_1 \neq b_2$, then $\|f_{b_1} - f_{b_2}\| = 1$. Therefore, $C(R)$ cannot be separable.

It is true, however, that given a separable metrizable space X, there is a metric d on X consistent with its topology such that $U_d(X)$ is separable. This is a consequence of the next proposition and the fact that separability implies the existence of a totally bounded metrization (Corollary 7.6.1). We prove this proposition with the aid of the following lemma.

Lemma 7.3 Let Y be a metrizable space, d a metric on Y consistent with its topology, and $X \subset Y$. If $g \in U_d(X)$, then g has a continuous extension to Y, i.e., there exists $\hat{g} \in C(Y)$ such that $g(x) = \hat{g}(x)$ for every $x \in X$, and the extension $\hat{g}$ can be chosen to satisfy $\|g\| = \|\hat{g}\|$. If X is dense in Y, $\hat{g}$ is unique.

Proof Since g is uniformly continuous on X, given $\varepsilon > 0$ there exists $\delta(\varepsilon) > 0$ such that if $x_1, x_2 \in X$ and $d(x_1, x_2) \leq \delta(\varepsilon)$, then $|g(x_1) - g(x_2)| \leq \varepsilon$. Suppose $y \in \bar{X}$. Then there exists a sequence $\{x_n\} \subset X$ for which $x_n \to y$. Given $\varepsilon > 0$, there exists $N(\varepsilon)$ such that $d(x_n, x_m) \leq \delta(\varepsilon)$ for all $n, m \geq N(\varepsilon)$, so $\{g(x_n)\}$ is Cauchy in R. Define $\hat{g}(y) = \lim_{n \to \infty} g(x_n)$. Note that $n \geq N(\varepsilon)$ implies $|g(x_n) - \hat{g}(y)| \leq \varepsilon$.

Suppose now that $x \in X$ and $d(x, y) \leq \delta(\varepsilon)/2$. Choose $n \geq N(\varepsilon)$ so that $d(x_n, y) \leq \delta(\varepsilon)/2$. Then $d(x, x_n) \leq \delta(\varepsilon)$ and

$$|g(x) - \hat{g}(y)| \leq |g(x) - g(x_n)| + |g(x_n) - \hat{g}(y)| \leq 2\varepsilon. \tag{16}$$

This shows that for any sequence $\{x_n'\} \subset X$ with $x_n' \to y$, we have $\hat{g}(y) = \lim_{n \to \infty} g(x_n')$, so the definition of $\hat{g}(y)$ is independent of the particular sequence $\{x_n\}$ chosen. If $y \in X$, we can take $x_n = y$, $n = 1, 2, \ldots$, and obtain $g(y) = \hat{g}(y)$, so $\hat{g}$ is an extension of g. If $\{y_m\}$ is a sequence in $\bar{X}$ which converges to $y \in \bar{X}$, then there exist sequences $\{x_{mn}\}$ in X with $y_m = \lim_{n \to \infty} x_{mn}$. Choose $n_1 < n_2 < \cdots$ so that $\lim_{m \to \infty} x_{mn_m} = y$ and $d(x_{mn_m}, y_m) \leq \delta(1/m)/2$. Then

$$\hat{g}(y) = \lim_{m \to \infty} g(x_{mn_m}), \tag{17}$$

and, by (16),

$$|g(x_{mn_m}) - \hat{g}(y_m)| \leq 2/m. \tag{18}$$

Letting $m \to \infty$ in (18) and using (17), we conclude that $\hat{g}(y) = \lim_{m \to \infty} \hat{g}(y_m)$ and $\hat{g}$ is continuous on $\bar{X}$. It is clear that

$$\sup_{x \in X} |g(x)| = \sup_{y \in \bar{X}} |\hat{g}(y)|.$$

If $\bar{X} = Y$, $\hat{g}$ is clearly unique and we are done. If $\bar{X}$ is a proper subset of Y, use the Tietze extension theorem (see, e.g., Ash [A1] or Dugundji [D7]) to extend $\hat{g}$ to all of Y so that

$$\|g\| = \sup_{y \in Y} |\hat{g}(y)|. \qquad \text{Q.E.D.}$$

Proposition 7.9 If (X, d) is a totally bounded metric space, then $U_d(X)$ is separable.

Proof Corollary 7.8.1 tells us that (X, d) can be isometrically embedded as a dense subset of a compact metric space (X_d, d_1). We regard X as a subspace of X_d. Given any $g \in U_d(X)$, by Lemma 7.3, g has a unique extension $\hat{g} \in C(X_d)$ such that $\|g\| = \|\hat{g}\|$. The mapping $g \to \hat{g}$ is linear and norm-preserving, thus an isometry from $U_d(X)$ to $C(X_d)$. The latter space is separable by Proposition 7.7, and the separability of $U_d(X)$ follows.   Q.E.D.

7.3 Borel Spaces

The constructions necessary for the subsequent theory of dynamic programming are impossible when the state space and control space are arbitrary sets or even when they are arbitrary measurable spaces. For this reason, we introduce the concept of a Borel space, and in this and subsequent sections we develop the properties of Borel spaces which permit these constructions.

Definition 7.6 If X is a topological space, the smallest σ-algebra of subsets of X which contains all open subsets of X is called the *Borel σ-algebra* and is denoted by $\mathscr{B}_X$. The members of $\mathscr{B}_X$ are called the *Borel subsets* of X.

If X is separable and metrizable and $\mathscr{F}$ is a σ-algebra on X containing a subbase $\mathscr{S}$ for its topology, then $\mathscr{F}$ contains $\mathscr{B}_X$. This is because, from Proposition 7.1, any open set in X can be written as a *countable* union of finite intersections of sets in $\mathscr{S}$. Thus we have $\mathscr{B}_X = \sigma(\mathscr{S})$ for any subbase $\mathscr{S}$.

We will often refer to the smallest σ-algebra containing a class of subsets as the *σ-algebra generated by the class*. Thus, $\mathscr{B}_X$ is the σ-algebra generated by the class of open subsets of X. Note that $\mathscr{B}_R$ is the class of Borel subsets of the real numbers in the usual sense, i.e., the σ-algebra generated by the intervals.

Given a class of real-valued functions on a topological space X, it is common to speak of the weakest topology with respect to which all functions in the class are continuous. In a similar vein, one can speak of the smallest σ-algebra with respect to which all functions in the class are measurable. If X is a metrizable space, it is easy to show that its topology is the weakest with respect to which all functions in $C(X)$ are continuous. The following proposition is the analogous result for $\mathscr{B}_X$. In the proof and in subsequent proofs, we will use the fact that for any two sets Ω, Ω', any collection $\mathscr{C}$ of subsets of Ω', and any function $f : \Omega \to \Omega'$, we have

$$\sigma[f^{-1}(\mathscr{C})] = f^{-1}[\sigma(\mathscr{C})].$$

Proposition 7.10 Let X be a metrizable space. Then $\mathscr{B}_X$ is the smallest σ-algebra with respect to which every function in $C(X)$ is measurable, i.e., $\mathscr{B}_X = \sigma[\bigcup_{f \in C(X)} f^{-1}(\mathscr{B}_R)]$.

Proof Denote $\mathscr{F} = \sigma[\bigcup_{f \in C(X)} f^{-1}(\mathscr{B}_R)]$ and let $\mathscr{T}_R$ be the topology of R. We have

$$\mathscr{F} = \sigma\left[\bigcup_{f \in C(X)} f^{-1}[\sigma(\mathscr{T}_R)]\right]$$

$$= \sigma\left[\bigcup_{f \in C(X)} \sigma[f^{-1}(\mathscr{T}_R)]\right] \subset \sigma\left[\bigcup_{f \in C(X)} \mathscr{B}_X\right] = \mathscr{B}_X.$$

To prove the reverse containment $\mathscr{B}_X \subset \mathscr{F}$ we need only establish that $\mathscr{F}$ contains every nonempty open set. By Lemma 7.2, it suffices to show that $\mathscr{F}$ contains every nonempty closed set. Let A be such a set. We may assume without loss of generality that $A \neq X$, so there exists $x \in X - A$. Let $B = \{x\}$, and let f be given by Lemma 7.1. Then $A = f^{-1}(\{0\}) \in \mathscr{F}$.   Q.E.D.

We use Lemma 7.2 to prove another useful characterization of the Borel σ-algebra in a metrizable space.

Proposition 7.11 Let X be a metrizable space. Then $\mathscr{B}_X$ is the smallest class of sets which is closed under countable unions and intersections and contains every closed (open) set.

Proof Let $\mathscr{D}$ be the smallest class of sets which contains every closed set and is closed under countable unions and intersections, i.e., $\mathscr{D}$ is the intersection of all such classes. Then $\mathscr{D} \subset \mathscr{B}_X$ and it suffices to prove that $\mathscr{D}$ is closed under complementation. Let $\mathscr{D}'$ be the class of complements of sets in $\mathscr{D}$. Then $\mathscr{D}'$ is also closed under countable unions and intersections. Lemma 7.2 implies that $\mathscr{D}$ contains every open set, so $\mathscr{D}'$ contains every closed set, and consequently $\mathscr{D} \subset \mathscr{D}'$. Given $D \in \mathscr{D}$, we have $D \in \mathscr{D}'$, so $D^c \in \mathscr{D}$.   Q.E.D.

Definition 7.7 Let X be a topological space. If there exists a complete separable metric space Y and a Borel subset $B \in \mathscr{B}_Y$ such that X is homeomorphic to B, then X is said to be a *Borel space*. The empty set will also be regarded as a Borel space.

Note that every Borel space is metrizable and separable. Also, every complete separable metrizable space is a Borel space. Examples of Borel spaces are R, R^n, and R^* with the weakest topology containing the intervals $[-\infty, \alpha)$, $(\beta, \infty]$, (α, β), $\alpha, \beta \in R$. (This is also the topology that makes the function φ defined by

$$\varphi(x) = \begin{cases} 1 & \text{if } x = \infty, \\ \operatorname{sgn}(x)(1 - e^{-|x|}) & \text{if } x \in R, \\ -1 & \text{if } x = -\infty, \end{cases}$$

a homeomorphism from R^* onto $[-1, 1]$). Any countable set X with the discrete topology (i.e., the topology consisting of all subsets of X) is also a Borel space. We will show that every Borel subset of a Borel space is itself a Borel space. For this we shall need the following two lemmas. The proof of the first is elementary and is left to the reader.

Lemma 7.4 If Y is a topological space and $E \subset Y$, then the σ-algebra $\mathscr{B}_E$ generated by the relative topology coincides with the relative σ-algebra, i.e., the collection $\{E \cap C \mid C \in \mathscr{B}_Y\}$. In particular, if $E \in \mathscr{B}_Y$, then $\mathscr{B}_E$ consists of the Borel subsets of Y contained in E.

Lemma 7.5 If X and Y are topological spaces and φ is a homeomorphism of X into Y, then $\varphi(\mathscr{B}_X) = \mathscr{B}_{\varphi(X)}$.

Proof If $\mathscr{T}_X$ is the topology of X, then $\varphi(\mathscr{T}_X)$ is the topology of $\varphi(X)$. Since φ is one-to-one, we have that φ is the inverse of a mapping, and

$$\varphi(\mathscr{B}_X) = \varphi[\sigma(\mathscr{T}_X)] = \sigma[\varphi(\mathscr{T}_X)] = \mathscr{B}_{\varphi(X)}. \qquad \text{Q.E.D.}$$

Proposition 7.12 If X is a Borel space and $B \in \mathscr{B}_X$, then B is a Borel space.

Proof Let φ be a homeomorphism of X into some complete separable metric space Y such that $\varphi(X) \in \mathscr{B}_Y$. From Lemma 7.5 and the fact that $B \in \mathscr{B}_X$, we obtain $\varphi(B) \in \mathscr{B}_{\varphi(X)}$. It follows from Lemma 7.4 that $\varphi(B) \in \mathscr{B}_Y$.
$$\text{Q.E.D.}$$

Like separability and completeness, the property of being a Borel space is preserved when taking countable Cartesian products.

Proposition 7.13 Let $X_1, X_2, \ldots$ be a sequence of Borel spaces and $Y_n = X_1 X_2 \cdots X_n$, $Y = X_1 X_2 \cdots$. Then Y and each Y_n with the product topology is a Borel space and the Borel σ-algebras coincide with the product σ-algebras, i.e., $\mathscr{B}_{Y_n} = \mathscr{B}_{X_1}\mathscr{B}_{X_2} \cdots \mathscr{B}_{X_n}$ and $\mathscr{B}_Y = \mathscr{B}_{X_1}\mathscr{B}_{X_2} \cdots$.

Proof As in Proposition 7.4, we focus our attention on the more difficult infinite product. Consider the last statement of the proposition. Each X_k has a countable base $\mathscr{G}_k$ for its topology, and the collection of sets of the form $G_1 G_2 \cdots G_n X_{n+1} X_{n+2} \cdots$, where G_k ranges over $\mathscr{G}_k$ and n ranges over the positive integers, is a base for the product topology on Y. The σ-algebra generated by this topology is $\mathscr{B}_Y$. Recall that the product σ-algebra $\mathscr{B}_{X_1}\mathscr{B}_{X_2} \cdots$ is the smallest σ-algebra containing all finite-dimensional measurable rectangles, i.e., all sets of the form $B_1 B_2 \cdots B_n X_{n+1}, X_{n+2} \cdots$, where $B_k \in \mathscr{B}_{X_k}$, $k = 1, \ldots, n$. It is clear that each basic set of the product topology on Y is a finite-dimensional measurable rectangle, and since each open subset of Y is a countable union of these basic open sets, every open subset of Y is $\mathscr{B}_{X_1}\mathscr{B}_{X_2} \cdots$ measurable. We conclude that $\mathscr{B}_Y \subset \mathscr{B}_{X_1}\mathscr{B}_{X_2} \cdots$. (Note that this argument relies only on the separability of the spaces $X_1, X_2, \ldots$. Without this separability assumption, the argument fails and the conclusion is false.) The reverse set containment follows from the observation that for each k and $B_k \in \mathscr{B}_{X_k}, X_1 X_2 \cdots X_{k-1} B_k X_{k+1} \cdots \in \mathscr{B}_Y$.

To prove that Y is a Borel space, note that X_k can be mapped by a homeomorphism φ_k onto a Borel subset of a separable topologically complete space $\tilde{X}_k$. The product $\tilde{Y} = \tilde{X}_1 \tilde{X}_2 \cdots$ is separable and topologically complete, and $\varphi : Y \to \tilde{Y}$ defined by

$$\varphi(x_1, x_2, \ldots) = (\varphi_1(x_1), \varphi_2(x_2), \ldots)$$

is a homeomorphism from Y onto $\varphi_1(X_1)\varphi_2(X_2)\cdots$. This last set is in $\mathscr{B}_{\tilde{X}_1}\mathscr{B}_{\tilde{X}_2}\cdots = \mathscr{B}_{\tilde{Y}}$, and the conclusion follows.   Q.E.D.

Definition 7.8   Let X and Y be topological spaces. A function $f : X \to Y$ is *Borel-measurable* if $f^{-1}(B) \in \mathscr{B}_X$ for every $B \in \mathscr{B}_Y$.

In many respects, Borel-measurable functions relate to Borel σ-algebras as continuous functions relate to topologies. We have already used the fact, for example, that if $f_k : X \to Y_k$ is continuous from a topological space X to a topological space Y_k, $k = 1, 2, \ldots$, then $F : X \to Y_1 Y_2 \cdots$ defined by $F(x) = (f_1(x), f_2(x), \ldots)$ is also continuous. This follows from the componentwise nature of convergence in product spaces. There is an analogous fact for Borel-measurable functions and Borel spaces.

Proposition 7.14   Let X be a Borel space, $Y_1, Y_2, \ldots$ a sequence of Borel spaces, and $f_k : X \to Y_k$ a sequence of functions. If each f_k is Borel-measurable, $k = 1, 2, \ldots$, then the function $F : X \to Y_1 Y_2 \cdots$ defined by

$$F(x) = (f_1(x), f_2(x), \ldots)$$

and the functions $F_n : X \to Y_1 Y_2 \cdots Y_n$ defined by

$$F_n(x) = (f_1(x), f_2(x), \ldots, f_n(x))$$

are Borel-measurable. Conversely, if F is Borel-measurable, then each f_k is Borel-measurable, $k = 1, 2, \ldots$, and if some F_n is Borel-measurable, then $f_1, f_2, \ldots, f_n$ are Borel-measurable.

Proof   Again we consider only the infinite product. The Borel σ-algebra in $Y_1 Y_2 \cdots$ is generated by sets of the form $B_1 B_2 \cdots$, where $B_k \in \mathscr{B}_{Y_k}$, $k = 1, 2, \ldots$. Now

$$F^{-1}(B_1 B_2 \cdots) = f_1^{-1}(B_1) \cap f_2^{-1}(B_2) \cap \cdots . \tag{19}$$

The left side of (19) is in $\mathscr{B}_X$ for each $B_k \in \mathscr{B}_{Y_k}$, $k = 1, 2, \ldots$, if and only if the sets $f_k^{-1}(B_k)$ are in $\mathscr{B}_X$ for each $B_k \in \mathscr{B}_{Y_k}$, $k = 1, 2, \ldots$, and the result follows.
                                                                        Q.E.D.

Corollary 7.14.1 Let X and Y be Borel spaces, D a Borel subset of X, and $f:D \to Y$ Borel-measurable. Then

$$\mathrm{Gr}(f) = \{(x, f(x)) \in XY | x \in D\}$$

is Borel-measurable.

Proof The mappings $(x, y) \to f(x)$ and $(x, y) \to y$ are Borel-measurable from DY to Y, so the mapping $F(x, y) = (f(x), y)$ is Borel-measurable from DY to YY. Then

$$\mathrm{Gr}(f) = F^{-1}(\{(y, y) | y \in Y\}).$$

Since $\{(y, y) | y \in Y\}$ is closed in YY, $\mathrm{Gr}(f)$ is Borel-measurable.   Q.E.D.

The concept of homeomorphism is instrumental in classifying topological spaces, since it allows us to identify those which are "topologically equivalent." We can also classify measurable spaces by identifying those which, when regarded only as sets with σ-algebras, are indistinguishable. We specialize this concept to Borel spaces.

Definition 7.9 Let X and Y be Borel spaces and $\varphi:X \to Y$ a Borel-measurable, one-to-one function such that φ^{-1} is Borel-measurable on $\varphi(X)$. Then φ is called a *Borel isomorphism*, and we say that X and $\varphi(X)$ are *Borel-isomorphic* (or simply *isomorphic*).

If X and Y are Borel spaces and $\varphi:X \to Y$ is a Borel isomorphism, it is tempting to think of X and $\varphi(X)$ as identical measurable spaces. The difficulty with this is that X is a Borel space, but $\varphi(X)$ is not required to be. This discrepancy is eliminated by the following intuitively plausible proposition, the rather lengthy proof of which can be found in Chapter I, Section 3 of Parthasarathy [P1]. We will not have occasion to use this result.

Proposition 7.15 (Kuratowski's theorem) Let X be a Borel space, Y a separable metrizable space, and $\varphi:X \to Y$ one-to-one and Borel-measurable. Then $\varphi(X)$ is a Borel subset of Y and φ^{-1} is Borel-measurable. In particular, if Y is a Borel space, then X and $\varphi(X)$ are isomorphic Borel spaces.

The advantage of classifying spaces by means of Borel isomorphisms is illustrated by the following result. We need this proposition for the subsequent development, but the proof is rather lengthy and is relegated to Appendix B, Section 2.

Proposition 7.16 Let X and Y be Borel spaces. Then X and Y are isomorphic if and only if they have the same cardinality.

Proposition 7.16 leads to a consideration of the possible cardinalities of Borel spaces. Of course, Borel spaces which are countably infinite are possible, as are Borel spaces which consist of a given finite number of elements. In both these cases, the Borel σ-algebra is the power set and the conclusion of Proposition 7.16 is trivial. Because every Borel space can be homeomorphically embedded in the Hilbert cube, every Borel space has cardinality less than or equal to c. Even if one were to admit the possibility of an uncountable cardinality strictly less than c, the proof of Proposition 7.16 as given in Appendix B shows that every uncountable Borel space has cardinality c. By combining this fact with Proposition 7.16, we obtain the following corollary.

Corollary 7.16.1   Every uncountable Borel space is Borel-isomorphic to every other uncountable Borel space. In particular, every uncountable Borel space is isomorphic to the unit interval $[0, 1]$ and the Baire null space $\mathcal{N}$.

7.4   Probability Measures on Borel Spaces

If X is a metrizable space, we shall refer to a probability measure p on the measurable space $(X, \mathscr{B}_X)$ as simply a probability measure on X. The set of all probability measures on X will be denoted by $P(X)$. A probability measure $p \in P(X)$ determines a linear functional $l_p : C(X) \to R$ defined by

$$l_p(f) = \int f \, dp. \tag{20}$$

On the other hand, a function $f \in C(X)$ determines a real-valued function $\theta_f : P(X) \to R$ defined by

$$\theta_f(p) = \int f \, dp. \tag{21}$$

These relationships and the metrizability of the underlying space X allow us to show several properties of $P(X)$. In particular, we will prove that there is a natural topology on $P(X)$, the weakest topology with respect to which every mapping of the form of (21) is continuous, under which $P(X)$ is a Borel space whenever X is a Borel space.

7.4.1   Characterization of Probability Measures

Definition 7.10   Let X be a metrizable space. A probability measure $p \in P(X)$ is said to be *regular* if for every $B \in \mathscr{B}_X$,

$$p(B) = \sup\{p(F)|F \subset B, F \text{ closed}\} = \inf\{p(G)|B \subset G, G \text{ open}\}. \tag{22}$$

Proposition 7.17   Let X be a metrizable space. Every probability measure in $P(X)$ is regular.

Proof Let $p \in P(X)$ be given and let $\mathscr{E}$ be the collection of $B \in \mathscr{B}_X$ for which (22) holds. If $H \subset X$ is open, then $H = \bigcup_{n=1}^{\infty} F_n$, where $\{F_n\}$ is an increasing sequence of closed sets (Lemma 7.2), so

$$\inf\{p(G) | H \subset G, G \text{ open}\} = p(H)$$
$$= \lim_{n \to \infty} p(F_n)$$
$$\leq \sup\{p(F) | F \subset H, F \text{ closed}\} \leq p(H).$$

Therefore $\mathscr{E}$ contains every open subset of X. We show that $\mathscr{E}$ is a σ-algebra and conclude that $\mathscr{E} = \mathscr{B}_X$.

If $B \in \mathscr{E}$, then

$$p(B^c) = 1 - p(B) = 1 - \sup\{p(F) | F \subset B, F \text{ closed}\}$$
$$= \inf\{p(G) | B^c \subset G, G \text{ open}\},$$

and similarly,

$$p(B^c) = \sup\{p(F) | F \subset B^c, F \text{ closed}\},$$

so $\mathscr{E}$ is closed under complementation. Now suppose $\{B_n\}$ is a sequence of sets in $\mathscr{E}$. Choose $\varepsilon > 0$ and $F_n \subset B_n \subset G_n$ such that F_n is closed, G_n is open, and $p(G_n - F_n) \leq \varepsilon/2^n$. Then

$$\bigcup_{n=1}^{\infty} B_n \subset \bigcup_{n=1}^{\infty} G_n = \left(\bigcup_{n=1}^{\infty} F_n\right) \cup \left[\bigcup_{n=1}^{\infty} (G_n - F_n)\right]$$
$$\subset \left(\bigcup_{n=1}^{\infty} B_n\right) \cup \left[\bigcup_{n=1}^{\infty} (G_n - F_n)\right], \qquad (23)$$

so

$$p\left(\bigcup_{n=1}^{\infty} G_n\right) \leq p\left(\bigcup_{n=1}^{\infty} B_n\right) + \varepsilon,$$

and since ε is arbitrary,

$$p\left(\bigcup_{n=1}^{\infty} B_n\right) = \inf\left\{p(G) \middle| \bigcup_{n=1}^{\infty} B_n \subset G, G \text{ open}\right\}.$$

It is also apparent from (23) that

$$p\left(\bigcup_{n=1}^{\infty} B_n\right) \leq p\left(\bigcup_{n=1}^{\infty} F_n\right) + \varepsilon,$$

so for N sufficiently large,

$$p\left(\bigcup_{n=1}^{\infty} B_n\right) \leq p\left(\bigcup_{n=1}^{N} F_n\right) + 2\varepsilon.$$

The finite union $\bigcup_{n=1}^{N} F_n$ is a closed subset of $\bigcup_{n=1}^{\infty} B_n$ and ε is arbitrary, so

$$p\left(\bigcup_{n=1}^{\infty} B_n\right) = \sup\left\{p(F) \,\middle|\, F \subset \bigcup_{n=1}^{\infty} B_n, \, F \text{ closed}\right\}.$$

This shows that $\mathscr{E}$ is closed under countable unions and completes the proof.   Q.E.D.

From Proposition 7.17 we conclude that a probability measure on a metrizable space is completely determined by its values on the open or closed sets. The following proposition is a similar result. It states that a probability measure p on a metric space (X, d) is completely determined by the values $\int g \, dp$, where g ranges over $U_d(X)$.

Proposition 7.18   Let X be a metrizable space and d a metric on X consistent with its topology. If $p_1, p_2 \in P(X)$ and

$$\int g \, dp_1 = \int g \, dp_2 \qquad \forall g \in U_d(X),$$

then $p_1 = p_2$.

Proof   Let F be any closed proper subset of X and let $G_n = \{x \in X \,|\, d(x, F) < 1/n\}$. For sufficiently large n, F and G_n^c are disjoint nonempty closed sets for which $\inf_{x \in F, y \in G_n^c} d(x, y) > 0$, so by Lemma 7.1, there exist functions $f_n \in U_d(X)$ such that $f_n(x) = 0$ for $x \in G_n^c$, $f_n(x) = 1$ for $x \in F$, and $0 \le f_n(x) \le 1$ for every $x \in X$. Then

$$p_1(F) \le \int f_n \, dp_1 = \int f_n \, dp_2 \le p_2(G_n),$$

and so

$$p_1(F) \le p_2\left(\bigcap_{n=1}^{\infty} G_n\right) = p_2(F).$$

Reversing the roles of p_1 and p_2, we obtain $p_1(F) = p_2(F)$. Proposition 7.17 implies $p_1(B) = p_2(B)$ for every $B \in \mathscr{B}_X$.   Q.E.D.

7.4.2   The Weak Topology

We turn now to a discussion of topologies on $P(X)$, where X is a metrizable space. Given $\varepsilon > 0$, $p \in P(X)$, and $f \in C(X)$, define the subset of $P(X)$:

$$V_\varepsilon(p; f) = \left\{q \in P(X) \,\middle|\, \left|\int f \, dq - \int f \, dp\right| < \varepsilon\right\}. \tag{24}$$

If $D \subset C(X)$, consider the collection of subsets of $P(X)$:

$$\mathscr{V}(D) = \left\{ V_\varepsilon(p;f) \middle| \varepsilon > 0, \, p \in P(X), \, f \in D \right\}.$$

Let $\mathscr{T}(D)$ be the weakest topology on $P(X)$ which contains the collection $\mathscr{V}(D)$, i.e., the topology for which $\mathscr{V}(D)$ is a subbase.

Lemma 7.6   Let X be a metrizable space and $D \subset C(X)$. Let $\{p_\alpha\}$ be a net in $P(X)$ and $p \in P(X)$. Then $p_\alpha \to p$ relative to the topology $\mathscr{T}(D)$ if and only if $\int f \, dp_\alpha \to \int f \, dp$ for every $f \in D$.

Proof   Suppose $p_\alpha \to p$ and $f \in D$. Then, given $\varepsilon > 0$, there exists β such that $\alpha \geq \beta$ implies $p_\alpha \in V_\varepsilon(p;f)$. Hence $\int f \, dp_\alpha \to \int f \, dp$. Conversely, if $\int f \, dp_\alpha \to \int f \, dp$ for every $f \in D$, and $G \in \mathscr{T}(D)$ contains p, then p is also contained in some basic open set $\bigcap_{k=1}^{n} V_{\varepsilon_k}(p;f_k) \subset G$, where $\varepsilon_k > 0$ and $f_k \in D$, $k = 1, \ldots, n$. Choose β such that for all $\alpha \geq \beta$ we have $\left| \int f_k \, dp_\alpha - \int f_k \, dp \right| < \varepsilon_k$, $k = 1, \ldots, n$. Then $p_\alpha \in G$ for $\alpha \geq \beta$, so $p_\alpha \to p$.   Q.E.D.

We are really interested in $\mathscr{T}[C(X)]$, the so-called *weak topology* on $P(X)$. The space $C(X)$ is too large to be manipulated easily, so we will need a countable set $D \subset C(X)$ such that $\mathscr{T}(D) = \mathscr{T}[C(X)]$. Such a set D is produced by the next three lemmas.

Lemma 7.7   Let X be a metrizable space and d a metric on X consistent with its topology. If $f \in C(X)$, then there exist sequences $\{g_n\}$ and $\{h_n\}$ in $U_d(X)$ such that $g_n \uparrow f$ and $h_n \downarrow f$.

Proof   We need only produce the sequence $\{g_n\}$, since the other case follows by considering $-f$. In Lemma 7.14 under weaker assumptions we will have occasion to utilize the construction about to be described, so we are careful to point out which assumptions are being used. If $f \in C(X)$, then f is bounded below by some $b \in R$, and for at least one $x_0 \in X$, $f(x_0) < \infty$. Define

$$g_n(x) = \inf_{y \in X} [f(y) + nd(x,y)]. \tag{25}$$

Note that for every $x \in X$,

$$b \leq g_n(x) \leq f(x) + nd(x,x) = f(x),$$

and

$$b \leq g_n(x) \leq f(x_0) + nd(x,x_0) < \infty.$$

Thus

$$b \leq g_1 \leq g_2 \leq \cdots \leq f, \tag{26}$$

and each g_n is finite-valued. For every $x, y, z \in X$,

$$f(y) + nd(x, y) \le f(y) + nd(z, y) + nd(x, z),$$

and infimizing first the left side and then the right over $y \in X$, we obtain

$$g_n(x) \le g_n(z) + nd(x, z).$$

Reverse the roles of x and z to show that

$$|g_n(x) - g_n(z)| \le nd(x, z),$$

so $g_n \in U_d(X)$ for each n. From (26) we have

$$\lim_{n \to \infty} g_n \le f. \tag{27}$$

We have so far used only the facts that f is bounded below and not identically ∞. To prove that equality holds in (27), we use the continuity of f. For $x \in X$, and $\varepsilon > 0$, let $\{y_n\} \subset X$ be such that

$$f(y_n) + nd(x, y_n) \le g_n(x) + \varepsilon.$$

As $n \to \infty$, either $g_n \uparrow \infty$, in which case equality must hold in (27), or else $y_n \to x$. In the latter case we have

$$f(x) = \lim_{n \to \infty} f(y_n) \le \lim_{n \to \infty} g_n(x) + \varepsilon, \tag{28}$$

and since x and ε are arbitrary, equality holds in (27).   Q.E.D.

Lemma 7.8   Let X be a metrizable space and d a metric on X consistent with its topology. Then $\mathcal{T}[C(X)] = \mathcal{T}[U_d(X)]$.

Proof   Since $U_d(X) \subset C(X)$, we have $\mathcal{V}[U_d(X)] \subset \mathcal{V}[C(X)]$ and $\mathcal{T}[U_d(X)] \subset \mathcal{T}[C(X)]$. To prove the reverse containment, we show that every set in $\mathcal{V}[C(X)]$ is open in the $\mathcal{T}[U_d(X)]$ topology. Thus, given any set $V_\varepsilon(p; f) \in \mathcal{V}[C(X)]$ and any point p_0 in this set, we will construct a set in $\mathcal{T}[U_d(X)]$ containing p_0 and contained in $V_\varepsilon(p; f)$. Given $V_\varepsilon(p; f)$ and $p_0 \in V_\varepsilon(p; f)$, there exists $\varepsilon_0 > 0$ for which $V_{\varepsilon_0}(p_0; f) \subset V_\varepsilon(p; f)$. By Lemma 7.7, there exist functions g and h in $U_d(X)$ such that $g \le f \le h$ and

$$\int f \, dp_0 < \int g \, dp_0 + (\varepsilon_0/2), \qquad \int h \, dp_0 < \int f \, dp_0 + (\varepsilon_0/2).$$

If $q \in V_{\varepsilon_0/2}(p_0; g) \cap V_{\varepsilon_0/2}(p_0; h)$, then

$$\int f \, dp_0 < \int g \, dp_0 + (\varepsilon_0/2) < \int g \, dq + \varepsilon_0 \le \int f \, dq + \varepsilon_0$$

and

$$\int f \, dq \le \int h \, dq < \int h \, dp_0 + (\varepsilon_0/2) < \int f \, dp_0 + \varepsilon_0,$$

so

$$\left| \int f \, dq - \int f \, dp_0 \right| < \varepsilon_0,$$

i.e., $q \in V_{\varepsilon_0}(p_0; f)$ and

$$V_{\varepsilon_0/2}(p_0; g) \cap V_{\varepsilon_0/2}(p_0; h) \subset V_\varepsilon(p; f). \qquad \text{Q.E.D.}$$

Lemma 7.9   Let X be a metrizable space and d a metric on X consistent with its topology. If D is dense in $U_d(X)$, then $\mathcal{T}[U_d(X)] = \mathcal{T}(D)$.

Proof   It is clear that $\mathcal{T}(D) \subset \mathcal{T}[U_d(X)]$. To prove the reverse set containment, we choose a set $V_\varepsilon(p; g) \in \mathcal{V}[U_d(X)]$, select a point p_0 in this set, and construct a set in $\mathcal{T}(D)$ containing p_0 and contained in $V_\varepsilon(p; g)$. Let

$$\varepsilon_0 = \varepsilon - \left| \int g \, dp_0 - \int g \, dp \right| > 0.$$

Let $h \in D$ be such that $\|g - h\| < \varepsilon_0/3$. Then for any $q \in V_{\varepsilon_0/3}(p_0; h)$, we have

$$\left| \int g \, dq - \int g \, dp \right| \leq \left| \int g \, dq - \int h \, dq \right| + \left| \int h \, dq - \int h \, dp_0 \right|$$

$$+ \left| \int h \, dp_0 - \int g \, dp_0 \right| + \left| \int g \, dp_0 - \int g \, dp \right|$$

$$< (\varepsilon_0/3) + (\varepsilon_0/3) + (\varepsilon_0/3) + \left| \int g \, dp_0 - \int g \, dp \right| = \varepsilon,$$

so $V_{\varepsilon_0/3}(p_0; h) \subset V_\varepsilon(p; g). \qquad \text{Q.E.D.}$

Proposition 7.19   Let X be a separable metrizable space. There is a metric d on X consistent with its topology and a countable dense subset D of $U_d(X)$ such that $\mathcal{T}(D)$ is the weak topology $\mathcal{T}[C(X)]$ on $P(X)$.

Proof   Corollary 7.6.1 states that the separable metrizable space X has a totally bounded metrization d. By Proposition 7.9, there exists a countable dense set D in $U_d(X)$. The conclusion follows from Lemmas 7.8 and 7.9.
$$\text{Q.E.D.}$$

From this point on, whenever X is a metrizable space, we will understand $P(X)$ to be a topological space with the weak topology $\mathcal{T}[C(X)]$. We will show that when X is separable and metrizable, $P(X)$ is separable and metrizable; when X is compact and metrizable, $P(X)$ is compact and metrizable; when X is separable and topologically complete, $P(X)$ is separable and topologically complete; and when X is a Borel space, $P(X)$ is a Borel space.

Proposition 7.20   If X is a separable metrizable space, then $P(X)$ is separable and metrizable.

Proof   Let d be a metric on X consistent with its topology and D a countable dense subset of $U_d(X)$ such that $\mathcal{T}(D)$ is the weak topology on $P(X)$ (Proposition 7.19). Let R^∞ be the product of countably many copies of the real line and let $\varphi : P(X) \to R^\infty$ be defined by

$$\varphi(p) = \left(\int g_1 \, dp, \int g_2 \, dp, \dots \right),$$

where $\{g_1, g_2, \dots\}$ is an enumeration of D. We will show that φ is a homeomorphism, and since R^∞ is metrizable and separable (Proposition 7.4), these properties for $P(X)$ will follow.

Suppose that $\varphi(p_1) = \varphi(p_2)$, so that $\int g_k \, dp_1 = \int g_k \, dp_2$ for every $g_k \in D$. If $g \in U_d(X)$, then there exists a sequence $\{g_{k_j}\} \subset D$ such that $\|g_{k_j} - g\| \to 0$ as $j \to \infty$. Then

$$\left| \int g \, dp_1 - \int g \, dp_2 \right| \le \limsup_{j \to \infty} \left| \int (g - g_{k_j}) \, dp_1 \right| + \limsup_{j \to \infty} \left| \int g_{k_j} \, dp_1 - \int g_{k_j} \, dp_2 \right|$$

$$+ \limsup_{j \to \infty} \left| \int (g_{k_j} - g) \, dp_2 \right|$$

$$\le 2 \limsup_{j \to \infty} \|g_{k_j} - g\| = 0,$$

so $\int g \, dp_1 = \int g \, dp_2$. Proposition 7.18 implies that $p_1 = p_2$, so φ is one-to-one. For each $g_k \in D$, the mapping $p \to \int g_k \, dp$ is continuous by Lemma 7.6, so φ is continuous. To show that φ^{-1} is continuous, let $\{p_\alpha\}$ be a net in $P(X)$ such that $\varphi(p_\alpha) \to \varphi(p)$ for some $p \in P(X)$. Then $\int g_k \, dp_\alpha \to \int g_k \, dp$ for every $g_k \in D$, and by Lemma 7.6, $p_\alpha \to p$.   Q.E.D.

Proposition 7.20 guarantees that when X is separable and metrizable, the topology on $P(X)$ can be characterized in terms of convergent sequences rather than nets. We give several conditions which are equivalent to convergence in $P(X)$.

Proposition 7.21   Let X be a separable metrizable space and let d be a metric on X consistent with its topology. Let $\{p_n\}$ be a sequence in $P(X)$ and $p \in P(X)$. The following statements are equivalent:

(a) $p_n \to p$;

(b) $\int f \, dp_n \to \int f \, dp$ for every $f \in C(X)$;

(c) $\int g \, dp_n \to \int g \, dp$ for every $g \in U_d(X)$;

(d) $\limsup_{n \to \infty} p_n(F) \le p(F)$ for every closed set $F \subset X$;

(e) $\liminf_{n \to \infty} p_n(G) \ge p(G)$ for every open set $G \subset X$.

Proof The equivalence of (a), (b), and (c) follows from Lemmas 7.6 and 7.8. The equivalence of (d) and (e) follows by complementation.

To show that (b) implies (d), let F be a closed proper nonempty subset of X and let $G_k = \{x \in X \mid d(x, F) < 1/k\}$. For k sufficiently large, F and G_k^c are disjoint nonempty sets, and there exist functions $f_k \in C(X)$ such that $f_k(x) = 1$ for $x \in F$, $f_k(x) = 0$ for $x \in G_k^c$, and $0 \le f(x) \le 1$ for every $x \in X$. Using (b) we have

$$\limsup_{n \to \infty} p_n(F) \le \lim_{n \to \infty} \int f_k \, dp_n = \int f_k \, dp \le p(G_k),$$

and letting $k \to \infty$, we obtain (d).

To show that (d) implies (b), choose $f \in C(X)$ and assume without loss of generality that $0 \le f \le 1$. Choose a positive integer K and define closed sets

$$F_k = \{x \in X \mid f(x) \ge k/K\}, \qquad k = 0, \ldots, K.$$

Define $\varphi : X \to [0,1]$ by

$$\varphi(x) = \sum_{k=0}^{K} (k/K) \chi_{F_k - F_{k+1}}(x),$$

where $F_{K+1} = \varnothing$. Then $f - (1/K) \le \varphi \le f$, and, for any $q \in P(X)$,

$$\int \varphi \, dq = \sum_{k=0}^{K} (k/K) q(F_k - F_{k+1}) = (1/K) \sum_{k=1}^{K} q(F_k).$$

Using (d) we have

$$\limsup_{n \to \infty} \int f \, dp_n - (1/K) \le \limsup_{n \to \infty} \int \varphi \, dp_n$$

$$= (1/K) \limsup_{n \to \infty} \sum_{k=1}^{K} p_n(F_k)$$

$$\le (1/K) \sum_{k=1}^{K} p(F_k) = \int \varphi \, dp \le \int f \, dp,$$

and since K is arbitrary, we obtain

$$\limsup_{n \to \infty} \int f \, dp_n \le \int f \, dp \tag{29}$$

for every $f \in C(X)$. In particular, (29) holds for $-f$, so

$$\liminf_{n \to \infty} \int f \, dp_n = -\limsup_{n \to \infty} \int (-f) \, dp_n \ge -\int (-f) \, dp = \int f \, dp. \tag{30}$$

Combine (29) and (30) to conclude (b).   Q.E.D.

When X is a metrizable space, we denote by p_x the probability measure on $p(X)$ which assigns unit point mass to x, i.e., $p_x(B) = 1$ if and only if $x \in B$.

Corollary 7.21.1 Let X be a metrizable space. The mapping $\delta: X \to P(X)$ defined by $\delta(x) = p_x$ is a homeomorphism.

Proof It is clear that δ is one-to-one. Suppose $\{x_n\}$ is a sequence in X and $x \in X$. If $x_n \to x$ and G is an open subset of X, then there are two possibilities. Either $x \in G$, in which case $x_n \in G$ for sufficiently large n, so $\liminf_{n \to \infty} p_{x_n}(G) = 1 = p_x(G)$, or else $x \notin G$, in which case $\liminf_{n \to \infty} p_{x_n}(G) \geq 0 = p_x(G)$. Proposition 7.21 implies $p_{x_n} \to p_x$, so δ is continuous. On the other hand, if $p_{x_n} \to p_x$ and G is an open neighborhood of x, then since $\liminf_{n \to \infty} p_{x_n}(G) \geq p_x(G) = 1$, we must have $x_n \in G$ for sufficiently large n, i.e., $x_n \to x$. This shows that δ is a homeomorphism.   Q.E.D.

From Corollary 7.21.1 we see that p_n can converge to p in such a way that strict inequality holds in (d) and (e) of Proposition 7.21. For example, let $G \subset X$ be open, let x be on the boundary of G, and let x_n converge to x through G. Then $p_{x_n}(G) = 1$ for every n, but $p_x(G) = 0$.

We now show that compactness of X is inherited by $P(X)$.

Proposition 7.22 If X is a compact metrizable space, then $P(X)$ is a compact metrizable space.

Proof If X is a compact metrizable space, it is separable (Corollary 7.6.2) and $C(X)$ is separable (Proposition 7.7). Let $\{f_k\}$ be a countable set in $C(X)$ such that $f_1 \equiv 1$, $\|f_k\| \leq 1$ for every k, and $\{f_k\}$ is dense in the unit sphere $\{f \in C(X) \mid \|f\| \leq 1\}$. Let $[-1,1]^\infty$ be the product of countably many copies of $[-1,1]$ and define $\varphi: P(X) \to [-1,1]^\infty$ by

$$\varphi(p) = \left(\int f_1 \, dp, \int f_2 \, dp, \ldots \right).$$

A trivial modification of the proof of Proposition 7.20 shows φ is a homeomorphism. We will show that $\varphi[P(X)]$ is closed in the compact space $[-1,1]^\infty$, and the compactness of $P(X)$ will follow.

Suppose $\{p_n\}$ is a sequence in $P(X)$ and $\varphi(p_n) \to (\alpha_1, \alpha_2, \ldots) \in [-1,1]^\infty$. Given $\varepsilon > 0$ and $f \in C(X)$ with $\|f\| \leq 1$, there is a function f_k with $\|f - f_k\| < \varepsilon/3$. There is a positive integer N such that $n, m \geq N$ implies $|\int f_k \, dp_n - \int f_k \, dp_m| < \varepsilon/3$. Then

$$\left| \int f \, dp_n - \int f \, dp_m \right| \leq \left| \int f \, dp_n - \int f_k \, dp_n \right| + \left| \int f_k \, dp_n - \int f_k \, dp_m \right|$$

$$+ \left| \int f_k \, dp_m - \int f \, dp_m \right| < \varepsilon,$$

so $\{\int f \, dp_n\}$ is Cauchy in $[-1, 1]$. Denote its limit by $E(f)$. If $\|f\| > 1$, define

$$E(f) = \|f\| E(f/\|f\|).$$

It is easily verified that E is a linear functional on $C(X)$, that $E(f) \geq 0$ whenever $f \geq 0$, $|E(f)| \leq \|f\|$ for every $f \in C(X)$, and $E(f_1) = 1$. Suppose $\{h_n\}$ is a sequence in $C(X)$ and $h_n(x) \downarrow 0$ for every $x \in X$. Then for each $\varepsilon > 0$, the set $K_n(\varepsilon) = \{x | h_n(x) \geq \varepsilon\}$ is compact, and $\bigcap_{n=1}^{\infty} K_n(\varepsilon) = \varnothing$. Therefore, for n sufficiently large, $K_n(\varepsilon) = \varnothing$, which implies $\|h_n\| \downarrow 0$. Consequently, $E(h_n) \downarrow 0$. This shows that the functional E is a *Daniell integral*, and by a classical theorem (see, e.g., Royden [R5, p. 299, Proposition 21]) there exists a unique probability measure on $\sigma[\bigcup_{f \in C(X)} f^{-1}(\mathscr{B}_R)]$ which satisfies $E(f) = \int f \, dp$ for every $f \in C(X)$. Proposition 7.10 implies $p \in P(X)$. We have

$$\alpha_k = \lim_{n \to \infty} \int f_k \, dp_n = E(f_k) = \int f_k \, dp, \qquad k = 1, 2, \ldots,$$

so $\varphi(p_n) \to \varphi(p)$. This proves $\varphi[P(X)]$ is closed.   Q.E.D.

In order to show that toplogical completeness and separability of X imply the same properties for $P(X)$, we need the following lemma.

Lemma 7.10   Let X and Y be separable metrizable spaces and $\varphi: X \to Y$ a homeomorphism. The mapping $\psi: P(X) \to P(Y)$ defined by

$$\psi(p)(B) = p[\varphi^{-1}(B)] \qquad \forall B \in \mathscr{B}_Y$$

is a homeomorphism.

Proof   Suppose $p_1, p_2 \in P(X)$ and $p_1 \neq p_2$. Since p_1 and p_2 are regular, there is an open set $G \subset X$ for which $p_1(G) \neq p_2(G)$. The image $\varphi(G)$ is relatively open in $\varphi(X)$, so $\varphi(G) = \varphi(X) \cap B$, where B is open in Y. It is clear that

$$\psi(p_1)(B) = p_1(G) \neq p_2(G) = \psi(p_2)(B),$$

so ψ is one-to-one. Let $\{p_n\}$ be a sequence in $P(X)$ and $p \in P(X)$. If $p_n \to p$, then since $\varphi^{-1}(H)$ is open in X for every open set $H \subset Y$, Proposition 7.21 implies

$$\liminf_{n \to \infty} \psi(p_n)(H) = \liminf_{n \to \infty} p_n[\varphi^{-1}(H)] \geq p[\varphi^{-1}(H)] = \psi(p)(H),$$

so $\psi(p_n) \to \psi(p)$ and ψ is continuous. If we are given $\{p_n\}$ and p such that $\psi(p_n) \to \psi(p)$, a reversal of this argument shows that $p_n \to p$ and ψ^{-1} is continuous.   Q.E.D.

Proposition 7.23   If X is a topologically complete separable space, then $P(X)$ is topologically complete and separable.

Proof By Urysohn's theorem (Proposition 7.2) there is a homeomorphism $\varphi: X \to \mathscr{H}$, and the mapping ψ obtained by replacing Y by $\mathscr{H}$ in Lemma 7.10 is a homeomorphism from $P(X)$ to $P(\mathscr{H})$. Alexandroff's theorem (Proposition 7.3) implies $\varphi(X)$ is a G_δ-subset of $\mathscr{H}$, and we see that

$$\psi[P(X)] = \{p \in P(\mathscr{H}) | p[\mathscr{H} - \varphi(X)] = 0\}. \tag{31}$$

We will show $\psi[P(X)]$ is a G_δ-subset of the compact space $P(\mathscr{H})$ (Proposition 7.22) and use Alexandroff's theorem again to conclude that $P(X)$ is topologically complete.

Since $\varphi(X)$ is a G_δ-subset of $\mathscr{H}$, we can find open sets $G_1 \supset G_2 \supset \cdots$ such that $\varphi(X) = \bigcap_{n=1}^\infty G_n$. It is clear from (31) that

$$\psi[P(X)] = \bigcap_{n=1}^\infty \{p \in P(\mathscr{H}) | p(\mathscr{H} - G_n) = 0\}$$

$$= \bigcap_{n=1}^\infty \bigcap_{k=1}^\infty \{p \in P(\mathscr{H}) | p(\mathscr{H} - G_n) < 1/k\}.$$

But for any closed set F and real number c, the set $\{p \in P(\mathscr{H}) | p(F) \geq c\}$ is closed by Proposition 7.21(d), and $\{p \in P(\mathscr{H}) | p(\mathscr{H} - G_n) < 1/k\}$ is the complement of such a set.   Q.E.D.

We turn now to characterizing the σ-algebra $\mathscr{B}_{P(X)}$ when X is metrizable and separable. From Lemma 7.6, we have that the mapping $\theta_f : P(X) \to R$ given by

$$\theta_f(p) = \int f \, dp$$

is continuous for every $f \in C(X)$. One can easily verify from Proposition 7.21 that the mapping $\theta_B : P(X) \to [0,1]$ defined by[†]

$$\theta_B(p) = p(B)$$

is Borel-measurable when B is a closed subset of X. (Indeed, in the final stage of the proof of Proposition 7.23, we used the fact that when B is closed the upper level sets $\{p \in P(X) | \theta_B(p) \geq c\}$ are closed.) Likewise, when B is open, θ_B is Borel-measurable. It is natural to ask if θ_B is also Borel-measurable when B is an arbitrary Borel set. The answer to this is yes, and in fact, $\mathscr{B}_{P(X)}$ is the smallest σ-algebra with respect to which θ_B is measurable for every $B \in \mathscr{B}_X$. A useful aid in proving this and several subsequent results is the concept of a Dynkin system.

[†] The use of the symbol θ_B here is a slight abuse of notation. In keeping with the definition of θ_f, the technically correct symbol would be θ_{χ_B}.

Definition 7.11   Let X be a set and $\mathscr{D}$ a class of subsets of X. We say $\mathscr{D}$ is a *Dynkin system* if the following conditions hold:

(a) $X \in \mathscr{D}$.
(b)   If $A, B \in \mathscr{D}$ and $B \subset A$, then $A - B \in \mathscr{D}$.
(c)   If $A_1, A_2, \ldots \in \mathscr{D}$ and $A_1 \subset A_2 \subset \cdots$, then $\bigcup_{n=1}^{\infty} A_n \in \mathscr{D}$.

Proposition 7.24   (Dynkin system theorem)   Let $\mathscr{F}$ be a class of subsets of a set X, and assume $\mathscr{F}$ is closed under finite intersections. If $\mathscr{D}$ is a Dynkin system containing $\mathscr{F}$, then $\mathscr{D}$ also contains $\sigma(\mathscr{F})$.

Proof   This is a standard result in measure theory. See, for example, Ash [A1, page 169].   Q.E.D.

Proposition 7.25   Let X be a separable metrizable space and $\mathscr{E}$ a collection of subsets of X which generates $\mathscr{B}_X$ and is closed under finite intersections. Then $\mathscr{B}_{P(X)}$ is the smallest σ-algebra with respect to which all functions of the form

$$\theta_E(p) = p(E), \qquad E \in \mathscr{E},$$

are measurable from $P(X)$ to $[0, 1]$, i.e.,

$$\mathscr{B}_{P(X)} = \sigma\left[\bigcup_{E \in \mathscr{E}} \theta_E^{-1}(\mathscr{B}_R) \right].$$

Proof   Let $\mathscr{F}$ be the smallest σ-algebra with respect to which θ_E is measurable for every $E \in \mathscr{E}$. To show $\mathscr{F} \subset \mathscr{B}_{P(X)}$, we show that θ_B is $\mathscr{B}_{P(X)}$-measurable for every $B \in \mathscr{B}_X$. Let $\mathscr{D} = \{B \in \mathscr{B}_X | \theta_B$ is $\mathscr{B}_{P(X)}$-measurable$\}$. It is easily verified that $\mathscr{D}$ is a Dynkin system. We have already seen that $\mathscr{D}$ contains every closed set, so the Dynkin system theorem (Proposition 7.24) implies $\mathscr{D} = \mathscr{B}_X$.

It remains to show that $\mathscr{B}_{P(X)} \subset \mathscr{F}$. Let $\mathscr{D}' = \{B \in \mathscr{B}_X | \theta_B$ is $\mathscr{F}$-measurable$\}$ As before, $\mathscr{D}'$ is a Dynkin system, and since $\mathscr{E} \subset \mathscr{D}'$, we have $\mathscr{D}' = \mathscr{B}_X$. Thus the function $\theta_f(p) = \int f \, dp$ is $\mathscr{F}$-measurable when f is the indicator of a Borel set. Therefore θ_f is $\mathscr{F}$-measurable when f is a Borel-measurable simple function. If $f \in C(X)$, then there is a sequence of simple functions f_n which are uniformly bounded below such that $f_n \uparrow f$. The monotone convergence theorem implies $\theta_{f_n} \uparrow \theta_f$, so θ_f is $\mathscr{F}$-measurable. It follows that for $\varepsilon > 0$, $p \in P(X)$, and $f \in C(X)$, the subbasic open set

$$V_\varepsilon(p; f) = \left\{ q \in P(X) \, \middle| \, \left| \int f \, dq - \int f \, dp \right| < \varepsilon \right\}$$

is $\mathscr{F}$-measurable. It follows that $\mathscr{B}_{P(X)} = \mathscr{F}$ (see the remark following Definition 7.6).   Q.E.D.

Corollary 7.25.1   If X is a Borel space, then $P(X)$ is a Borel space.

Proof   Let φ be a homeomorphism mapping X onto a Borel subset of a topologically complete separable space Y. Then, by Lemma 7.10, $P(X)$ is homeomorphic to the Borel set $\{p \in P(Y) \mid p[\varphi(X)] = 1\}$. Since $P(Y)$ is topologically complete and separable (Proposition 7.23), the result follows.   Q.E.D.

7.4.3   Stochastic Kernels

We now consider probability measures on a separable metrizable space parameterized by the elements of another separable metrizable space.

Definition 7.12   Let X and Y be separable metrizable spaces. A *stochastic kernel* $q(dy|x)$ on Y given X is a collection of probability measures in $P(Y)$ parameterized by $x \in X$. If $\mathscr{F}$ is a σ-algebra on X and $\gamma^{-1}[\mathscr{B}_{P(Y)}] \subset \mathscr{F}$, where $\gamma : X \to P(Y)$ is defined by

$$\gamma(x) = q(dy|x), \tag{32}$$

then $q(dy|x)$ is said to be $\mathscr{F}$-*measurable*. If γ is continuous, $q(dy|x)$ is said to be *continuous*.

Proposition 7.26   Let X and Y be Borel spaces, $\mathscr{E}$ a collection of subsets of Y which generates $\mathscr{B}_Y$ and is closed under finite intersections, and $q(dy|x)$ a stochastic kernel on Y given X. Then $q(dy|x)$ is Borel-measurable if and only if the mapping $\lambda_E : X \to [0, 1]$ defined by

$$\lambda_E(x) = q(E|x)$$

is Borel-measurable for every $E \in \mathscr{E}$.

Proof   Let $\gamma : X \to P(Y)$ be defined by $\gamma(x) = q(dy|x)$. Then for $E \in \mathscr{E}$, we have $\lambda_E = \theta_E \circ \gamma$. If $q(dy|x)$ is Borel-measurable (i.e., γ is Borel-measurable), then Proposition 7.25 implies λ_E is Borel-measurable for every $E \in \mathscr{E}$. Conversely, if λ_E is Borel-measurable for every $E \in \mathscr{E}$, then $\sigma[\bigcup_{E \in \mathscr{E}} \lambda_E^{-1}(\mathscr{B}_R)] \subset \mathscr{B}_X$. Proposition 7.25 implies

$$\gamma^{-1}[\mathscr{B}_{P(Y)}] = \gamma^{-1}\left[\sigma\left(\bigcup_{E \in \mathscr{E}} \theta_E^{-1}(\mathscr{B}_R)\right)\right]$$

$$= \sigma\left[\bigcup_{E \in \mathscr{E}} \gamma^{-1}(\theta_E^{-1}(\mathscr{B}_R))\right] = \sigma\left[\bigcup_{E \in \mathscr{E}} \lambda_E^{-1}(\mathscr{B}_R)\right] \subset \mathscr{B}_X,$$

so $q(dy|x)$ is Borel-measurable.   Q.E.D.

Corollary 7.26.1   Let X and Y be Borel spaces and $q(dy|x)$ a Borel-measurable stochastic kernel on Y given X. If $B \in \mathscr{B}_{XY}$, then the mapping

$\Lambda_B : X \to [0, 1]$ defined by

$$\Lambda_B(x) = q(B_x|x), \tag{33}$$

where $B_x = \{ y \in Y | (x, y) \in B \}$, is Borel-measurable.

Proof If $B \in \mathscr{B}_{XY}$ and $x \in X$, then $B_x \subset Y$ is homeomorphic to $B \cap [\{x\} Y] \in \mathscr{B}_{XY}$. It follows that $B_x \in \mathscr{B}_Y$, so $q(B_x|x)$ is defined. It is easy to show that the collection $\mathscr{D} = \{ B \in \mathscr{B}_{XY} | \Lambda_B$ is Borel-measurable$\}$ is a Dynkin system. Proposition 7.26 implies that $\mathscr{D}$ contains the measurable rectangles, so $\mathscr{D} = \mathscr{B}_{XY}$.   Q.E.D.

We now show that one can decompose a probability measure on a product of Borel spaces into a marginal and a Borel-measurable stochastic kernel. This decomposition is possible even when a measurable dependence on a parameter is admitted, and, as we shall see in Chapter 10, this result is essential to the filtering algorithm for imperfect state information dynamic programming models.

As a notational convenience, we use $\underline{X}$ to denote a typical Borel subset of a Borel space X.

Proposition 7.27   Let $(X, \mathscr{F})$ be a measurable space, let Y and Z be Borel spaces, and let $q(d(y, z)|x)$ be a stochastic kernel on YZ given X. Assume that $q(B|x)$ is $\mathscr{F}$-measurable in x for every $B \in \mathscr{B}_{YZ}$. Then there exists a stochastic kernel $r(dz|x, y)$ on Z given XY and a stochastic kernel $s(dy|x)$ on Y given X such that $r(\underline{Z}|x, y)$ is $\mathscr{F}\mathscr{B}_Y$-measurable in (x, y) for every $\underline{Z} \in \mathscr{B}_Z$, $s(\underline{Y}|x)$ is $\mathscr{F}$-measurable in x for every $\underline{Y} \in \mathscr{B}_Y$, and

$$q(\underline{YZ}|x) = \int_{\underline{Y}} r(\underline{Z}|x, y) s(dy|x) \qquad \forall \underline{Y} \in \mathscr{B}_Y, \quad \underline{Z} \in \mathscr{B}_Z. \tag{34}$$

Proof   We prove this proposition under the assumption that Y and Z are uncountable. If either Y or Z or both are countable, slight modifications (actually simplifications) of this proof are necessary. From Corollary 7.16.1, we may assume without loss of generality that $Y = Z = (0, 1]$.

Let $s(dy|x)$ be the marginal of $q(d(y, z)|x)$ on Y, i.e., $s(\underline{Y}|x) = q(\underline{Y}Z|x)$ for every $\underline{Y} \in \mathscr{B}_Y$. For each positive integer n, define subsets of Y

$$M(j, n) = ((j - 1)/2^n, j/2^n], \qquad j = 1, \ldots, 2^n.$$

Then each $M(j, n + 1)$ is a subset of some $M(k, n)$, and the collection $\{ M(j, n) | n = 1, 2, \ldots ; j = 1, \ldots, 2^n \}$ generates $\mathscr{B}_Y$. For $z \in Q \cap Z$, define $q(dy(0, z]|x)$ to be the measure on Y whose value at $\underline{Y} \in \mathscr{B}_Y$ is $q(\underline{Y}(0, z]|x)$. Then $q(dy(0, z]|x)$ is absolutely continuous with respect to $s(dy|x)$ for every

$z \in Q \cap Z$ and $x \in X$. Define for $z \in Q \cap Z$

$$G_n(z|x, y) = \begin{cases} q[M(j, n)(0, z]|x]/s[M(j, n)|x] \\ \quad \text{if} \quad y \in M(j, n) \quad \text{and} \quad s[M(j, n)|x] > 0, \\ 0 \quad \text{if} \quad y \in M(j, n) \quad \text{and} \quad s[M(j, n)|x] = 0. \end{cases}$$

The functions $G_n(z|x, y)$ can be regarded as generalized difference quotients of $q(dy(0, z]|x)$ relative to $s(dy|x)$. For each z, the set

$$B(z) = \left\{ (x, y) \in XY \Big| \lim_{n \to \infty} G_n(z|x, y) \text{ exists in } R \right\}$$

$$= \{ (x, y) \in XY \,|\, \{G_n(z|x, y)\} \text{ is Cauchy} \}$$

$$= \bigcap_{k=1}^{\infty} \bigcup_{N=1}^{\infty} \bigcap_{n, m \geq N} \{ (x, y) \in XY \,|\, |G_n(z|x, y) - G_m(z|x, y)| < 1/k \}$$

is $\mathscr{F}\mathscr{B}_Y$-measurable. Theorem 2.5, page 612 of Doob [D4] states that

$$s[B(z)_x|x] = 1 \qquad \forall x \in X, \quad z \in Q \cap Z,$$

and if we define

$$G(z|x, y) = \begin{cases} \lim_{n \to \infty} G_n(z|x, y) & \text{if} \quad (x, y) \in B(z), \\ z & \text{otherwise,} \end{cases}$$

then

$$q[\underline{Y}(0, z]|x] = \int_{\underline{Y}} G(z|x, y)s(dy|x) \qquad \forall x \in X, \quad z \in Q \cap Z, \quad \underline{Y} \in \mathscr{B}_Y. \quad (35)$$

It is clear that for any z, $G(z|x, y)$ is $\mathscr{F}\mathscr{B}_Y$-measurable in (x, y).[†]

A comparison of (34) and (35) suggests that we should try to extend $G(z|x, y)$ in such a way that for fixed (x, y), $G(z|x, y)$ is a distribution function.

[†] For the reader familiar with martingales, we give the proof of the theorem just referenced. Fix x and y and observe that for $m \geq n$,

$$q[M(j, n)(0, z]|x] = \int_{M(j, n)} G_m(z|x, y)s(dy|x). \qquad (*)$$

Since $\{M(j, n)|j = 1, \ldots, 2^n\}$ is the σ-algebra generated by $G_n(z|x, y)$ regarded as a function of y, we conclude that $G_n(z|x, y), n = 1, 2, \ldots$ is a martingale on Y under the measure $s(dy|x)$. Each $G_n(z|x, y)$ is bounded above by 1, so by the martingale convergence theorem (see, e.g., Ash [A1, p. 292]), $G_n(z|x, y)$ converges for $s(dy|x)$ almost every y. Thus $s[B(z)_x|x] = 1$ and the definition of $G(z|x, y)$ given above is possible. Let $m \to \infty$ in ($*$) to see that (35) holds whenever $\underline{Y} = M(j, n)$ for some j and n. The collection of sets $\underline{Y}$ for which (35) holds is a Dynkin system, and it follows from Proposition 7.24 that (35) holds for every $\underline{Y} \in \mathscr{B}_Y$.

Toward this end, for each $z_0 \in Q \cap Z$, we define

$$C(z_0) = \{(x, y) \in XY \mid \exists z \in Q \cap Z \text{ with } z \le z_0 \text{ and } G(z|x, y) > G(z_0|x, y)\},$$

$$= \bigcup_{\substack{z \in Q \cap Z \\ z \le z_0}} \{(x, y) \in XY \mid G(z|x, y) > G(z_0|x, y)\},$$

$$C = \bigcup_{z_0 \in Q \cap Z} C(z_0),$$

$$D(z_0) = \{(x, y) \in XY \mid G(\cdot|x, y) \text{ is not right-continuous at } z_0\}$$

$$= \bigcup_{n=1}^{\infty} \bigcap_{k=1}^{\infty} \bigcup_{\substack{z \in Q \cap Z \\ z_0 \le z < z_0 + 1/k}} \{(x, y) \in XY \mid |G(z|x, y) - G(z_0|x, y)| \ge 1/n\},$$

$$D = \bigcup_{z_0 \in Q \cap Z} D(z_0),$$

$$E = \{(x, y) \in XY \mid G(z|x, y) \text{ does not converge to zero as } z \downarrow 0\}$$

$$= \bigcup_{n=1}^{\infty} \bigcap_{k=1}^{\infty} \bigcup_{\substack{z \in Q \cap Z \\ z < 1/k}} \{(x, y) \in XY \mid |G(z|x, y)| \ge 1/n\},$$

and

$$F = \{(x, y) \in XY \mid G(1|x, y) \ne 1\}.$$

For fixed $x \in X$ and $z_0 \in Q \cap Z$, (35) implies that whenever $z \in Q \cap Z$, $z \le z_0$, then

$$\int_{\underline{Y}} G(z|x, y) s(dy|x) \le \int_{\underline{Y}} G(z_0|x, y) s(dy|x) \qquad \forall \underline{Y} \in \mathcal{B}_Y.$$

Therefore $G(z|x, y) \le G(z_0|x, y)$ for $s(dy|x)$ almost all y, so $s[C(z_0)_x|x] = 0$ and

$$s(C_x|x) = 0. \tag{36}$$

Equation (36) implies that $G(z|x, y)$ is nondecreasing in z for $s(dy|x)$ almost all y. This fact and (35) imply that if $z \downarrow z_0$ ($z \in Q \cap Z$), then

$$\int_Y G(z|x, y) s(dy|x) \downarrow \int_Y G(z_0|x, y) s(dy|x),$$

and

$$G(z|x, y) \downarrow G(z_0|x, y)$$

for $s(dy|x)$ almost all y. Therefore $s[D(z_0)_x|x] = 0$ and

$$s(D_x|x) = 0. \tag{37}$$

Equation (35) also implies that as $z \downarrow 0$ $(z \in Q \cap Z)$

$$\int_{\underline{Y}} G(z|x, y)s(dy|x) \downarrow 0 \qquad \forall \underline{Y} \in \mathscr{B}_Y.$$

Since $G(z|x, y)$ is nondecreasing in z for $s(dy|x)$ almost all y, we must have $G(z|x, y) \downarrow 0$ for $s(dy|x)$ almost all y, i.e.,

$$s(E_x|x) = 0. \tag{38}$$

Substituting $z = 1$ in (35), we see that

$$\int_{\underline{Y}} G(1|x, y)s(dy|x) = s(\underline{Y}|x) \qquad \forall \underline{Y} \in \mathscr{B}_Y,$$

so $G(1|x, y) = 1$ for $s(dy|x)$ almost all y, i.e.,

$$s(F_x|x) = 0. \tag{39}$$

For $z \in Z$, let $\{z_n\}$ be a sequence in $Q \cap Z$ such that $z_n \downarrow z$ and define, for every $x \in X$, $y \in Y$,

$$F(z|x, y) = \begin{cases} \lim_{n \to \infty} G(z_n|x, y) & \text{if } (x, y) \in XY - (C \cup D \cup E \cup F), \\ z & \text{otherwise.} \end{cases} \tag{40}$$

For $(x, y) \in XY - (C \cup D \cup E \cup F)$, $G(z|x, y)$ is a nondecreasing right-continuous function of $z \in Q \cap Z$, so $F(z|x, y)$ is well defined, nondecreasing, and right-continuous. It also satisfies for every $(x, y) \in XY$,

$$0 \le F(z|x, y) \le 1 \qquad \forall z \in Z,$$
$$F(1|x, y) = 1,$$

and

$$\lim_{z \downarrow 0} F(z|x, y) = 0.$$

It is a standard result of probability theory (Ash [A1, p. 24]) that for each (x, y) there is a probability measure $r(dz|x, y)$ on Z such that

$$r((0, z]|x, y) = F(z|x, y) \qquad \forall z \in (0, 1].$$

The collection of subsets $\underline{Z} \in \mathscr{B}_Z$ for which $r(\underline{Z}|x, y)$ is $\mathscr{F}\mathscr{B}_Y$-measurable in (x, y) forms a Dynkin system which contains $\{(0, z]|z \in Z\}$, so $r(\underline{Z}|x, y)$ is $\mathscr{F}\mathscr{B}_Y$-measurable for every $\underline{Z} \in \mathscr{B}_Z$. Relations (35)–(40) and the monotone convergence theorem imply

$$q[\underline{Y}(0, z]|x] = \int_{\underline{Y}} F(z|x, y)s(dy|x)$$
$$= \int_{\underline{Y}} r((0, z]|x, y)s(dy|x) \qquad \forall x \in X, \quad z \in Z, \quad \underline{Y} \in \mathscr{B}_Y. \tag{41}$$

The collection of subsets $\underline{Z} \in \mathcal{B}_Z$ for which (34) holds forms a Dynkin system which contains $\{(0, z] \mid z \in Z\}$, so (34) holds for every $\underline{Z} \in \mathcal{B}_Z$.   Q.E.D.

If $\mathcal{F} = \mathcal{B}_X$, an application of Proposition 7.26 reduces Proposition 7.27 to the following form.

Corollary 7.27.1   Let X, Y, and Z be Borel spaces and let $q(d(y, z) \mid x)$ be a Borel-measurable stochastic kernel on YZ given X. Then there exist Borel-measurable stochastic kernels $r(dz \mid x, y)$ and $s(dy \mid x)$ on Z given XY and on Y given X, respectively, such that (34) holds.

If there is no dependence on the parameter x in Corollary 7.27.1, we have the following well-known result for Borel spaces.

Corollary 7.27.2   Let Y and Z be Borel spaces and $q \in P(YZ)$. Then there exists a Borel-measurable stochastic kernel $r(dz \mid y)$ on Z given Y such that

$$q(\underline{Y}\underline{Z}) = \int_{\underline{Y}} r(\underline{Z} \mid y) s(dy) \qquad \forall \underline{Y} \in \mathcal{B}_Y, \quad \underline{Z} \in \mathcal{B}_Z,$$

where s is the marginal of q on Y.

7.4.4   Integration

As in Section 2.1, we adopt the convention

$$-\infty + \infty = +\infty - \infty = \infty. \tag{42}$$

With this convention, for a, b, $c \in R^*$ the associative law

$$(a + b) + c = a + (b + c)$$

still holds, since if either a, b, or c is ∞, then both sides of (42) are ∞, while if neither a, b, nor c is ∞, the usual arithmetic involving finite numbers and $-\infty$ applies. Also, if $a, b, c \in R^*$ and $a + b = c$, then $a = c - b$, provided $b \neq \pm\infty$. It is always true however that if $a + b \leq c$, then $a \leq c - b$.

We use convention (42) to extend the definition of the integral. If X is a metrizable space, $p \in P(X)$, and $f : X \to R^*$ is Borel-measurable, we define

$$\int f \, dp = \int f^+ \, dp - \int f^- \, dp. \tag{43}$$

Note that if $\int f^+ \, dp < \infty$ or if $\int f^- \, dp < \infty$, (43) reduces to the classical definition of $\int f \, dp$. We collect some of the properties of integration in this extended sense in the following lemma.

Lemma 7.11   Let X be a metrizable space and let $p \in P(X)$ be given. Let f, g and f_n, $n = 1, 2, \ldots$, be Borel-measurable, extended real-valued functions on X.

(a)   Using (42) to define $f + g$, we have

$$\int (f + g)\, dp \leq \int f\, dp + \int g\, dp. \tag{44}$$

(b)   If either
    (b1) $\int f^+\, dp < \infty$ and $\int g^+\, dp < \infty$, or
    (b2) $\int f^-\, dp < \infty$ and $\int g^-\, dp < \infty$, or
    (b3) $\int g^+\, dp < \infty$ and $\int g^-\, dp < \infty$, then

$$\int (f + g)\, dp = \int f\, dp + \int g\, dp. \tag{45}$$

(c)   If $0 < \alpha < \infty$, then $\int (\alpha f)\, dp = \alpha \int f\, dp$.
(d)   If $f \leq g$, then $\int f\, dp \leq \int g\, dp$.
(e)   If $f_n \uparrow f$ and $\int f_1\, dp > -\infty$, then $\int f_n\, dp \uparrow \int f\, dp$.
(f)   If $f_n \downarrow f$ and $\int f_1\, dp < \infty$, then $\int f_n\, dp \downarrow \int f\, dp$.

Proof   We prove (b) first and then return to (a). Under assumption (b1), we have $f(x) < \infty$ and $g(x) < \infty$ for p almost every x, so the sum $f(x) + g(x)$ can be defined without resort to the convention (42) for p almost every x. Furthermore, $\int f\, dp < \infty$ and $\int g\, dp < \infty$, so (45) follows from the additivity theorem for classical integration theory (Ash [A1, p. 45]). The proof of (45) under assumption (b2) is similar. Under assumption (b3), either $\int f^+\, dp = \infty$, in which case both sides of (45) are ∞, or else $\int f^+\, dp < \infty$, in which case assumption (b1) holds. Returning to (a), we note that if assumption (b1) holds, then (45) implies (44). If assumption (b1) fails to hold, then

$$\int f\, dp + \int g\, dp = \infty,$$

so (44) is still valid. Statements (c) and (d) are simple consequences of (42) and (43). Statement (e) follows from the extended monotone convergence theorem (Ash [A1, p. 47]) if $\int f_1^-\, dp < \infty$. If $\int f_1^-\, dp = \infty$, then $\int f_1\, dp > -\infty$ implies $\int f_1^+\, dp = \int f_1\, dp = \infty$, and the conclusion follows from (d). Statement (f) follows from the extended monotone convergence theorem.   Q.E.D.

We saw in Corollary 7.27.2 that a probability measure on a product of Borel spaces can be decomposed into a stochastic kernel and a marginal. This process can be reversed, that is, given a probability measure and one or more Borel-measurable stochastic kernels on Borel spaces, a unique probability measure on the product space can be constructed.

Proposition 7.28   Let $X_1, X_2, \ldots$ be a sequence of Borel spaces, $Y_n = X_1 X_2 \cdots X_n$ and $Y = X_1 X_2 \cdots$. Let $p \in P(X_1)$ be given, and, for $n = 1, 2, \ldots$, let $q_n(dx_{n+1} | y_n)$ be a Borel-measurable stochastic kernel on X_{n+1} given Y_n.

Then for $n = 2, 3, \ldots$, there exist unique probability measures $r_n \in P(Y_n)$ such that

$$r_n(\underline{X}_1 \underline{X}_2 \cdots \underline{X}_n) = \int_{\underline{X}_1} \int_{\underline{X}_2} \cdots \int_{\underline{X}_{n-1}} q_{n-1}(\underline{X}_n | x_1, x_2, \ldots, x_{n-1})$$
$$\times q_{n-2}(dx_{n-1} | x_1, x_2, \ldots, x_{n-2}) \cdots$$
$$\times q_1(dx_2 | x_1) p(dx_1) \qquad \forall \underline{X}_1 \in \mathscr{B}_{X_1}, \ldots, \underline{X}_n \in \mathscr{B}_{X_n}. \quad (46)$$

If $f: Y_n \to R^*$ is Borel-measurable and either $\int f^+ \, dr_n < \infty$ or $\int f^- \, dr_n < \infty$, then

$$\int_{Y_n} f \, dr_n = \int_{X_1} \int_{X_2} \cdots \int_{X_n} f(x_1, x_2, \ldots, x_n) q_{n-1}(dx_n | x_1, x_2, \ldots, x_{n-1}) \cdots$$
$$\times q_1(dx_2 | x_1) p(dx_1). \quad (47)$$

Furthermore, there exists a unique probability measure r on $Y = X_1 X_2 \cdots$ such that for each n the marginal of r on Y_n is r_n.

Proof   The spaces Y_n, $n = 2, 3, \ldots$, and Y are Borel by Proposition 7.13. If there exists $r_n \in P(Y_n)$ satisfying (46), it must be unique. To see this, suppose $r'_n \in P(Y_n)$ also satisfies (46). The collection $\mathscr{D} = \{B \in \mathscr{B}_{Y_n} | r_n(B) = r'_n(B)\}$ is a Dynkin system containing the measurable rectangles, so $\mathscr{D} = \mathscr{B}_{Y_n}$ and $r_n = r'_n$. We establish the existence of r_n by induction, considering first the case $n = 2$. For $B \in \mathscr{B}_{Y_2}$, use Corollary 7.26.1 to define

$$r_2(B) = \int_{X_1} q_1(B_{x_1} | x_1) p(dx_1). \quad (48)$$

It is easily verified that $r_2 \in P(Y_2)$ and r_2 satisfies (46). If f is the indicator of $B \in \mathscr{B}_{Y_2}$, the $\int_{X_2} f(x_1, x_2) q_1(dx_2 | x_1)$ is Borel-measurable and, by (48),

$$\int_{Y_2} f \, dr_2 = \int_{X_1} \int_{X_2} f(x_1, x_2) q_1(dx_2 | x_1) p(dx_1). \quad (49)$$

Linearity of the integral implies that (49) holds for Borel-measurable simple functions as well. If $f: Y_2 \to [0, \infty]$ is Borel-measurable, then there exists an increasing sequence of simple functions such that $f_n \uparrow f$. By the monotone convergence theorem,

$$\lim_{n \to \infty} \int_{X_2} f_n(x_1, x_2) q_1(dx_2 | x_1) = \int_{X_2} f(x_1, x_2) q_1(dx_2 | x_1) \qquad \forall x_1 \in X_1,$$

so $\int_{X_2} f(x_1, x_2) q_1(dx_2 | x_1)$ is Borel-measurable and

$$\lim_{n \to \infty} \int_{Y_2} f_n \, dr_2 = \lim_{n \to \infty} \int_{X_1} \int_{X_2} f_n(x_1, x_2) q_1(dx_2 | x_1) p(dx_1)$$
$$= \int_{X_1} \int_{X_2} f(x_1, x_2) q_1(dx_2 | x_1) p(dx_1).$$

But $\int_{Y_2} f_n \, dr_2 \uparrow \int_{Y_2} f \, dr_2$, so (49) holds for any Borel-measurable nonnegative f. For a Borel-measurable $f: Y_2 \to R^*$ satisfying $\int f^+ \, dr_n < \infty$ or $\int f^- \, dr_n < \infty$, we have

$$\int_{Y_2} f^+ \, dr_2 = \int_{X_1} \int_{X_2} f^+(x_1, x_2) q_1(dx_2|x_1) p(dx_1),$$

and

$$\int_{Y_2} f^- \, dr_2 = \int_{X_1} \int_{X_2} f^-(x_1, x_2) q_1(dx_2|x_1) p(dx_1).$$

Assume for specificity that $\int_{Y_2} f^- \, dr_2 < \infty$. Then the functions

$$\int_{X_2} f^+(x_1, x_2) q_1(dx_2|x_1)$$

and

$$-\int_{X_2} f^-(x_1, x_2) q_1(dx_2|x_1)$$

satisfy condition (b2) of Lemma 7.11, so

$$\int_{Y_2} f \, dr_2 = \int_{Y_2} f^+ \, dr_2 - \int_{Y_2} f^- \, dr_2$$

$$= \int_{X_1} \left[\int_{X_2} f^+(x_1, x_2) q_1(dx_2|x_1) - \int_{X_2} f^-(x_1, x_2) q_1(dx_2|x_1) \right] p(dx_1)$$

$$= \int_{X_1} \int_{X_2} f(x_1, x_2) q_1(dx_2|x_1) p(dx_1),$$

where the last step is a direct result of the definition of $\int_{X_2} f(x_1, x_2) q_1(dx_2|x_1)$.

Assume now that $r_k \in P(Y_k)$ exists for which (46) and (47) hold when $n = k$. For $B \in Y_{k+1}$, let

$$r_{k+1}(B) = \int_{Y_k} q_k(B_{y_k}|y_k) r_k(dy_k).$$

Then $r_{k+1} \in P(Y_{k+1})$. If $B = \underline{X}_1 \underline{X}_2 \cdots \underline{X}_k \underline{X}_{k+1}$, where $\underline{X}_j \in \mathscr{B}_{X_j}$, then

$$r_{k+1}(B) = \int \chi_{\underline{X}_1 \underline{X}_2 \cdots \underline{X}_k}(y_k) q_k(\underline{X}_{k+1}|y_k) r_k(dy_k)$$

$$= \int_{\underline{X}_1} \int_{\underline{X}_2} \cdots \int_{\underline{X}_k} q_k(\underline{X}_{k+1}|x_1, x_2, \ldots, x_k) q_{k-1}(dx_k|x_{k-1}) \cdots$$

$$\times \, q_1(dx_2|x_1) p(dx_1) \qquad (50)$$

by (47) when $n = k$. This proves (46) for $n = k + 1$. Now use (50) to prove (47) when $n = k + 1$ and f is an indicator function. As before, extend this to the case of $f: Y_{k+1} \to [0, \infty]$. If $f: Y_{k+1} \to R^*$ is Borel-measurable and either $\int f^+ \, dr_{k+1} < \infty$ or $\int f^- \, dr_{k+1} < \infty$, then the validity of (47) for nonnegative functions and the induction hypothesis imply

$$\int_{Y_{k+1}} f^+ \, dr_{k+1} = \int_{X_1} \int_{X_2} \cdots \int_{X_{k+1}} f^+(x_1, \ldots, x_{k+1}) q_k(dx_{k+1}|x_1, x_2, \ldots, x_k) \cdots$$
$$\times \, q_1(dx_2|x_1) p(dx_1)$$
$$= \int_{X_1 \cdots X_k} \int_{X_{k+1}} f^+(x_1, \ldots, x_{k+1}) q_k(dx_{k+1}|x_1, x_2, \ldots, x_k) \, dr_k,$$

and likewise

$$\int_{Y_{k+1}} f^- \, dr_{k+1} = \int_{X_1 \ldots X_k} \int_{X_{k+1}} f^-(x_1, \ldots, x_{k+1}) q_k(dx_{k+1}|x_1, x_2, \ldots, x_k) \, dr_k.$$

Assume for specificity that $\int_{Y_{k+1}} f^- \, dr_{k+1} < \infty$. Then the functions

$$\int_{X_{k+1}} f^+(x_1, \ldots, x_{k+1}) q_k(dx_{k+1}|x_1, x_2, \ldots, x_k)$$

and

$$-\int_{X_{k+1}} f^-(x_1, \ldots, x_{k+1}) q_k(dx_{k+1}|x_1, x_2, \ldots, x_k)$$

satisfy condition (b2) of Lemma 7.11, so as before

$$\int_{Y_{k+1}} f \, dr_{k+1} = \int_{X_1 \cdots X_k} \int_{X_{k+1}} f(x_1, \ldots, x_{k+1}) q_k(dx_{k+1}|x_1, x_2, \ldots, x_k) \, dr_k.$$

$$(51)$$

Since

$$\left[\int_{X_{k+1}} f(x_1, \ldots, x_{k+1}) q_k(dx_{k+1}|x_1, x_2, \ldots, x_k) \right]^-$$
$$\leq \int_{X_{k+1}} f^-(x_1, \ldots, x_{k+1}) q_k(dx_{k+1}|x_1, x_2, \ldots, x_k),$$

we can apply the induction hypothesis to the right-hand side of (51) to conclude that (47) holds in the generality stated in the proposition.

To establish the existence of a unique probability measure $r \in P(Y)$ whose marginal on Y_n is r_n, $n = 2, 3, \ldots$, we note that the measures r_n are consistent, i.e., if $m \geq n$, then the marginal of r_m on Y_n is r_n. If each X_k is complete, the Kolmogorov extension theorem (see, e.g., Ash [A1, p. 191]) guarantees the existence of a unique $r \in P(Y)$ whose marginal on each Y_n is r_n. If X_k is not complete, it can be homeomorphically embedded as a Borel subset in a complete separable metric space $\tilde{X}_k$. As in Proposition 7.13, each Y_n is homeomorphic to a Borel subset of the complete separable metric space $\tilde{Y}_n = \tilde{X}_1 \tilde{X}_2 \cdots \tilde{X}_n$ and Y is homeomorphic to a Borel subset of the complete separable metric space $\tilde{Y} = \tilde{X}_1 \tilde{X}_2 \cdots$. Each $r_n \in P(Y_n)$ can be identified with

$\tilde{r}_n \in P(\tilde{Y}_n)$ in the manner of Lemma 7.10, and, invoking the Kolmogorov extension theorem, we establish the existence of a unique $\tilde{r} \in P(\tilde{Y})$ whose marginal on each $\tilde{Y}_n$ is $\tilde{r}_n$. It is straightforward to show that $\tilde{r}$ assigns probability one to the image of Y in $\tilde{Y}$, so $\tilde{r}$ corresponds to some $r \in P(Y)$ whose marginal on each Y_n is r_n. The uniqueness of $\tilde{r}$ implies the uniqueness of r.   Q.E.D.

In the course of proving Proposition 7.28, we have also proved the following.

Proposition 7.29 Let X and Y be Borel spaces and $q(dy|x)$ a Borel-measurable stochastic kernel on Y given X. If $f: XY \to R^*$ is Borel-measurable, then the function $\lambda: X \to R^*$ defined by

$$\lambda(x) = \int f(x, y)q(dy|x) \tag{52}$$

is Borel-measurable.

Corollary 7.29.1 Let X be a Borel space and let $f: X \to R^*$ be Borel-measurable. Then the function $\theta_f: P(X) \to R^*$ defined by

$$\theta_f(p) = \int f \, dp$$

is Borel-measurable.

Proof Define a Borel-measurable stochastic kernel on X given $P(X)$ by $q(dx|p) = p(dx)$. Define $\tilde{f}: P(X)X \to R^*$ by $\tilde{f}(p, x) = f(x)$. Then

$$\theta_f(p) = \int f(x)p(dx) = \int \tilde{f}(p, x)q(dx|p)$$

is Borel-measurable by Proposition 7.29.   Q.E.D.

If $f \in C(XY)$ and $q(dy|x)$ is continuous, then the mapping λ of (52) is also continuous. We prove this with the aid of the following lemma.

Lemma 7.12 Let X and Y be separable metrizable spaces. Then the mapping $\sigma: P(X)P(Y) \to P(XY)$ defined by

$$\sigma(p, q) = pq,$$

where pq is the product of the measures p and q, is continuous.

Proof We use Urysohn's theorem (Proposition 7.2) to homeomorphically embed X and Y into the Hilbert cube $\mathcal{H}$, and, for simplicity of notation, we treat X and Y as subsets of $\mathcal{H}$. Let d be a metric on $\mathcal{H}\mathcal{H}$ consistent with its topology. If $g \in U_d(XY)$, then Lemma 7.3 implies that g can be extended to a function $\hat{g} \in C(\mathcal{H}\mathcal{H})$. The set of finite linear combinations of the form $\sum_{j=1}^k \hat{f}_j(x)\hat{h}_j(y)$, where $\hat{f}_j$ and $\hat{h}_j$ range over $C(\mathcal{H})$ and k ranges over the positive integers, is an algebra which separates points in $\mathcal{H}\mathcal{H}$, so given $\varepsilon > 0$, the Stone–Weierstrass theorem implies that such a linear combination can be found satisfying $\|\sum_{j=1}^{k} \hat{f}_j \hat{h}_j - \hat{g}\| < \varepsilon$. If $\{p_n\}$ is a sequence in $P(X)$ converging to $p \in P(X)$, $\{q_n\}$ a sequence in $P(Y)$ converging to $q \in P(Y)$, and f_j and h_j the restrictions of $\hat{f}_j$ and $\hat{h}_j$ to X and Y, respectively, then

$$\limsup_{n \to \infty} \left| \int_{XY} g \, d(p_n q_n) - \int_{XY} g \, d(pq) \right|$$

$$\leq \limsup_{n \to \infty} \left| \int_{XY} \left(g - \sum_{j=1}^{k} f_j h_j \right) d(p_n q_n) \right|$$

$$+ \sum_{j=1}^{k} \lim_{n \to \infty} \left| \int_X f_j \, dp_n \int_Y h_j \, dq_n - \int_X f_j \, dp \int_Y h_j \, dq \right|$$

$$+ \limsup_{n \to \infty} \left| \int_{XY} \left(\sum_{j=1}^{k} f_j h_j - g \right) d(pq) \right|$$

$$\leq 2\varepsilon.$$

The continuity of σ follows from the equivalence of (a) and (c) of Proposition 7.21.   Q.E.D.

Proposition 7.30 Let X and Y be separable metrizable spaces and let $q(dy|x)$ be a continuous stochastic kernel on Y given X. If $f \in C(XY)$, then the function $\lambda: X \to R$ defined by

$$\lambda(x) = \int f(x, y) q(dy|x)$$

is continuous.

Proof The mapping $v: X \to P(XY)$ defined by $v(x) = p_x q(dy|x)$ is continuous by Corollary 7.21.1 and Lemma 7.12. We have $\lambda(x) = (\theta_f \circ v)(x)$, where $\theta_f: P(XY) \to R$ is defined by $\theta_f(r) = \int f \, dr$. By Proposition 7.21, θ_f is continuous. Hence, λ is continuous.   Q.E.D.

7.5 Semicontinuous Functions and Borel-Measurable Selection

In the dynamic programming algorithm given by (17) and (18) of Chapter 1, three operations are performed repetitively. First, there is the evaluation of a conditional expectation. Second, an extended real-valued function in two variables (state and control) is infimized over one of these variables (control). Finally, if an optimal or nearly optimal policy is to be constructed, a "selector" which maps each state to a control which achieves or nearly achieves the infimum in the second step must be chosen. In this section, we give results which will enable us to show that, under certain conditions, the extended real-valued functions involved are semicontinuous and the selectors can be chosen to be Borel-measurable. The results are applied to dynamic programming in Propositions 8.6–8.7 and Corollaries 9.17.2–9.17.3.

Definition 7.13 Let X be a metrizable space and f an extended real-valued function on X. If $\{x \in X \,|\, f(x) \leq c\}$ is closed for every $c \in R$, f is said to be *lower semicontinuous*. If $\{x \in X \,|\, f(x) \geq c\}$ is closed for every $c \in R$, f is said to be *upper semicontinuous*.

Note that f is lower semicontinuous if and only if $-f$ is upper semi-continuous. We will use this duality in the proofs of the following propositions to assert facts about upper semicontinuous functions given analogous facts about lower semicontinuous functions. Note also that if f is lower semi-continuous, the sets $\{x \in X \,|\, f(x) = -\infty\}$ and $\{x \in X \,|\, f(x) \leq \infty\}$ are closed, since the former is equal to $\bigcap_{n=1}^{\infty} \{x \in X \,|\, f(x) \leq -n\}$ and the latter is X. There is a similar result for upper semicontinuous functions. The following lemma provides an alternative characterization of lower and upper semi-continuous functions.

Lemma 7.13 Let X be a metrizable space and $f : X \to R^*$.

(a) The function f is lower semicontinuous if and only if for each sequence $\{x_n\} \subset X$ converging to $x \in X$

$$\liminf_{n \to \infty} f(x_n) \geq f(x). \tag{53}$$

(b) The function f is upper semicontinuous if and only if for each sequence $\{x_n\} \subset X$ converging to $x \in X$

$$\limsup_{n \to \infty} f(x_n) \leq f(x). \tag{54}$$

Proof Suppose f is lower semicontinuous and $x_n \to x$. We can extract a subsequence $\{x_{n_k}\}$ such that $x_{n_k} \to x$ as $k \to \infty$ and

$$\lim_{k \to \infty} f(x_{n_k}) = \liminf_{n \to \infty} f(x_n).$$

Given $\varepsilon > 0$, define

$$\theta(\varepsilon) = \begin{cases} \liminf_{n \to \infty} f(x_n) + \varepsilon & \text{if } \liminf_{n \to \infty} f(x_n) > -\infty, \\ -1/\varepsilon & \text{otherwise.} \end{cases}$$

There exists a positive integer $k(\varepsilon)$ such that $f(x_{n_k}) \leq \theta(\varepsilon)$ for all $k \geq k(\varepsilon)$. The set $\{y \in X \mid f(y) \leq \theta(\varepsilon)\}$ is closed, and hence it contains x. Inequality (53) follows. Conversely, if (53) holds and for some $c \in R$, $\{x_n\}$ is a sequence in $\{y \in X \mid f(y) \leq c\}$ converging to x, then $f(x) \leq c$, so f is lower semicontinuous.

Part (b) of the proposition follows from part (a) by the duality mentioned earlier.   Q.E.D.

If f and g are lower semicontinuous and bounded below on X and if $x_n \to x$, then

$$\liminf_{n \to \infty} [f(x_n) + g(x_n)] \geq \liminf_{n \to \infty} f(x_n) + \liminf_{n \to \infty} g(x_n)$$

$$\geq f(x) + g(x),$$

so $f + g$ is lower semicontinuous. If f is lower semicontinuous and $\alpha > 0$, then αf is lower semicontinuous as well. Upper semicontinuous functions have similar properties.

It is clear from (53) and (54) that $f : X \to R^*$ is continuous if and only if it is both lower and upper semicontinuous. We can often infer properties of semicontinuous functions from properties of continuous functions by means of the next lemma.

Lemma 7.14   Let X be a metrizable space and $f : X \to R^*$.

(a)   The function f is lower semicontinuous and bounded below if and only if there exists a sequence $\{f_n\} \subset C(X)$ such that $f_n \uparrow f$.

(b)   The function f is upper semicontinuous and bounded above if and only if there exists a sequence $\{f_n\} \subset C(X)$ such that $f_n \downarrow f$.

Proof   We prove only part (a) of the proposition and appeal to duality for part (b). Assume f is lower semicontinuous and bounded below by $b \in R$, and let d be a metric on X consistent with its topology. We may assume without loss of generality that for some $x_0 \in X$, $f(x_0) < \infty$, since the result is trivial otherwise. (Take $f_n(x) = n$ for every $x \in X$.) As in Lemma 7.7, define

$$g_n(x) = \inf_{y \in X} [f(y) + nd(x, y)].$$

Exactly as in the proof of Lemma 7.7, we show that $\{g_n\}$ is an increasing sequence of continuous functions bounded below by b and above by f. The characterization (53) of lower semicontinuous functions can be used in place of continuity to prove $g_n \uparrow f$. In particular, (28) becomes

$$f(x) \leq \liminf_{n \to \infty} f(y_n) \leq \lim_{n \to \infty} g_n(x) + \varepsilon.$$

Now define $f_n = \min\{n, g_n\}$. Then each f_n is continuous and bounded and $f_n \uparrow f$. This concludes the proof of the direct part of the proposition. For the converse part, suppose $\{f_n\} \subset C(X)$ and $f_n \uparrow f$. For $c \in R$,

$$\{x \in X \,|\, f(x) \le c\} = \bigcap_{n=1}^{\infty} \{x \in X \,|\, f_n(x) \le c\}$$

is closed.   Q.E.D.

The following proposition shows that the semicontinuity of a function of two variables is preserved when one of the variables is integrated out via a continuous stochastic kernel.

Proposition 7.31   Let X and Y be separable metrizable spaces, let $q(dy|x)$ be a continuous stochastic kernel on Y given X, and let $f : XY \to R^*$ be Borel-measurable. Define

$$\lambda(x) = \int f(x, y) q(dy|x).$$

(a)   If f is lower semicontinuous and bounded below, then λ is lower semicontinuous and bounded below.

(b)   If f is upper semicontinuous and bounded above, then λ is upper semicontinuous and bounded above.

Proof   We prove part (a) of the proposition and appeal to duality for part (b). If $f : XY \to R^*$ is lower semicontinuous and bounded below, then by Lemma 7.14 there exists a sequence $\{f_n\} \subset C(XY)$ such that $f_n \uparrow f$. Define $\lambda_n(x) = \int f_n(x, y) q(dy|x)$. By Proposition 7.30, we have that λ_n is continuous, and by the monotone convergence theorem $\lambda_n \uparrow \lambda$. By Lemma 7.14, λ is lower semicontinuous.   Q.E.D.

An important operation in the execution of the dynamic programming algorithm is the infimization over one of the variables of a bivariate function. In the context of semicontinuity, we have the following result related to this operation.

Proposition 7.32   Let X and Y be metrizable spaces and let $f : XY \to R^*$ be given. Define

$$f^*(x) = \inf_{y \in Y} f(x, y). \tag{55}$$

(a)   If f is lower semicontinuous and Y is compact, then f^* is lower semicontinuous and for every $x \in X$ the infimum in (55) is attained by some $y \in Y$.

(b)   If f is upper semicontinuous, then f^* is upper semicontinuous.

Proof   (a)   Fix x and let $\{y_n\} \subset Y$ be such that $f(x, y_n) \downarrow f^*(x)$. Then $\{y_n\}$ accumulates at some $y_0 \in Y$, and part (a) of Lemma 7.13 implies that $f(x, y_0) = f^*(x)$. To show that f^* is lower semicontinuous, let $\{x_n\} \subset X$ be such that $x_n \to x_0$. Choose a sequence $\{y_n\} \subset Y$ such that

$$f(x_n, y_n) = f^*(x_n), \qquad n = 1, 2, \dots .$$

There is a subsequence of $\{(x_n, y_n)\}$, call it $\{(x_{n_k}, y_{n_k})\}$, such that $\liminf_{n \to \infty} f(x_n, y_n) = \lim_{k \to \infty} f(x_{n_k}, y_{n_k})$. The sequence $\{y_{n_k}\}$ accumulates at some $y_0 \in Y$, and, by Lemma 7.13(a),

$$\liminf_{n \to \infty} f^*(x_n) = \liminf_{n \to \infty} f(x_n, y_n) = \lim_{k \to \infty} f(x_{n_k}, y_{n_k}) \geq f(x_0, y_0) \geq f^*(x_0),$$

so f^* is lower semicontinuous.

(b) Let d_1 be a metric on X and d_2 a metric on Y consistent with their topologies. If $G \subset XY$ is open and $x_0 \in \text{proj}_X(G)$, then there is some $y_0 \in Y$ for which $(x_0, y_0) \in G$, and there is some $\varepsilon > 0$ such that

$$N_\varepsilon(x_0, y_0) = \{(x, y) \in XY \,|\, d_1(x, x_0) < \varepsilon, d_2(y, y_0) < \varepsilon\} \subset G.$$

Then

$$x_0 \in \text{proj}_X[N_\varepsilon(x_0, y_0)] = \{x \in X \,|\, d_1(x, x_0) < \varepsilon\} \subset \text{proj}_X(G),$$

so $\text{proj}_X(G)$ is open in X. For $c \in R$,

$$\{x \in X \,|\, f^*(x) < c\} = \text{proj}_X(\{(x, y) \in XY \,|\, f(x, y) < c\}).$$

The upper semicontinuity of f implies that $\{(x, y) \,|\, f(x, y) < c\}$ is open, so $\{x \in X \,|\, f^*(x) < c\}$ is open and f^* is upper semicontinuous.   Q.E.D.

Another important operation in the dynamic programming algorithm is the choice of a measurable "selector" which assigns to each $x \in X$ a $y \in Y$ which attains or nearly attains the infimum in (55). We first discuss Borel-measurable selection in case (a) of Proposition 7.32. For this we will need the Hausdorff metric and the corresponding topology on the set 2^Y of closed subsets of a compact metric space Y (Appendix C). The space 2^Y under this topology is compact (Proposition C.2) and, therefore, complete and separable. Several preliminary lemmas are required.

Lemma 7.15   Let Y be a compact metrizable space and let $g: Y \to R^*$ be lower semicontinuous. Define $g^*: 2^Y \to R^*$ by

$$g^*(A) = \begin{cases} \min_{y \in A} g(y) & \text{if} \quad A \neq \varnothing, \\ \infty & \text{if} \quad A = \varnothing. \end{cases} \tag{56}$$

Then g^* is lower semicontinuous.

Proof   Since the empty set is an isolated point in 2^Y, we need only prove that g^* is lower semicontinuous on $2^Y - \{\varnothing\}$. We have already shown [Proposition 7.32(a)] that, given a nonempty set $A \in 2^Y$, there exists $y \in A$

such that $g^*(A) = g(y)$. Let $\{A_n\} \subset 2^Y$ be a sequence of nonempty sets with limit $A \in 2^Y$, and let $y_n \in A_n$ be such that $g^*(A_n) = g(y_n)$, $n = 1, 2, \ldots$. Choose a subsequence $\{y_{n_k}\}$ such that

$$\lim_{k \to \infty} g(y_{n_k}) = \liminf_{n \to \infty} g(y_n) = \liminf_{n \to \infty} g^*(A_n).$$

The subsequence $\{y_{n_k}\}$ accumulates at some $y_0 \in Y$, and, by Lemma 7.13(a),

$$g(y_0) \leq \lim_{k \to \infty} g(y_{n_k}) = \liminf_{n \to \infty} g^*(A_n).$$

From (14) of Appendix C and from Proposition C.3, we have (in the notation of Appendix C)

$$y_0 \in \overline{\lim_{n \to \infty}} A_n = A,$$

so

$$g^*(A) \leq g(y_0) \leq \liminf_{n \to \infty} g^*(A_n).$$

The result follows from Lemma 7.13(a).   Q.E.D.

Lemma 7.16   Let Y be a compact metrizable space and let $g: Y \to R^*$ be lower semicontinuous. Define $G: 2^Y R^* \to 2^Y$ by

$$G(A, c) = A \cap \{y \in Y \mid g(y) \leq c\}. \tag{57}$$

Then G is Borel-measurable.

Proof   We show that G is upper semicontinuous (K) (Definition C.2) and apply Proposition C.4. Let $\{(A_n, c_n)\} \subset 2^Y R^*$ be a sequence with limit (A, c). If $\overline{\lim}_{n \to \infty} G(A_n, c_n) = \varnothing$, then

$$\overline{\lim_{n \to \infty}} G(A_n, c_n) \subset G(A, c). \tag{58}$$

Otherwise, choose $y \in \overline{\lim}_{n \to \infty} G(A_n, c_n)$. There is a sequence $n_1 < n_2 < \cdots$ of positive integers and a sequence $y_{n_k} \in G(A_{n_k}, c_{n_k})$, $k = 1, 2, \ldots$, such that $y_{n_k} \to y$. By definition, $y_{n_k} \in A_{n_k}$ for every k, so $y \in \overline{\lim}_{n \to \infty} A_n = A$. We also have $g(y_{n_k}) \leq c_{n_k}$, $k = 1, 2, \ldots$, and using the lower semicontinuity of g, we obtain

$$g(y) \leq \liminf_{k \to \infty} g(y_{n_k}) \leq \lim_{k \to \infty} c_{n_k} = c.$$

Therefore $y \in G(A, c)$, (58) holds, and G is upper semicontinuous (K).

Q.E.D.

Lemma 7.17 Let Y be a compact metrizable space and let $g: Y \to R^*$ be lower semicontinuous. Let $g^*: 2^Y \to R^*$ be defined by (56) and define $G^*: 2^Y \to 2^Y$ by

$$G^*(A) = A \cap \{y \in Y | g(y) \le g^*(A)\}. \tag{59}$$

Then G^* is Borel-measurable.

Proof Let G be the Borel-measurable function given by (57). Lemma 7.15 implies g^* is Borel-measurable. A comparison of (57) and (59) shows that

$$G^*(A) = G[A, g^*(A)].$$

It follows that G^* is also Borel-measurable.   Q.E.D.

Lemma 7.18 Let Y be a compact metrizable space. There is a Borel-measurable function $\sigma: 2^Y - \{\varnothing\} \to Y$ such that $\sigma(A) \in A$ for every $A \in 2^Y - \{\varnothing\}$.

Proof Let $\{g_n | n = 1, 2, \ldots\}$ be a subset of $C(Y)$ which separates points in Y (for example, the one constructed in the proof of Proposition 7.7). As in Lemma 7.15, define $g_n^*: 2^Y \to R^*$ by

$$g_n^*(A) = \begin{cases} \min_{y \in A} g_n(y) & \text{if } A \ne \varnothing, \\ \infty & \text{if } A = \varnothing, \end{cases}$$

and, as in Lemma 7.17, define $G_n^*: 2^Y \to 2^Y$ by

$$G_n^*(A) = A \cap \{y \in Y | g_n(y) \le g_n^*(A)\} = \{y \in A | g_n(y) = g_n^*(A)\}.$$

Let $H_n: 2^Y \to 2^Y$ be defined recursively by

$$H_0(A) = A,$$
$$H_n(A) = G_n^*[H_{n-1}(A)], \qquad n = 1, 2, \ldots.$$

Then for $A \ne \varnothing$, each $H_n(A)$ is nonempty and compact, and

$$A = H_0(A) \supset H_1(A) \supset H_2(A) \supset \cdots.$$

Therefore, $\bigcap_{n=0}^{\infty} H_n(A) \ne \varnothing$. If $y, y' \in \bigcap_{n=0}^{\infty} H_n(A)$, then for $n = 1, 2, \ldots$, we have

$$g_n(y) = g_n^*[H_{n-1}(A)] = g_n(y').$$

Since $\{g_n | n = 1, 2, \ldots\}$ separates points in Y, we have $y = y'$, and $\bigcap_{n=0}^{\infty} H_n(A)$ must consist of a single point, which we denote by $\sigma(A)$.

We show that for $A \neq \emptyset$

$$\lim_{n \to \infty} H_n(A) = \bigcap_{n=0}^{\infty} H_n(A) = \{\sigma(A)\}. \tag{60}$$

Since the sequence $\{H_n(A)\}$ is nonincreasing, we have from (14) and (15) of Appendix C that

$$\bigcap_{n=0}^{\infty} H_n(A) \subset \varliminf_{n \to \infty} H_n(A) \subset \varlimsup_{n \to \infty} H_n(A). \tag{61}$$

If $y \in \varlimsup_{n \to \infty} H_n(A)$, then there exist positive integers $n_1 < n_2 < \cdots$ and a sequence $y_{n_k} \in H_{n_k}(A)$, $k = 1, 2, \ldots$, such that $y_{n_k} \to y$. For fixed k, $y_{n_j} \in H_{n_k}$ for all $j \geq k$, and since $H_{n_k}(A)$ is closed, we have $y \in H_{n_k}(A)$. Therefore, $y \in \bigcap_{n=0}^{\infty} H_n(A)$ and

$$\varlimsup_{n \to \infty} H_n(A) \subset \bigcap_{n=0}^{\infty} H_n(A). \tag{62}$$

From relations (61) and (62), we obtain (60).

Since G_n^* and H_n are Borel-measurable for every n, the mapping $v: 2^Y - \{\emptyset\} \to 2^Y$ defined by $v(A) = \{\sigma(A)\}$ is Borel-measurable. It is easily seen that the mapping $\tau: Y \to 2^Y$ defined by $\tau(y) = \{y\}$ is a homeomorphism. Since Y is compact, $\tau(Y)$ is compact, thus closed in 2^Y, and $\tau^{-1}: \tau(Y) \to Y$ is Borel-measurable. Since $\sigma = \tau^{-1} \circ v$, it follows that σ is Borel-measurable.

Q.E.D.

Lemma 7.19   Let X be a metrizable space, Y a compact metrizable space, and let $f: XY \to R^*$ be lower semicontinuous. Define $F: XR^* \to 2^Y$ by

$$F(x, c) = \{y \in Y \mid f(x, y) \leq c\}. \tag{63}$$

Then F is Borel-measurable.

Proof   The proof is very similar to that of Lemma 7.16. We show that F is upper semicontinuous (K) and apply Proposition C.4. Let $(x_n, c_n) \to (x, c)$ in XR^* and let y be an element of $\varlimsup_{n \to \infty} F(x_n, c_n)$, provided this set is nonempty. There exist positive integers $n_1 < n_2 < \cdots$ and $y_{n_k} \in F(x_{n_k}, c_{n_k})$ such that $y_{n_k} \to y$. Since $f(x_{n_k}, y_{n_k}) \leq c_{n_k}$ and f is lower semicontinuous, we conclude that $f(x, y) \leq c$, so that $\varlimsup_{n \to \infty} F(x_n, c_n) \subset F(x, c)$. The result follows.

Q.E.D.

Lemma 7.20   Let X be a metrizable space, Y a compact metrizable space, and let $f: XY \to R^*$ be lower semicontinuous. Let $f^*: X \to R^*$ be given by $f^*(x) = \min_{y \in Y} f(x, y)$, and define $F^*: X \to 2^Y$ by

$$F^*(x) = \{y \in Y \mid f(x, y) \leq f^*(x)\}. \tag{64}$$

Then F^* is Borel-measurable.

Proof Let F be the Borel-measurable function defined by (63). Proposition 7.32(a) implies that f^* is Borel-measurable. From (63) and (64) we have

$$F^*(x) = F[x, f^*(x)].$$

It follows that F^* is also Borel-measurable.   Q.E.D.

We are now ready to prove the selection theorem for lower semicontinuous functions.

Proposition 7.33 Let X be a metrizable space, Y a compact metrizable space, D a closed subset of XY, and let $f:D \to R^*$ be lower semicontinuous. Let $f^*:\mathrm{proj}_X(D) \to R^*$ be given by

$$f^*(x) = \min_{y \in D_x} f(x, y). \tag{65}$$

Then $\mathrm{proj}_X(D)$ is closed in X, f^* is lower semicontinuous, and there exists a Borel-measurable function $\varphi:\mathrm{proj}_X(D) \to Y$ such that $\mathrm{Gr}(\varphi) \subset D$ and

$$f[x, \varphi(x)] = f^*(x) \qquad \forall x \in \mathrm{proj}_X(D). \tag{66}$$

Proof We first prove the result for the case where $D = XY$. As in Lemma 7.18, let $\sigma:2^Y - \{\varnothing\} \to Y$ be a Borel-measurable function satisfying $\sigma(A) \in A$ for every $A \in 2^Y - \{\varnothing\}$. As in Lemma 7.20, let $F^*:X \to 2^Y$ be the Borel-measurable function defined by

$$F^*(x) = \{y \in Y \mid f(x, y) = f^*(x)\}.$$

Proposition 7.32(a) implies that f^* is lower semicontinuous and $F^*(x) \neq \varnothing$ for every $x \in X$. The composition $\varphi = \sigma \circ F^*$ satisfies (66).

Suppose now that D is not necessarily XY. To see that $\mathrm{proj}_X(D)$ is closed, note that the function $g = -\chi_D$ is lower semicontinuous and

$$\mathrm{proj}_X(D) = \{x \in X \mid g^*(x) \leq -1\},$$

where $g^*(x) = \min_{y \in Y} g(x, y)$. By the special case of the proposition already proved, g^* is lower semicontinuous, $\mathrm{proj}_X(D)$ is closed, and there is a Borel-measurable function $\varphi_1:X \to Y$ such that $g[x, \varphi_1(x)] = g^*(x)$ for every $x \in X$ or, equivalently, $(x, \varphi_1(x)) \in D$ for every $x \in \mathrm{proj}_X(D)$.

Define now the lower semicontinuous function $\hat{f}:XY \to R^*$ by

$$\hat{f}(x, y) = \begin{cases} f(x, y) & \text{if } (x, y) \in D, \\ \infty & \text{otherwise.} \end{cases}$$

For all $c \in R$,

$$\{x \in \mathrm{proj}_X(D) \mid f^*(x) \leq c\} = \left\{x \in X \mid \min_{y \in Y} \hat{f}(x, y) \leq c\right\}.$$

Since $\min_{y \in Y} \hat{f}(x, y)$ is lower semicontinuous, it follows that f^* is also lower semicontinuous. Let $\varphi_2 : X \to Y$ be a Borel-measurable function satisfying

$$\hat{f}[x, \varphi_2(x)] = \min_{y \in Y} \hat{f}(x, y) \qquad \forall x \in X.$$

Clearly $(x, \varphi_2(x)) \in D$ for all x in the Borel set

$$\left\{ x \in X \middle| \min_{y \in Y} \hat{f}(x, y) < \infty \right\}.$$

Define $\varphi : \operatorname{proj}_X(D) \to Y$ by

$$\varphi(x) = \begin{cases} \varphi_1(x) & \text{if } \min_{y \in Y} \hat{f}(x, y) = \infty, \\ \varphi_2(x) & \text{if } \min_{y \in Y} \hat{f}(x, y) < \infty. \end{cases}$$

The function φ is Borel-measurable and satisfies (66).　　Q.E.D.

We turn our attention to selection in the case of an upper semicontinuous function. The analysis is considerably simpler, but in contrast to the "exact selector" of (66) we will obtain only an approximate selector for this case.

Lemma 7.21　Let X be a metrizable space, Y a separable metrizable space, and G an open subset of XY. Then $\operatorname{proj}_X(G)$ is open and there exists a Borel-measurable function $\varphi : \operatorname{proj}_X(G) \to Y$ such that $\operatorname{Gr}(\varphi) \subset G$.

Proof　Let $\{ y_n | n = 1, 2, \ldots \}$ be a countable dense subset of Y. For fixed $y \in Y$, the mapping $x \to (x, y)$ is continuous, so $\{ x \in X | (x, y) \in G \}$ is open. Let $G_n = \{ x \in X | (x, y_n) \in G \}$, and note that $\operatorname{proj}_X(G) = \bigcup_{n=1}^{\infty} G_n$ is open. Define $\varphi : \operatorname{proj}_X(G) \to Y$ by

$$\varphi(x) = \begin{cases} y_1 & \text{if } x \in G_1, \\ y_n & \text{if } x \in G_n - \bigcup_{k=1}^{n-1} G_k, \quad n = 2, 3, \ldots. \end{cases}$$

Then φ is Borel-measurable and $\operatorname{Gr}(\varphi) \subset G$.　　Q.E.D.

Proposition 7.34　Let X be a metrizable space, Y a separable metrizable space, D an open subset of XY, and let $f : D \to R^*$ be upper semicontinuous. Let $f^* : \operatorname{proj}_X(D) \to R^*$ be given by

$$f^*(x) = \inf_{y \in D_x} f(x, y). \qquad (67)$$

Then $\operatorname{proj}_X(D)$ is open in X, f^* is upper semicontinuous, and for every $\varepsilon > 0$, there exists a Borel-measurable function $\varphi_\varepsilon : \operatorname{proj}_X(D) \to Y$ such that

$Gr(\varphi_\varepsilon) \subset D$ and for all $x \in \text{proj}_X(D)$

$$f[x, \varphi_\varepsilon(x)] \leq \begin{cases} f^*(x) + \varepsilon & \text{if } f^*(x) > -\infty, \\ -1/\varepsilon & \text{if } f^*(x) = -\infty. \end{cases} \tag{68}$$

Proof The set $\text{proj}_X(D)$ is open in X by Lemma 7.21. To show that f^* is upper semicontinuous, define an upper semicontinuous function $\hat{f} : XY \to R^*$ by

$$\hat{f}(x, y) = \begin{cases} f(x, y) & \text{if } (x, y) \in D, \\ \infty & \text{otherwise.} \end{cases}$$

For $c \in R$, we have

$$\{x \in \text{proj}_X(D) \mid f^*(x) < c\} = \left\{ x \in X \mid \inf_{y \in Y} \hat{f}(x, y) < c \right\},$$

and this set is open by Proposition 7.32(b).

Let $\varepsilon > 0$ be given. For $k = 0, \pm 1, \pm 2, \ldots$, define (see Fig. 7.1)

$$A(k) = \{(x, y) \in D \mid f(x, y) < k\varepsilon\},$$
$$B(k) = \{x \in \text{proj}_X(D) \mid (k - 1)\varepsilon \leq f^*(x) < k\varepsilon\},$$
$$B(-\infty) = \{x \in \text{proj}_X(D) \mid f^*(x) = -\infty\},$$
$$B(\infty) = \{x \in \text{proj}_X(D) \mid f^*(x) = \infty\}.$$

The sets $A(k)$, $k = 0, \pm 1, \pm 2, \ldots$, are open, while the sets $B(k)$, $k = 0, \pm 1, \pm 2, \ldots, B(-\infty)$, and $B(\infty)$ are Borel-measurable. By Lemma 7.21,

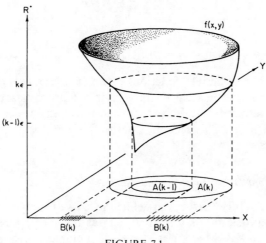

FIGURE 7.1

there exists for each $k = 0, \pm 1, \pm 2, \ldots$ a Borel-measurable $\varphi_k : \text{proj}_X(A_k) \to Y$ such that $\text{Gr}(\varphi_k) \subset A_k$, and there exists a Borel-measurable $\bar{\varphi} : \text{proj}_X(D) \to Y$ such that $\text{Gr}(\bar{\varphi}) \subset D$. Let k^* be an integer such that $k^* \leq -1/\varepsilon^2$. Define $\varphi_\varepsilon : \text{proj}_X(D) \to Y$ by

$$\varphi_\varepsilon(x) = \begin{cases} \varphi_k(x) & \text{if} \quad x \in B(k), \quad k = 0, \pm 1, \pm 2, \ldots, \\ \bar{\varphi}(x) & \text{if} \quad x \in B(\infty), \\ \varphi_{k^*}(x) & \text{if} \quad x \in B(-\infty). \end{cases}$$

Since $B(k) \subset \text{proj}_X[A(k)]$ and $B(-\infty) \subset \text{proj}_X[A(k)]$ for all k, this definition is possible. It is clear that φ_ε is Borel-measurable and $\text{Gr}(\varphi_\varepsilon) \subset D$. If $x \in B(k)$, then, since $(x, \varphi_k(x)) \in A(k)$, we have

$$f[x, \varphi_\varepsilon(x)] = f[x, \varphi_k(x)] < k\varepsilon \leq f^*(x) + \varepsilon.$$

If $x \in B(\infty)$, then $f(x, y) = \infty$ for all $y \in D_x$ and $f[x, \varphi_\varepsilon(x)] = \infty = f^*(x)$. If $x \in B(-\infty)$, we have

$$f[x, \varphi_\varepsilon(x)] = f[x, \varphi_{k^*}(x)] < k^*\varepsilon \leq -1/\varepsilon.$$

Hence φ_ε has the required properties.   Q.E.D.

7.6   Analytic Sets

The dynamic programming algorithm is centered around infimization of functions, and this is intimately connected with projections of sets. More specifically, if $f : XY \to R^*$ is given and $f^* : X \to R^*$ is defined by

$$f^*(x) = \inf_{y \in Y} f(x, y),$$

then for each $c \in R$

$$\{x \in X \,|\, f^*(x) < c\} = \text{proj}_X(\{(x, y) \in XY \,|\, f(x, y) < c\}).$$

If f is a Borel-measurable function, then $\{(x, y) \,|\, f(x, y) < c\}$ is a Borel-measurable set. Unfortunately, the projection of a Borel-measurable set need not be Borel-measurable. In Borel spaces, however, the projection of a Borel set is an analytic set. This section is devoted to development of properties of analytic sets.

7.6.1   Equivalent Definitions of Analytic Sets

There are a number of ways to define the class of analytic sets in a Borel space X. One possibility is to define them as the projections on X of the Borel subsets of XY, where Y is some uncountable Borel space. Another possibility is to define them as the images of the Baire null space $\mathcal{N}$ under continuous functions from $\mathcal{N}$ into X. Still another possibility is to define them as all sets of the form

$$\bigcup_{(\sigma_1, \sigma_2, \ldots) \in \mathcal{N}} \bigcap_{n=1}^{\infty} S(\sigma_1, \sigma_2, \ldots, \sigma_n),$$

where $\mathcal{N}$ is the set of all sequences of positive integers (the Baire null space) and the sets $S(\sigma_1, \sigma_2, \ldots, \sigma_n)$ are closed in X. All these definitions are equivalent, as we show in Proposition 7.41. We will take the third definition as our starting point, since this is the most convenient analytically. We first formalize the set operation just given in terms of the notion of a Suslin scheme in a paved space.

Definition 7.14   Let X be a set. A *paving* $\mathcal{P}$ of X is a nonempty collection of subsets of X. The pair $(X, \mathcal{P})$ is called a *paved space*.

If $(X, \mathcal{P})$ is a paved space, we denote by $\sigma(\mathcal{P})$ the σ-algebra generated by $\mathcal{P}$, we denote by $\mathcal{P}_\delta$ the collection of all intersections of countably many members of $\mathcal{P}$, and we denote by $\mathcal{P}_\sigma$ the collection of all unions of countably many members of $\mathcal{P}$. Recall that N is the set of positive integers, $\mathcal{N}$ is the set of all infinite sequences of positive integers, and Σ is the set of all finite sequences of positive integers.

Definition 7.15   Let $(X, \mathcal{P})$ be a paved space. A *Suslin scheme* for $\mathcal{P}$ is a mapping from Σ into $\mathcal{P}$. The *nucleus of a Suslin scheme* $S: \Sigma \to \mathcal{P}$ is

$$N(S) = \bigcup_{(\sigma_1, \sigma_2, \ldots) \in \mathcal{N}} \bigcap_{n=1}^{\infty} S(\sigma_1, \sigma_2, \ldots, \sigma_n). \tag{69}$$

The set of all nuclei of Suslin schemes for a paving $\mathcal{P}$ will be denoted by $\mathcal{S}(\mathcal{P})$.

In order to simplify notation, we write, for $s = (\sigma_1, \sigma_2, \ldots, \sigma_n) \in \Sigma$ and $z = (\zeta_1, \zeta_2, \ldots) \in \mathcal{N}$,

$$s < z \quad \text{if} \quad \sigma_1 = \zeta_1, \quad \sigma_2 = \zeta_2, \ldots, \sigma_n = \zeta_n.$$

With this notation, (69) can also be written as

$$N(S) = \bigcup_{z \in \mathcal{N}} \bigcap_{s < z} S(s).$$

We will use both expressions interchangeably.

Note that the union in (69) is uncountable, so if $\mathcal{P}$ is a σ-algebra and S is a Suslin scheme for $\mathcal{P}$, $N(S)$ may be outside $\mathcal{P}$. Several properties of $\mathcal{S}(\mathcal{P})$ are given below.

Proposition 7.35   Let X be a space with pavings $\mathscr{P}$ and $\mathscr{Q}$ such that $\mathscr{P} \subset \mathscr{Q}$. Then

(a) $\mathscr{S}(\mathscr{P}) \subset \mathscr{S}(\mathscr{Q})$,
(b) $\mathscr{S}(\mathscr{P})_\delta = \mathscr{S}(\mathscr{P})$,
(c) $\mathscr{S}(\mathscr{P})_\sigma = \mathscr{S}(\mathscr{P})$,
(d) $\mathscr{P} \subset \mathscr{S}(\mathscr{P})$,
(e) $\mathscr{S}(\mathscr{P}) = \mathscr{S}[\mathscr{S}(\mathscr{P})]$.

Proof   (a)   Obvious.

(b)   It is clear that $\mathscr{S}(\mathscr{P})_\delta \supset \mathscr{S}(\mathscr{P})$. Now choose $\bigcap_{k=1}^\infty N(S_k) \in \mathscr{S}(\mathscr{P})_\delta$, where S_k is a Suslin scheme for $\mathscr{P}$, $k = 1, 2, \ldots$. It suffices to construct a Suslin scheme S for $\mathscr{P}$ such that

$$N(S) = \bigcap_{k=1}^\infty N(S_k). \tag{70}$$

For $k = 1, 2, \ldots$, let $\Pi_k = \{(2j-1)2^{k-1} \mid j = 1, 2, \ldots\}$. Then $\Pi_1, \Pi_2, \ldots$ is a partition of N into infinitely many infinite sets. For each positive integer k, let $\varphi_k : \mathscr{N} \to \mathscr{N}$ be defined by

$$\varphi_k(\zeta_1, \zeta_2, \ldots) = (\zeta_{2^{k-1}}, \zeta_{3 \cdot 2^{k-1}}, \zeta_{5 \cdot 2^{k-1}}, \ldots),$$

i.e., φ_k picks out the components of $(\zeta_1, \zeta_2, \ldots)$ with indices in Π_k. We want to construct a Suslin scheme S for which

$$\bigcap_{s < z} S(s) = \bigcap_{k=1}^\infty \bigcap_{s < \varphi_k(z)} S_k(s) \qquad \forall z \in \mathscr{N}. \tag{71}$$

We may rewrite (71) as

$$\bigcap_{n=1}^\infty S(\zeta_1, \zeta_2, \ldots, \zeta_n)$$

$$= \bigcap_{k=1}^\infty \bigcap_{j=1}^\infty S_k(\zeta_{2^{k-1}}, \zeta_{3 \cdot 2^{k-1}}, \ldots, \zeta_{(2j-1)2^{k-1}}) \qquad \forall(\zeta_1, \zeta_2, \ldots) \in \mathscr{N}. \tag{72}$$

Given $(\zeta_1, \zeta_2, \ldots, \zeta_n) \in \Sigma$, we have $n = (2j-1)2^{k-1}$ for exactly one pair of positive integers j and k. Define

$$S(\zeta_1, \zeta_2, \ldots, \zeta_n) = S_k(\zeta_{2^{k-1}}, \zeta_{3 \cdot 2^{k-1}}, \ldots, \zeta_{(2j-1)2^{k-1}}). \tag{73}$$

This defines a Suslin scheme S for which (72), and hence (71), is easily verified.

We now use (71) to prove (70). Choose

$$x \in N(S) = \bigcup_{z \in \mathscr{N}} \bigcap_{s < z} S(s).$$

For some $z_0 \in \mathcal{N}$, we have

$$x \in \bigcap_{s < z_0} S(s) = \bigcap_{k=1}^{\infty} \bigcap_{s < \varphi_k(z_0)} S_k(s).$$

Thus, for every k,

$$x \in \bigcap_{s < \varphi_k(z_0)} S_k(s) \subset \bigcup_{z \in \mathcal{N}} \bigcap_{s < z} S_k(s) = N(S_k).$$

It follows that $x \in \bigcap_{k=1}^{\infty} N(S_k)$ and

$$N(S) \subset \bigcap_{k=1}^{\infty} N(S_k). \tag{74}$$

If we are given $x \in \bigcap_{k=1}^{\infty} N(S_k)$, then $x \in \bigcup_{z \in \mathcal{N}} \bigcap_{s < z} S_k(s)$ for every k, so for every k, there exists $z_k \in \mathcal{N}$ such that $x \in \bigcap_{s < z_k} S_k(s)$. Let $z_0 \in \mathcal{N}$ be such that $\varphi_k(z_0) = z_k$, $k = 1, 2, \ldots$. Then $x \in \bigcap_{k=1}^{\infty} \bigcap_{s < \varphi_k(z_0)} S_k(s)$. An application of (71) shows that

$$x \in \bigcap_{s < z_0} S(s) \subset \bigcup_{z \in \mathcal{N}} \bigcap_{s < z} S(s) = N(S)$$

and

$$N(S) \supset \bigcap_{k=1}^{\infty} N(S_k). \tag{75}$$

Relation (70) follows from (74) and (75).

(c) It is clear that $\mathcal{S}(\mathcal{P})_\sigma \supset \mathcal{S}(\mathcal{P})$. Choose $\bigcup_{k=1}^{\infty} N(S_k) \in \mathcal{S}(\mathcal{P})_\sigma$, where S_k is a Suslin scheme for $\mathcal{P}$, $k = 1, 2, \ldots$. It suffices to construct a Suslin scheme S for $\mathcal{P}$ for which

$$N(S) = \bigcup_{k=1}^{\infty} N(S_k). \tag{76}$$

Given $(\zeta_1, \zeta_2, \ldots, \zeta_n) \in \Sigma$, we have $\zeta_1 = (2j - 1)2^{k-1}$ for exactly one pair of positive integers j and k. Define

$$S(\zeta_1, \zeta_2, \ldots, \zeta_n) = S((2j - 1)2^{k-1}, \zeta_2, \ldots, \zeta_n) = S_k(j, \zeta_2, \ldots, \zeta_n).$$

This defines a Suslin scheme S for which

$$\bigcap_{n=1}^{\infty} S((2j - 1)2^{k-1}, \zeta_2, \ldots, \zeta_n) = \bigcap_{n=1}^{\infty} S_k(j, \zeta_2, \ldots, \zeta_n)$$

$$\forall k \in N, \forall (j, \zeta_2, \ldots) \in \mathcal{N}. \tag{77}$$

Returning to (76), we choose $x \in N(S) = \bigcup_{z \in \mathcal{N}} \bigcap_{s < z} S(s)$. For some $(\zeta_1, \zeta_2, \ldots) \in \mathcal{N}$, we have $x \in \bigcap_{n=1}^{\infty} S(\zeta_1, \zeta_2, \ldots, \zeta_n)$, and choosing j and k so that $\zeta_1 = (2j - 1)2^{k-1}$, we have, from (77),

$$x \in \bigcap_{n=1}^{\infty} S_k(j, \zeta_2, \dots, \zeta_n) \subset N(S_k) \subset \bigcup_{k=1}^{\infty} N(S_k),$$

so

$$N(S) \subset \bigcup_{k=1}^{\infty} N(S_k). \tag{78}$$

If, on the other hand, we choose

$$x \in \bigcup_{k=1}^{\infty} N(S_k) = \bigcup_{k=1}^{\infty} \bigcup_{z \in \mathcal{N}} \bigcap_{s < z} S_k(s),$$

then for some $k \in N$ and $(j, \zeta_2, \dots) \in \mathcal{N}$, we have $x \in \bigcap_{n=1}^{\infty} S_k(j, \zeta_2, \dots, \zeta_n)$. Equation (77) implies

$$x \in \bigcap_{n=1}^{\infty} S((2j - 1)2^{k-1}, \zeta_2, \dots, \zeta_n) \subset N(S),$$

so

$$N(S) \supset \bigcup_{k=1}^{\infty} N(S_k). \tag{79}$$

Relation (76) follows from (78) and (79).

(d)   For $P \in \mathscr{P}$, define $S(s) = P$ for every $s \in \Sigma$. Then $N(S) = P$.

(e)   The proof of this takes us somewhat far afield, so is given as Proposition B.2 of Appendix B.   Q.E.D.

It is not in general true that $\mathscr{S}(\mathscr{P})$ is closed under complementation, so $\mathscr{S}(\mathscr{P})$ is generally not a σ-algebra. In order for $\mathscr{S}(\mathscr{P})$ to contain $\sigma(\mathscr{P})$, we need one additional assumption.

Corollary 7.35.1   Let $(X, \mathscr{P})$ be a paved space and assume that the complement of each set in $\mathscr{P}$ is in $\mathscr{S}(\mathscr{P})$. Then $\sigma(\mathscr{P}) \subset \mathscr{S}(\mathscr{P})$.

Proof   The smallest algebra containing $\mathscr{P}$ consists of the finite intersections of finite unions of sets in $\mathscr{P}$ and complements of sets in $\mathscr{P}$. By Proposition 7.35, these sets are contained in $\mathscr{S}[\mathscr{S}[\mathscr{S}(\mathscr{P})]] = \mathscr{S}(\mathscr{P})$. Since $\mathscr{S}(\mathscr{P})$ is a monotone class, it contains the σ-algebra generated by $\mathscr{P}$ as well (Ash [A1, p. 19]).   Q.E.D.

Definition 7.16   Let X be a Borel space. Denote by $\mathscr{F}_X$ the collection of closed subsets of X. The *analytic subsets* of X are the members of $\mathscr{S}(\mathscr{F}_X)$.

Corollary 7.35.2   Let X be a Borel space. The countable intersections and unions of analytic subsets of X are analytic.

Proof This follows from Proposition 7.35(b) and (c).   Q.E.D.

Proposition 7.36 Let X be a Borel space. Then every Borel subset of X is analytic. Indeed, the class of analytic sets $\mathscr{S}(\mathscr{F}_X)$ is equal to $\mathscr{S}(\mathscr{B}_X)$.

Proof Every open subset of X is an F_σ (Lemma 7.2), so every open set is analytic. Corollary 7.35.1 implies $\mathscr{B}_X \subset \mathscr{S}(\mathscr{F}_X)$. Proposition 7.35(a) and (e) implies

$$\mathscr{S}(\mathscr{F}_X) \subset \mathscr{S}(\mathscr{B}_X) \subset \mathscr{S}[\mathscr{S}(\mathscr{F}_X)] = \mathscr{S}(\mathscr{F}_X). \qquad \text{Q.E.D.}$$

If the Borel space X is countable, then every subset of X is both analytic and Borel-measurable. If X is uncountable, however, the class of analytic subsets of X is strictly larger than $\mathscr{B}_X$. This is shown in Appendix B, where we prove the existence of an analytic set whose complement is not analytic.

Note that an immediate consequence of Proposition 7.36 is that if Y is a Borel subset of the Borel space X, then the analytic subsets of Y are the analytic subsets of X contained in Y. A generalization of this fact is the following.

Corollary 7.36.1 Let X and Y be Borel spaces and $\varphi : X \to Y$ a Borel isomorphism. Then $A \subset X$ is analytic if and only if $\varphi(A) \subset Y$ is analytic.

Proof If $\varphi : X \to Y$ is a Borel isomorphism and $A \subset X$ is analytic, then $A = N(S)$, S is a Suslin scheme for $\mathscr{F}_X$. It is easily seen that $\varphi(A) = N(\varphi \circ S)$, where $\varphi \circ S$ is the Suslin scheme for $\mathscr{B}_Y$ defined by $(\varphi \circ S)(s) = \varphi[S(s)]$, so $\varphi(A)$ is analytic by Proposition 7.36. If $\varphi(A) \subset Y$ is analytic, $A \subset X$ is analytic by a similar argument.   Q.E.D.

We proceed to the development of several equivalent characterizations of analytic sets. The general definition of a Suslin scheme is unrestrictive with respect to the form of the mapping $S : \Sigma \to \mathscr{P}$. In the event that X is a separable metric space and $\mathscr{P} = \mathscr{F}_X$, one can assume without loss of generality that S has more structure.

Definition 7.17 Let $(X, \mathscr{P})$ be a paved space and S a Suslin scheme for $\mathscr{P}$. The Suslin scheme S is *regular* if for each $n \in N$ and $(\sigma_1, \sigma_2, \ldots, \sigma_{n+1}) \in \Sigma$, we have

$$S(\sigma_1, \sigma_2, \ldots, \sigma_n) \supset S(\sigma_1, \sigma_2, \ldots, \sigma_n, \sigma_{n+1}).$$

Lemma 7.22 Let (X, d) be a separable metric space and S a Suslin scheme for $\mathscr{F}_X$. Then there exists a regular Suslin scheme R for $\mathscr{F}_X$ such that $N(R) = N(S)$ and, for every $z = (\zeta_1, \zeta_2, \ldots) \in \mathscr{N}$,

$$\lim_{n \to \infty} \operatorname{diam} R(\zeta_1, \zeta_2, \ldots, \zeta_n) = 0 \qquad \text{if} \quad R(\zeta_1, \zeta_2, \ldots, \zeta_n) \neq \varnothing \quad \forall n. \quad (80)$$

Proof By the Lindelöf property, for each positive integer k, X can be covered by a countable collection of open spheres of the form $B_{kj} = \{x \in X \mid d(x, x_{kj}) < 1/k\}, j = 1, 2, \ldots$. For $(\bar{\zeta}_1, \zeta_1, \bar{\zeta}_2, \zeta_2, \ldots) \in \mathcal{N}$, define

$$R(\bar{\zeta}_1) = \bar{B}_{1\bar{\zeta}_1},$$
$$R(\bar{\zeta}_1, \zeta_1) = R(\bar{\zeta}_1) \cap S(\zeta_1),$$
$$R(\bar{\zeta}_1, \zeta_1, \bar{\zeta}_2) = R(\bar{\zeta}_1, \zeta_1) \cap \bar{B}_{2\bar{\zeta}_2},$$
$$R(\bar{\zeta}_1, \zeta_1, \bar{\zeta}_2, \zeta_2) = R(\bar{\zeta}_1, \zeta_1, \bar{\zeta}_2) \cap S(\zeta_1, \zeta_2),$$

etc. Thus

$$\bigcap_{s < (\bar{\zeta}_1, \zeta_1, \bar{\zeta}_2, \zeta_2, \ldots)} R(s) = \left[\bigcap_{k=1}^{\infty} \bar{B}_{k\bar{\zeta}_k} \right] \cap \left[\bigcap_{s < z} S(s) \right], \tag{81}$$

where $z = (\zeta_1, \zeta_2, \ldots)$. It is clear that R is a regular Suslin scheme for $\mathscr{F}_X$ and (80) holds. If $x \in N(R)$, then there exists $(\bar{\zeta}_1, \zeta_1, \bar{\zeta}_2, \zeta_2, \ldots) \in \mathcal{N}$ such that $x \in \bigcap_{s < (\bar{\zeta}_1, \zeta_1, \bar{\zeta}_2, \zeta_2, \ldots)} R(s)$, so by (81) $x \in \bigcap_{s < (\zeta_1, \zeta_2, \ldots)} S(s) \subset N(S)$, and therefore $N(R) \subset N(S)$. If $x \in N(S)$, then there exists $(\zeta_1, \zeta_2, \ldots) \in \mathcal{N}$ such that $x \in \bigcap_{s < (\zeta_1, \zeta_2, \ldots)} S(s)$. Since for each positive integer k, the collection $\{B_{kj} \mid j = 1, 2, \ldots\}$ covers X, there exists for each k a positive integer $\bar{\zeta}_k$ for which $x \in B_{k\bar{\zeta}_k}$. Then $x \in \bigcap_{k=1}^{\infty} B_{k\bar{\zeta}_k}$ and, by (81), $x \in \bigcap_{s < (\bar{\zeta}_1, \zeta_1, \bar{\zeta}_2, \zeta_2, \ldots)} R(s) \subset N(R)$, so $N(R) \supset N(S)$. It follows that $N(R) = N(S)$.   Q.E.D.

Note that if a regular Suslin scheme R satisfies (80), then for all z in

$$\mathcal{N}_1 = \left\{ z \in \mathcal{N} \mid \bigcap_{s < z} R(s) \neq \varnothing \right\},$$

the set $\bigcap_{s < z} R(s)$ consists of a single point, say $f(z)$. Thus we have

$$N(R) = f(\mathcal{N}_1),$$

and this relation provides the basis for an alternative way of characterizing analytic sets. We have the following lemma.

Lemma 7.23 Let (X, d) be a complete separable metric space. If $A \subset X$ is a nonempty analytic set, then there exist a closed subset $\mathcal{N}_1$ of $\mathcal{N}$ and a continuous function $f: \mathcal{N}_1 \to X$ such that $A = f(\mathcal{N}_1)$. Conversely, if $\mathcal{N}_1 \subset \mathcal{N}$ is closed and $f: \mathcal{N}_1 \to X$ is continuous, then $f(\mathcal{N}_1)$ is analytic.

Proof Let $A = N(R)$ be nonempty, where R is a regular Suslin scheme for $\mathscr{F}_X$ satisfying (80). Define

$$\mathcal{N}_1 = \left\{ z \in \mathcal{N} \mid \bigcap_{s < z} R(s) \neq \varnothing \right\}.$$

Let $z = (\zeta_1, \zeta_2, \ldots)$ be in $\mathcal{N}$. If for each n we have $R(\zeta_1, \zeta_2, \ldots, \zeta_n) \neq \varnothing$, then it is possible to chose $x_n \in R(\zeta_1, \zeta_2, \ldots, \zeta_n)$. The sequence $\{x_n\}$ is Cauchy by (80), and since (X, d) is complete, $\{x_n\}$ has a limit $x \in X$. But for each n the regularity of R implies $\{x_m | m \geq n\} \subset R(\zeta_1, \zeta_2, \ldots, \zeta_n)$, so $x \in R(\zeta_1, \zeta_2, \ldots, \zeta_n)$. Therefore $x \in \bigcap_{s < z} R(s)$. Now suppose $z \in \mathcal{N} - \mathcal{N}_1$. The preceding argument shows that for some $s_n < z$, we have $R(s_n) = \varnothing$. The open neighborhood $\{w \in \mathcal{N} | s_n < w\}$ contains z and is contained in $\mathcal{N} - \mathcal{N}_1$, so $\mathcal{N} - \mathcal{N}_1$ is open and $\mathcal{N}_1$ is closed.

For $z \in \mathcal{N}_1$, define $f(z)$ to be the unique point in $\bigcap_{s < z} R(s)$. If $\{z_k\}$ is a sequence in $\mathcal{N}_1$ converging to $z_0 = (\zeta_1, \zeta_2, \ldots) \in \mathcal{N}_1$, then given $\varepsilon > 0$, (80) implies that there exists $s_n < z_0$ for which diam $R(s_n) < \varepsilon$. For k sufficiently large, $z_k \in \{z \in \mathcal{N} | s_n < z\}$, so $f(z_k) \in R(s_n)$. Therefore $d(f(z_k), f(z_0)) \leq$ diam $R(s_n) < \varepsilon$, which shows that f is continuous.

For the converse, suppose $\mathcal{N}_1 \subset \mathcal{N}$ is closed and $f : \mathcal{N}_1 \to X$ is continuous. Define a regular Suslin scheme R for $\mathscr{F}_X$ by

$$R(s) = \overline{f(\{z \in \mathcal{N}_1 | s < z\})},$$

where $R(s) = \varnothing$ if $\{z \in \mathcal{N}_1 | s < z\} = \varnothing$. If $z \in \mathcal{N}_1$, then $f(z) \in \bigcap_{s < z} R(s) \subset N(R)$, so $f(\mathcal{N}_1) \subset N(R)$. If $x \in N(R)$, then for some $z_0 = (\zeta_1, \zeta_2, \ldots) \in \mathcal{N}$ we have $x \in \bigcap_{s < z_0} R(s)$. Then for each n,

$$x \in \overline{f(\{z \in \mathcal{N}_1 | (\zeta_1, \zeta_2, \ldots, \zeta_n) < z\})},$$

so given $\varepsilon > 0$, there exists a $z_n \in \mathcal{N}_1$ with $(\zeta_1, \zeta_2, \ldots, \zeta_n) < z_n$ and $d(x, f(z_n)) < \varepsilon$. But as $n \to \infty$, z_n must converge to z_0. The closedness of $\mathcal{N}_1$ implies $z_0 \in \mathcal{N}_1$, and the continuity of f implies $d(x, f(z_0)) \leq \varepsilon$. Since $\varepsilon > 0$ is arbitrary, we have $f(z_0) = x$, $x \in f(\mathcal{N}_1)$, and $N(R) \subset f(\mathcal{N}_1)$. Q.E.D.

We have thus characterized analytic sets as the continuous images of closed subsets of $\mathcal{N}$. We will obtain an even sharper characterization, for which we need the following lemma.

Lemma 7.24 If $\mathcal{N}_1$ is a nonempty closed subset of $\mathcal{N}$, then there exists a continuous function $g : \mathcal{N} \to \mathcal{N}$ such that $\mathcal{N}_1 = g(\mathcal{N})$.

Proof Use the Lindelöf property to cover $\mathcal{N}_1$ with a countable collection of nonempty closed sets $\{S(\zeta_1) | \zeta_1 \in N\}$ which satisfy

$$\mathcal{N}_1 \supset S(\zeta_1), \qquad \text{diam } S(\zeta_1) \leq 1, \quad \zeta_1 = 1, 2, \ldots,$$

where d is a metric on $\mathcal{N}$ consistent with its topology and diam $S(\zeta_1)$ is given by (9). Cover each $S(\zeta_1)$ with a countable collection of nonempty closed sets $\{S(\zeta_1, \zeta_2) | \zeta_2 \in N\}$ which satisfy

$$S(\zeta_1) \supset S(\zeta_1, \zeta_2), \qquad \text{diam } S(\zeta_1, \zeta_2) \leq \tfrac{1}{2}, \quad \zeta_2 = 1, 2, \ldots.$$

Continue in this manner so that, for any $(\zeta_1, \zeta_2, \ldots, \zeta_{n-1})$,

$$S(\zeta_1, \zeta_2, \ldots, \zeta_n) \neq \varnothing, \qquad \zeta_n = 1, 2, \ldots, \tag{82}$$

$$S(\zeta_1, \zeta_2, \ldots, \zeta_{n-1}) = \bigcup_{\zeta_n = 1}^{\infty} S(\zeta_1, \zeta_2, \ldots, \zeta_n), \tag{83}$$

$$S(\zeta_1, \zeta_2, \ldots, \zeta_{n-1}) \supset S(\zeta_1, \zeta_2, \ldots, \zeta_{n-1}, \zeta_n), \qquad \zeta_n = 1, 2, \ldots, \tag{84}$$

$$\operatorname{diam} S(\zeta_1, \zeta_2, \ldots, \zeta_n) \leq 1/n, \qquad \zeta_n = 1, 2, \ldots. \tag{85}$$

The completeness of $\mathcal{N}$ and (82)–(85) imply that for each $z \in \mathcal{N}$, $\bigcap_{s < z} S(s)$ consists of a single point. Define $g(z)$ to be this point. Then

$$N(S) = g(\mathcal{N}) = \mathcal{N}_1.$$

The continuity of g follows by an argument similar to the one used in the proof of Lemma 7.23.   Q.E.D.

Proposition 7.37   Let X be a Borel space. A nonempty set $A \subset X$ is analytic if and only if $A = f(\mathcal{N})$ for some continuous function $f : \mathcal{N} \to X$.

Proof   If X is complete, the proposition follows from Lemmas 7.23 and 7.24. If X is not complete, it is still homeomorphic to a Borel subset of a complete separable space, and the result follows from Corollary 7.36.1.
                                                                    Q.E.D.

Proposition 7.37 gives a very useful characterization of nonempty analytic sets in terms of continuous functions and the Baire null space $\mathcal{N}$. The Baire null space has a simple description and its topology allows considerable flexibility. We have already shown, for example, that it is homeomorphic to $\mathcal{N}_0$, the space of irrationals in $[0, 1]$. Another important homeomorphism is the following.

Lemma 7.25   The space $\mathcal{N}$ is homeomorphic to any finite or countably infinite product of copies of itself.

Proof   We prove the lemma for the case of a countably infinite product. Let $\Pi_1, \Pi_2, \ldots$ be a partition of N, the set of positive integers, into infinitely many infinite sets. Define $\varphi : \mathcal{N} \to \mathcal{N} \mathcal{N} \mathcal{N} \cdots$ by

$$\varphi(z) = (z_1, z_2, \ldots), \tag{86}$$

where z_k consists of the components of z with indices in Π_k. Then φ is one-to-one and onto and, because convergence in a product space is componentwise, φ is a homeomorphism.   Q.E.D.

Combination of Lemma 7.25 with Proposition 7.37 gives the following.

Proposition 7.38 Let $X_1, X_2, \ldots$ be a sequence of Borel spaces and A_k an analytic subset of $X_k, k = 1, 2, \ldots$. Then the sets $A_1 A_2 \cdots$ and $A_1 A_2 \cdots A_n$, $n = 1, 2, \ldots$, are analytic subsets of $X_1 X_2 \cdots$ and $X_1 X_2 \cdots X_n$, respectively.

Proof Let $f_k : \mathcal{N} \to X_k$ be continuous such that $A_k = f_k(\mathcal{N}), k = 1, 2, \ldots$. Let φ be given by (86) and $F : \mathcal{N} \mathcal{N} \cdots \to X_1 X_2 \cdots$ be given by

$$F(z_1, z_2, \ldots) = (f_1(z_1), f_2(z_2), \ldots).$$

Then $F \circ \varphi$ is continuous and maps $\mathcal{N}$ onto $A_1 A_2 \cdots$. The finite products are handled similarly.   Q.E.D.

Another consequence of Proposition 7.37 is that the continuous image of an analytic set, in particular, the projection of an analytic set, is analytic. As discussed at the beginning of this section, this property motivated our inquiry into analytic sets. We formalize this and a related fact to obtain another characterization of analytic sets.

Proposition 7.39 Let X and Y be Borel spaces and A an analytic subset of XY. Then $\text{proj}_X(A)$ is analytic. Conversely, given any analytic set $C \subset X$ and any uncountable Borel space Y, there is a Borel set $B \subset XY$ such that $C = \text{proj}_X(B)$. If $Y = \mathcal{N}$, B can be chosen to be closed.

Proof If $A = f(\mathcal{N}) \subset XY$ is analytic, where f is continuous, then $\text{proj}_X(A) = (\text{proj}_X \circ f)(\mathcal{N})$ is analytic by Proposition 7.37. If $C = f(\mathcal{N}) \subset X$ is nonempty and analytic, then

$$C = \text{proj}_X[\tilde{\text{Gr}}(f)],$$

where $\tilde{\text{Gr}}(f) = \{(f(z), z) \in X\mathcal{N} \mid z \in \mathcal{N}\}$ is closed because f is continuous. If Y is any uncountable Borel space, then there exists a Borel isomorphism φ from $\mathcal{N}$ onto Y (Corollary 7.16.1). The mapping Φ defined by

$$\Phi(x, z) = (x, \varphi(z))$$

is a Borel isomorphism from $X\mathcal{N}$ onto XY, and

$$C = \text{proj}_X(\Phi[\tilde{\text{Gr}}(f)]).   \text{Q.E.D.}$$

So far we have treated only the continuous images of analytic sets. With the aid of Proposition 7.39, we can consider their images under Borel-measurable functions as well.

Proposition 7.40 Let X and Y be Borel spaces and $f : X \to Y$ a Borel-measurable function. If $A \subset X$ is analytic, then $f(A)$ is analytic. If $B \subset Y$ is analytic, then $f^{-1}(B)$ is analytic.

Proof Suppose $A \subset X$ is analytic. By Proposition 7.39, there exists a Borel set $B \subset X\mathcal{N}$ such that $A = \text{proj}_X(B)$. Define $\psi : B \to Y$ by $\psi(x, z) = f(x)$.

Then ψ is Borel-measurable, and Corollary 7.14.1 implies that $\mathrm{Gr}(\psi) \in \mathscr{B}_{X\mathscr{N}Y}$. Finally, $f(A) = \mathrm{proj}_Y[\mathrm{Gr}(\psi)]$ is analytic by Proposition 7.39.

If $B \subset Y$ is analytic, then $B = N(S)$, where S is some Suslin scheme for $\mathscr{F}_Y$. Then $f^{-1}(B) = N(f^{-1} \circ S)$, where $f^{-1} \circ S$ is the Suslin scheme for $\mathscr{B}_X$ defined by

$$(f^{-1} \circ S)(s) = f^{-1}[S(s)] \qquad \forall s \in \Sigma.$$

The analyticity of $f^{-1}(B)$ follows from Proposition 7.36. Q.E.D.

We summarize the equivalent definitions of analytic sets in Borel spaces.

Proposition 7.41 Let X be a Borel space. The following definitions of the collection of analytic subsets of X are equivalent:

(a) $\mathscr{S}(\mathscr{F}_X)$;

(b) $\mathscr{S}(\mathscr{B}_X)$;

(c) the empty set and the images of $\mathscr{N}$ under continuous functions from $\mathscr{N}$ into X;

(d) the projections into X of the closed subsets of $X\mathscr{N}$;

(e) the projections into X of the Borel subsets of XY, where Y is an uncountable Borel space;

(f) the images of Borel subsets of Y under Borel-measurable functions from Y into X, where Y is an uncountable Borel space.

Proof The only new characterization here is (f). If Y is an uncountable Borel space and $f: Y \to X$ is Borel-measurable, then for every $B \in \mathscr{B}_Y$, $f(B)$ is analytic in X by Proposition 7.40. To show that every nonempty analytic set $A \subset X$ can be obtained this way, let φ be a Borel isomorphism from Y onto $X\mathscr{N}$ and let $F \subset X\mathscr{N}$ be closed and satisfy $\mathrm{proj}_X(F) = A$. Define $B = \varphi^{-1}(F) \in \mathscr{B}_Y$. Then $(\mathrm{proj}_X \circ \varphi)(B) = A$. If $A = \varnothing$, then $f(\varnothing) = A$ for any Borel-measurable $f: Y \to X$. Q.E.D.

7.6.2 Measurability Properties of Analytic Sets

At the beginning of this section we indicated that extended real-valued functions on a Borel space X whose lower level sets are analytic arise naturally via partial infimization. Because the collection of analytic subsets of an uncountable Borel space is strictly larger than the Borel σ-algebra (Appendix B), such functions need not be Borel-measurable. Nonetheless, they can be integrated with respect to any probability measure on $(X, \mathscr{B}_X)$. To show this, we must discuss the measurability properties of analytic sets.

If X is a Borel space and $p \in P(X)$, we define *p-outer measure*, denoted by p^*, on the set of all subsets of X by

$$p^*(E) = \inf\{p(B) | E \subset B, B \in \mathscr{B}_X\}. \tag{87}$$

Outer measure on an increasing sequence of sets has a convergence property, namely, $p^*(E_n) \uparrow p^*(\bigcup_{n=1}^{\infty} E_n)$ if $E_1 \subset E_2 \subset \cdots$. This is easy to verify from (87) and also follows from Eq. (5) and Proposition A.1 of Appendix A (see also Ash [A1, Lemma 1.3.3(d)]). The collection of sets $\mathscr{B}_X(p)$ defined by

$$\mathscr{B}_X(p) = \{E \subset X | p^*(E) + p^*(E^c) = 1\}$$

is a σ-algebra (Ash [A1, Theorem 1.3.5]), called the *completion of $\mathscr{B}_X$ with respect to p*. It can be described as the class of sets of the form $B \cup N$ as B ranges over $\mathscr{B}_X$ and N ranges over all subsets of sets of p-measure zero in $\mathscr{B}_X$ (Ash [A1, p. 18]), and we have

$$p^*(B \cup N) = p(B).$$

Furthermore, p^* restricted to $\mathscr{B}_X(p)$ is a probability measure, and is the only extension of p to $\mathscr{B}_X(p)$ that is a probability measure. In what follows, we denote this measure also by p and write $p(E)$ in place of $p^*(E)$ for all $E \in \mathscr{B}_X(p)$.

Definition 7.18 Let X be a Borel space. The *universal σ-algebra $\mathscr{U}_X$* is defined by $\mathscr{U}_X = \bigcap_{p \in P(X)} \mathscr{B}_X(p)$. If $E \in \mathscr{U}_X$, we say E is *universally measurable*.

The usefulness of analytic sets in measure theory is in large degree derived from the following proposition.

Proposition 7.42 (Lusin's theorem) Let X be a Borel space and S a Suslin scheme for $\mathscr{U}_X$. Then $N(S)$ is universally measurable. In other words, $\mathscr{S}(\mathscr{U}_X) = \mathscr{U}_X$.

Proof Denote $A = N(S)$, where S is a Suslin scheme for $\mathscr{U}_X$. For $(\sigma_1, \ldots, \sigma_k) \in \Sigma$, define

$$N(\sigma_1, \ldots, \sigma_k) = \{(\zeta_1, \zeta_2, \ldots) \in \mathscr{N} | \zeta_1 = \sigma_1, \ldots, \zeta_k = \sigma_k\} \tag{88}$$

and

$$M(\sigma_1, \ldots, \sigma_k) = \{(\zeta_1, \zeta_2, \ldots) \in \mathscr{N} | \zeta_1 \leq \sigma_1, \ldots, \zeta_k \leq \sigma_k\}$$
$$= \bigcup_{\tau_1 \leq \sigma_1, \ldots, \tau_k \leq \sigma_k} N(\tau_1, \ldots, \tau_k). \tag{89}$$

Define also

$$R(\sigma_1, \ldots, \sigma_k) = \bigcup_{z \in M(\sigma_1, \ldots, \sigma_k)} \bigcap_{s < z} S(s). \tag{90}$$

Then

$$R(\sigma_1, \ldots, \sigma_k) \subset K(\sigma_1, \ldots, \sigma_k), \tag{91}$$

where

$$K(\sigma_1, \ldots, \sigma_k) = \bigcup_{\tau_1 \leq \sigma_1, \ldots, \tau_k \leq \sigma_k} \bigcap_{j=1}^{k} S(\tau_1, \ldots, \tau_j). \tag{92}$$

As $\sigma_1 \uparrow \infty$, $M(\sigma_1) \uparrow \mathcal{N}$, so $R(\sigma_1) \uparrow A$. Likewise, as $\sigma_k \uparrow \infty$, $M(\sigma_1, \ldots, \sigma_{k-1}, \sigma_k) \uparrow M(\sigma_1, \ldots, \sigma_{k-1})$, so $R(\sigma_1, \ldots, \sigma_{k-1}, \sigma_k) \uparrow R(\sigma_1, \ldots, \sigma_{k-1})$. Given $p \in P(X)$ and $\varepsilon > 0$, choose $\bar{\zeta}_1, \bar{\zeta}_2, \ldots$ such that

$$p^*(A) \le p^*[R(\bar{\zeta}_1)] + (\varepsilon/2),$$
$$p^*[R(\bar{\zeta}_1, \ldots, \bar{\zeta}_{k-1})] \le p^*[R(\bar{\zeta}_1, \ldots, \bar{\zeta}_{k-1}, \bar{\zeta}_k)] + (\varepsilon/2^k), \qquad k = 2, 3, \ldots.$$

Then

$$p^*(A) \le p^*[R(\bar{\zeta}_1, \ldots, \bar{\zeta}_k)] + \varepsilon, \qquad k = 1, 2, \ldots. \tag{93}$$

The set $K(\bar{\zeta}_1, \ldots, \bar{\zeta}_k)$ is universally measurable, so (91) and (93) imply

$$\begin{aligned}
1 &= p[K(\bar{\zeta}_1, \ldots, \bar{\zeta}_k)] + p[X - K(\bar{\zeta}_1, \ldots, \bar{\zeta}_k)] \\
&\ge p^*[R(\bar{\zeta}_1, \ldots, \bar{\zeta}_k)] + p[X - K(\bar{\zeta}_1, \ldots, \bar{\zeta}_k)] \\
&\ge p^*(A) - \varepsilon + p[X - K(\bar{\zeta}_1, \ldots, \bar{\zeta}_k)].
\end{aligned} \tag{94}$$

We show that

$$\bigcap_{k=1}^{\infty} K(\zeta_1, \ldots, \zeta_k) \subset A \qquad \forall (\zeta_1, \zeta_2, \ldots) \in \mathcal{N}. \tag{95}$$

If

$$x \in \bigcap_{k=1}^{\infty} K(\zeta_1, \ldots, \zeta_k) = \bigcap_{k=1}^{\infty} \bigcup_{\tau_1 \le \zeta_1, \ldots, \tau_k \le \zeta_k} \bigcap_{j=1}^{k} S(\tau_1, \ldots, \tau_j), \tag{96}$$

an argument by contradiction will be used to show that for some $\bar{\tau}_1 \le \zeta_1$, we have

$$x \in S(\bar{\tau}_1) \cap \left[\bigcap_{k=2}^{\infty} \bigcup_{\tau_2 \le \zeta_2, \ldots, \tau_k \le \zeta_k} \bigcap_{j=2}^{k} S(\bar{\tau}_1, \tau_2, \ldots, \tau_j) \right]. \tag{97}$$

If no such $\bar{\tau}_1$ existed, then for every $\tau_1 \le \zeta_1$, there would exist a positive integer $k(\tau_1)$ such that

$$x \notin S(\tau_1) \cap \left[\bigcap_{k=2}^{k(\tau_1)} \bigcup_{\tau_2 \le \zeta_2, \ldots, \tau_k \le \zeta_k} \bigcap_{j=2}^{k} S(\tau_1, \tau_2, \ldots, \tau_j) \right].$$

If $\bar{k} = \max_{\tau_1 \le \zeta_1} k(\tau_1)$, then

$$x \notin \bigcup_{\tau_1 \le \zeta_1} \left\{ S(\tau_1) \cap \left[\bigcap_{k=2}^{\bar{k}} \bigcup_{\tau_2 \le \zeta_2, \ldots, \tau_k \le \zeta_k} \bigcap_{j=2}^{k} S(\tau_1, \tau_2, \ldots, \tau_j) \right] \right\}$$

$$\supset \bigcup_{\tau_1 \le \zeta_1, \ldots, \tau_{\bar{k}} \le \zeta_{\bar{k}}} \bigcap_{j=1}^{\bar{k}} S(\tau_1, \tau_2, \ldots, \tau_j) = K(\zeta_1, \ldots, \zeta_{\bar{k}})$$

and a contradiction is reached. Replace (96) by (97) and apply the same argument to establish the existence of $\bar{\tau}_2 \leq \zeta_2$ such that

$$x \in S(\bar{\tau}_1) \cap S(\bar{\tau}_1, \bar{\tau}_2) \cap \left[\bigcap_{k=3}^{\infty} \bigcup_{\tau_3 \leq \zeta_3, \ldots, \tau_k \leq \zeta_k} \bigcap_{j=3}^{k} S(\bar{\tau}_1, \bar{\tau}_2, \tau_3, \ldots, \tau_j) \right].$$

Continuing this process, construct a sequence $\bar{\tau}_1 \leq \zeta_1, \bar{\tau}_2 \leq \zeta_2, \ldots$ such that

$$x \in \bigcap_{k=1}^{\infty} S(\bar{\tau}_1, \ldots, \bar{\tau}_k) \subset N(S) = A.$$

This proves (95), i.e., as $k \to \infty$, $K(\zeta_1, \ldots, \zeta_k)$ decreases to a set contained in A, and $X - K(\zeta_1, \ldots, \zeta_k)$ increases to a set containing $X - A$. Letting $k \to \infty$ in (94), we obtain

$$1 \geq p^*(A) - \varepsilon + p^*(X - A).$$

Since $\varepsilon > 0$ is arbitrary, this implies that

$$1 \geq p^*(A) + p^*(X - A).$$

It is true for any $E \subset X$ that $p^*(E) + p^*(X - E) \geq 1$, so

$$p^*(A) + p^*(X - A) = 1,$$

and A is measurable with respect to p.   Q.E.D.

Corollary 7.42.1   Let X be a Borel space. Every analytic subset of X is universally measurable.

Proof   The closed subsets of X are universally measurable, so $\mathscr{S}(\mathscr{F}_X) \subset \mathscr{U}_X$ by Proposition 7.42.   Q.E.D.

As remarked earlier, the class of analytic subsets of an uncountable Borel space is not a σ-algebra, so there are universally measurable sets which are not analytic. In fact, we show in Appendix B that in any uncountable Borel space, the universal σ-algebra is strictly larger than the σ-algebra generated by the analytic subsets.

7.6.3   An Analytic Set of Probability Measures

In Proposition 7.25 we saw that when X is a Borel space, the function $\theta_A : P(X) \to [0, 1]$ defined by $\theta_A(p) = p(A)$ is Borel-measurable for every Borel-measurable $A \subset X$. We now investigate the properties of this function when A is analytic. The main result is that the set $\{p \in P(X) | p(A) \geq c\}$ is analytic for each real c.

Proposition 7.43   Let X be a Borel space and A an analytic subset of X. For each $c \in R$, the set $\{p \in P(X) | p(A) \geq c\}$ is analytic.

Proof   Let S be a Suslin scheme for $\mathscr{F}_X$, the class of closed subsets of X, such that $A = N(S)$. For $s \in \Sigma$, let $N(s)$, $M(s)$, $R(s)$, and $K(s)$ be defined by (88)–(90) and (92). Then (91) and (95) hold and each $K(s)$ is closed. We show that for $c \in R$

$$\{p \in P(X)|p(A) \geq c\} = \bigcap_{n=1}^{\infty} \bigcup_{z \in \mathscr{N}} \bigcap_{s < z} \{p \in P(X)|p[K(s)] \geq c - (1/n)\}. \quad (98)$$

If $\bar{p}(A) \geq c$, then for any $n \geq 1$, there exists $(\bar{\zeta}_1, \bar{\zeta}_2, \ldots) \in \mathscr{N}$ such that (93) is satisfied with $p = \bar{p}$ and $\varepsilon = 1/n$. Then by (91), for $k = 1, 2, \ldots$

$$\bar{p}[K(\bar{\zeta}_1, \ldots, \bar{\zeta}_k)] \geq \bar{p}[R(\bar{\zeta}_1, \ldots, \bar{\zeta}_k)] \geq \bar{p}(A) - (1/n) \geq c - (1/n),$$

so

$$\bar{p} \in \bigcap_{n=1}^{\infty} \bigcup_{z \in \mathscr{N}} \bigcap_{s < z} \{p \in P(X)|p[K(s)] \geq c - (1/n)\}.$$

On the other hand, given

$$\bar{p} \in \bigcap_{n=1}^{\infty} \bigcup_{z \in \mathscr{N}} \bigcap_{s < z} \{p \in P(X)|p[K(s)] \geq c - (1/n)\},$$

we have that for each n there exists $(\zeta_1, \zeta_2, \ldots) \in \mathscr{N}$ for which

$$\bar{p}\left[\bigcap_{k=1}^{\infty} K(\zeta_1, \ldots, \zeta_k)\right] = \lim_{k \to \infty} \bar{p}[K(\zeta_1, \ldots, \zeta_k)] \geq c - (1/n).$$

We have then from (95) that

$$\bar{p}(A) \geq c - (1/n), \qquad n = 1, 2, \ldots,$$

so $\bar{p}(A) \geq c$, and (98) is proved.

Proposition 7.25 guarantees that for each $n \geq 1$ and $s \in \Sigma$, the set

$$T_n(s) = \{p \in P(X)|p[K(s)] \geq c - (1/n)\}$$

is Borel-measurable in $P(X)$. We have from (98) that

$$\{p \in P(X)|p(A) \geq c\} = \bigcap_{n=1}^{\infty} N(T_n),$$

and the proposition follows from Proposition 7.36 and Corollary 7.35.2.

Q.E.D.

Corollary 7.43.1   Let X be a Borel space and A an analytic subset of X. For each $c \in R$, the set $\{p \in P(X)|p(A) > c\}$ is analytic.

Proof For each $c \in R$,

$$\{p \in P(X) | p(A) > c\} = \bigcup_{n=1}^{\infty} \{p \in P(X) | p(A) \geq c + (1/n)\}.$$

The result follows from Corollary 7.35.2 and Proposition 7.43.   Q.E.D.

7.7 Lower Semianalytic Functions and Universally Measurable Selection

In a Borel space X, there are at least three σ-algebras which arise naturally[†]: the Borel σ-algebra $\mathscr{B}_X$ of Definition 7.6, the universal σ-algebra $\mathscr{U}_X$ of Definition 7.18, and the analytic σ-algebra $\mathscr{A}_X$, which we define now.

Definition 7.19 Let X be a Borel space. The *analytic σ-algebra* $\mathscr{A}_X$ is the smallest σ-algebra containing the analytic subsets of X. In symbols, $\mathscr{A}_X = \sigma[\mathscr{S}(\mathscr{F}_X)]$. If $E \in \mathscr{A}_X$, we say that E is *analytically measurable*.

From Proposition 7.36 and Lusin's theorem (Proposition 7.42), we have that for any Borel space X

$$\mathscr{B}_X \subset \mathscr{S}(\mathscr{F}_X) \subset \mathscr{A}_X \subset \mathscr{U}_X.$$

If X is countable, each of these collections of sets is equal to the power set of X (the collection of all subsets of X). We show in Appendix B that if X is uncountable, each set containment above is strict. This fact will not be used in the constructive part of the theory, but only to give examples showing that results cannot be strengthened.

Corresponding to the three σ-algebras just discussed, we will treat three types of measurability of functions. Borel-measurable functions were defined in Definition 7.8. The other two types are defined next.

Definition 7.20 Let X and Y be Borel spaces and f a function mapping $D \subset X$ into Y. If $D \in \mathscr{A}_X$ and $f^{-1}(B) \in \mathscr{A}_X$ for every $B \in \mathscr{B}_Y$, f is said to be *analytically measurable*. If $D \in \mathscr{U}_X$ and $f^{-1}(B) \in \mathscr{U}_X$ for every $B \in \mathscr{B}_Y$, f is said to be *universally measurable*.

From the preceding discussion, we see that every Borel-measurable function is analytically measurable, and every analytically measurable function is universally measurable. The converses of these statements are false.

We begin by stating for future reference the following characterization of the universal σ-algebra.

[†] A fourth σ-algebra, the *limit σ-algebra* $\mathscr{L}_X$, which lies between $\mathscr{A}_X$ and $\mathscr{U}_X$, is defined in Appendix B, and treated there and in Section 11.1.

Lemma 7.26   Let X be a Borel space and $E \subset X$. Then $E \in \mathcal{U}_X$ if and only if, given any $p \in P(X)$, there exists $B \in \mathcal{B}_X$ such that $p(E \triangle B) = 0$.

We turn now to the question of composition of measurable functions. If Borel-measurable functions are composed, the result is again Borel-measurable. Unfortunately, the composition of analytically measurable functions need not be analytically measurable (Appendix B). We have the following result for universally measurable functions.

Proposition 7.44   Let X, Y, and Z be Borel spaces, $D \in \mathcal{U}_X$, and $E \in \mathcal{U}_Y$. Suppose $f : D \to Y$ and $g : E \to Z$ are universally measurable and $f(D) \subset E$. Then the composition $g \circ f$ is universally measurable.

Proof   We must show that given $B \in \mathcal{B}_Z$, the set $f^{-1}[g^{-1}(B)]$ is universally measurable. Since $g^{-1}(B) \in \mathcal{U}_Y$, it suffices to prove that $f^{-1}(U) \in \mathcal{U}_X$ for every $U \in \mathcal{U}_Y$. For $p \in P(X)$, define $p' \in P(Y)$ by

$$p'(C) = p[f^{-1}(C)] \qquad \forall C \in \mathcal{B}_Y.$$

Let $V \in \mathcal{B}_Y$ be such that

$$p[f^{-1}(V) \triangle f^{-1}(U)] = p'(V \triangle U) = 0.$$

The set $f^{-1}(V)$ is in $\mathcal{U}_X$, so there exists $W \in \mathcal{B}_X$ for which $p[W \triangle f^{-1}(V)] = 0$. Then $p[W \triangle f^{-1}(U)] = 0$. The result follows from Lemma 7.26.   Q.E.D.

The proof of Proposition 7.44 also establishes the following fact.

Corollary 7.44.1   Let X and Y be Borel spaces, $D \in \mathcal{U}_X$, and $f : D \to Y$ a universally measurable function. If $U \in \mathcal{U}_Y$, then $f^{-1}(U) \in \mathcal{U}_X$.

Since $\mathcal{A}_X \subset \mathcal{U}_X$, we can specialize these results to analytically measurable sets and functions.

Corollary 7.44.2   Let X, Y, and Z be Borel spaces, $D \in \mathcal{A}_X$, and $E \in \mathcal{A}_Y$. Suppose $f : D \to Y$ and $g : E \to Z$ are analytically measurable and $f(D) \subset E$. Then the composition $g \circ f$ is universally measurable. If $A \in \mathcal{A}_Y$, then $f^{-1}(A) \in \mathcal{U}_X$.

We remind the reader that if X and Y are Borel spaces, a stochastic kernel $q(dy|x)$ on Y given X is said to be universally measurable if the mapping $\gamma(x) = q(dy|x)$ is universally measurable from X to $P(Y)$ (Definition 7.12).

Corollary 7.44.3   Let X and Y be Borel spaces, let $f : X \to Y$ be a function, and let $q(dy|x)$ be a stochastic kernel on Y given X such that, for each x, $q(dy|x)$ assigns probability one to the point $f(x) \in Y$. Then $q(dy|x)$ is universally measurable if and only if f is universally measurable.

Proof Let $\delta: Y \to P(Y)$ be the homeomorphism defined by $\delta(y) = p_y$ (Corollary 7.21.1). Let $\gamma: X \to P(Y)$ be the mapping $\gamma(x) = q(dy|x)$. Then $\gamma = \delta \circ f$ and $f = \delta^{-1} \circ \gamma$. The result follows from Proposition 7.44.   Q.E.D.

If X is a Borel space and $f: X \to R^*$ is universally measurable, then given any $p \in P(X)$, f is measurable with respect to the completed Borel σ-algebra $\mathscr{B}_X(p)$, and $\int f \, dp$ is defined by

$$\int f \, dp = \int f^+ \, dp - \int f^- \, dp,$$

where the convention $\infty - \infty = \infty$ is used and the integrations are performed on the measure space $(X, \mathscr{B}_X(p), p)$. If $D \in \mathscr{U}_X$, the integral $\int_D f \, dp$ is defined similarly. Having thus defined $\int f \, dp$ without resort to p-outer measure, we have all the classical integration theorems at our disposal, provided that we take care with the addition of infinities.

We proceed now to show that universally measurable stochastic kernels can be used to define probability measures on product spaces in the manner of Proposition 7.28. For this we need some preparatory lemmas.

Lemma 7.27 Let X be a Borel space and $f: X \to R^*$. The function f is universally measurable if and only if, for every $p \in P(X)$, there is a Borel-measurable function $f_p: X \to R^*$ such that $f(x) = f_p(x)$ for p almost every x.

Proof Suppose f is universally measurable and let $p \in P(X)$ be given. For $r \in Q^*$, let $U(r) = \{x | f(x) \leq r\}$. Then $f(x) = \inf\{r \in Q^* | x \in U(r)\}$. Let $B(r) \in \mathscr{B}_X$ be such that $p[B(r) \triangle U(r)] = 0$. Define

$$f_p(x) = \inf\{r \in Q^* | x \in B(r)\} = \inf_{r \in Q^*} \psi_r(x),$$

where $\psi_r(x) = r$ if $x \in B(r)$ and $\psi_r(x) = \infty$ otherwise. Then $f_p: X \to R^*$ is Borel-measurable, and

$$\{x | f(x) \neq f_p(x)\} \subset \bigcup_{r \in Q^*} [B(r) \triangle U(r)]$$

has p-measure zero.

Conversely, if, given $p \in P(X)$, there is a Borel-measurable f_p such that $f(x) = f_p(x)$ for p almost every x, then

$$p(\{x | f(x) \leq c\} \triangle \{x | f_p(x) \leq c\}) = 0$$

for every $c \in R^*$, and the universal measurability of f follows.   Q.E.D.

Lemma 7.27 can be used to give an alternative definition of $\int f \, dp$ when f is a universally measurable, extended real-valued function on a Borel space X and $p \in P(X)$. Letting f_p be as in the proof of that lemma, we can define $\int f \, dp = \int f_p \, dp$. It is easy to show that this definition is equivalent to the one which precedes Lemma 7.27.

Lemma 7.28   Let X and Y be Borel spaces and let $q(dy|x)$ be a stochastic kernel on Y given X. The following statements are equivalent:

(a)   The stochastic kernel $q(dy|x)$ is universally measurable.

(b)   For any $B \in \mathscr{B}_Y$, the mapping $\lambda_B : X \to R$ defined by $\lambda_B(x) = q(B|x)$ is universally measurable.

(c)   For any $p \in P(X)$, there exists a Borel-measurable stochastic kernel $q_p(dy|x)$ on Y given X such that $q(dy|x) = q_p(dy|x)$ for p almost every x.

Proof   We show (a) $\Rightarrow$ (b) $\Rightarrow$ (c) $\Rightarrow$ (a). Assume (a) holds. Then the function $\gamma : X \to P(Y)$ given by $\gamma(x) = q(dy|x)$ is universally measurable. If $B \in \mathscr{B}_X$, λ_B is defined as in (b), and $\theta_B : P(Y) \to R$ is given by $\theta_B(p) = p(B)$, then $\lambda_B = \theta_B \circ \gamma$, which is universally measurable by Propositions 7.25 and 7.44. Therefore (a) $\Rightarrow$ (b).

Assume (b) holds and choose $p \in P(X)$. Since Y is separable and metrizable, there exists a countable base $\mathscr{B}$ for the topology in Y. Let $\mathscr{F}$ be the collection of sets in $\mathscr{B}$ and their finite intersections. For $F \in \mathscr{F}$, let f_F be a Borel-measurable function for which

$$f_F(x) = q(F|x) \quad \forall x \in B_F,$$

where $B_F \in \mathscr{B}_X$ and $p(B_F) = 1$. Such an f_F and B_F exist by assumption (b) and Lemma 7.27. For $x \in \bigcap_{F \in \mathscr{F}} B_F$, let $q_p(dy|x) = q(dy|x)$. For $x \notin \bigcap_{F \in \mathscr{F}} B_F$, let $q_p(dy|x)$ be some fixed probability measure in $P(Y)$. Then $q(dy|x) = q_p(dy|x)$ for p almost every x. The class of sets $\underline{Y}$ in $\mathscr{B}_Y$ for which $q_p(\underline{Y}|x)$ is Borel-measurable in x is a Dynkin system containing $\mathscr{F}$. The class $\mathscr{F}$ is closed under finite intersections and generates $\mathscr{B}_Y$, so statement (c) follows from the Dynkin system theorem (Proposition 7.24). Therefore (b) $\Rightarrow$ (c).

Assume (c) holds and choose $p \in P(X)$. Let $q_p(dy|x)$ be as in assumption (c) and define $\gamma, \gamma_p : X \to P(Y)$ by $\gamma(x) = q(dy|x)$, $\gamma_p(x) = q_p(dy|x)$. If $B \in \mathscr{B}_{P(X)}$, then $p[\gamma^{-1}(B) \triangle \gamma_p^{-1}(B)] = 0$. Lemma 7.26 implies that $\gamma^{-1}(B)$ is universally measurable. Therefore (c) $\Rightarrow$ (a).   Q.E.D.

Lemma 7.29   Let X, Y, and Z be Borel spaces and let $f : XY \to Z$ be a universally measurable function. For fixed $x \in X$, define $g_x : Y \to Z$ by

$$g_x(y) = f(x, y).$$

Then g_x is universally measurable for every $x \in X$.

Proof   For fixed $x_0 \in X$, let $\varphi : Y \to XY$ be the continuous function defined by $\varphi(y) = (x_0, y)$. For $\underline{Z} \in \mathscr{B}_Z$,

$$\{y \in Y | g_{x_0}(y) \in \underline{Z}\} = \varphi^{-1}(\{(x, y) \in XY | f(x, y) \in \underline{Z}\}),$$

and this set is universally measurable by Corollary 7.44.1.   Q.E.D.

It is worth noting that if $(\Omega_1, \mathscr{F}_1, p)$ and $(\Omega_2, \mathscr{F}_2, q)$ are probability spaces, then there are two natural σ-algebras on $\Omega_1\Omega_2$, namely, $\mathscr{F}_1\mathscr{F}_2$ and the completion $\overline{\mathscr{F}_1\mathscr{F}_2}$ of $\mathscr{F}_1\mathscr{F}_2$ with respect to pq. If $f: \Omega_1\Omega_2 \to R$ is $\mathscr{F}_1\mathscr{F}_2$-measurable, then for every $\omega_1 \in \Omega_1$, the function $g_{\omega_1}(\omega_2) = f(\omega_1, \omega_2)$ is $\mathscr{F}_2$-measurable. However, if f is only $\overline{\mathscr{F}_1\mathscr{F}_2}$-measurable, then $g_{\omega_1}(\omega_2)$ can be guaranteed to be $\mathscr{F}_2$-measurable only for p almost all ω_1. The case treated by Lemma 7.29 is intermediate to these two, since $\mathscr{U}_X\mathscr{U}_Y \subset \mathscr{U}_{XY}$, and if $p \in P(X)$, $q \in P(Y)$, and $\overline{\mathscr{U}_X\mathscr{U}_Y}$ denotes the completion of $\mathscr{U}_X\mathscr{U}_Y$ with respect to pq, then $\mathscr{U}_{XY} \subset \overline{\mathscr{U}_X\mathscr{U}_Y}$. Note that the stronger result that $g_x(y)$ is $\mathscr{U}_Y$-measurable for every $x \in X$ holds, although the assumption that f is $\mathscr{U}_{XY}$-measurable may be weaker than the assumption that f is $\mathscr{U}_X\mathscr{U}_Y$-measurable.

We now use the properties of universally measurable functions and stochastic kernels to extend Proposition 7.28.

Proposition 7.45 Let $X_1, X_2, \ldots$ be a sequence of Borel spaces, $Y_n = X_1 X_2 \cdots X_n$ and $Y = X_1 X_2 \cdots$. Let $p \in P(X_1)$ be given and, for $n = 1, 2, \ldots$, let $q_n(dx_{n+1}|y_n)$ be a universally measurable stochastic kernel on X_{n+1} given Y_n. Then for $n = 2, 3, \ldots$, there exist unique probability measures $r_n \in P(Y_n)$, such that

$$r_n(\underline{X}_1\underline{X}_2\cdots\underline{X}_n) = \int_{\underline{X}_1}\int_{\underline{X}_2}\cdots\int_{\underline{X}_{n-1}} q_{n-1}(\underline{X}_n|x_1, x_2, \ldots, x_{n-1})$$
$$\times q_{n-2}(dx_{n-1}|x_1, x_2, \ldots, x_{n-2})\cdots$$
$$\times q_1(dx_2|x_1)p(dx_1) \qquad \forall \underline{X}_1 \in \mathscr{B}_{X_1}, \ldots, \underline{X}_n \in \mathscr{B}_{X_n}. \quad (99)$$

If $f: Y_n \to R^*$ is universally measurable and either $\int f^+ \, dr_n < \infty$ or $\int f^- \, dr_n < \infty$, then

$$\int_{Y_n} f \, dr_n = \int_{X_1}\int_{X_2}\cdots\int_{X_n} f(x_1, x_2, \ldots, x_n)q_{n-1}(dx_n|x_1, x_2, \ldots, x_{n-1})\cdots$$
$$\times q_1(dx_2|x_1)p(dx_1). \quad (100)$$

Furthermore, there exists a unique probability measure $r \in P(Y)$ such that for each n the marginal of r on Y_n is r_n.

Proof There is a Borel-measurable stochastic kernel $\bar{q}_1(dx_2|x_1)$ which agrees with $q(dx_2|x_1)$ for p almost every x_1. Define $r_2 \in P(Y_2)$ by specifying it on measurable rectangles to be (Proposition 7.28)

$$r_2(\underline{X}_1\underline{X}_2) = \int_{\underline{X}_1} \bar{q}_1(\underline{X}_2|x_1)p(dx_1) \qquad \forall \underline{X}_1 \in \mathscr{B}_{X_1}, \quad \underline{X}_2 \in \mathscr{B}_{X_2}.$$

Assume $f: Y_2 \to [0, \infty]$ is universally measurable and let $\bar{f}: Y_2 \to [0, \infty]$ be Borel-measurable and agree with f on $Y_2 - N$, where $N \in \mathscr{B}_{Y_2}$ and $r_2(N) = 0$.

By Proposition 7.28,

$$0 = r_2(N) = \int_{X_1}\int_{X_2} \chi_N(x_1, x_2)\bar{q}_1(dx_2|x_1)p(dx_1)$$

$$= \int_{X_1} \bar{q}_1(N_{x_1}|x_1)p(dx_1),$$

so $\bar{q}_1(N_{x_1}|x_1) = 0$ for p almost every x_1. Now $f(x_1, x_2) = \bar{f}(x_1, x_2)$ for $x_2 \notin N_{x_1}$ so

$$\left| \int_{X_2} [f(x_1, x_2) - \bar{f}(x_1, x_2)]\bar{q}_1(dx_2|x_1) \right|$$

$$\leq \int_{N_{x_1}} |f(x_1, x_2) - \bar{f}(x_1, x_2)|\bar{q}_1(dx_2|x_1) = 0$$

for p almost every x_1. It follows that

$$\int_{X_1} \bar{f}(x_1, x_2)\bar{q}_1(dx_2|x_1) = \int_{X_2} f(x_1, x_2)\bar{q}_1(dx_2|x_1)$$

$$= \int_{X_2} f(x_1, x_2)q_1(dx_2|x_1)$$

for p almost every x_1. The left-hand side is Borel-measurable by Proposition 7.29, so the right-hand side is universally measurable by Lemma 7.27. Furthermore,

$$\int_{Y_2} f\, dr_2 = \int_{Y_2} \bar{f}\, dr_2 = \int_{X_1}\int_{X_2} \bar{f}(x_1, x_2)\bar{q}_1(dx_2|x_1)p(dx_1)$$

$$= \int_{X_1}\int_{X_2} f(x_1, x_2)q_1(dx_2|x_1)p(dx_1).$$

This proves (100) for $n = 2$ and $f \geq 0$. If $f: Y_2 \to R^*$ is universally measurable and satisfies $\int f^+ dr_2 < \infty$ or $\int f^- dr_2 < \infty$, then (100) holds for f^+ and f^-, so it holds for f as well. Take $f = \chi_{X_1 X_2}$ to obtain (99).

Now assume the proposition holds for $n = k$. Let $\bar{q}_k(dx_{k+1}|y_k)$ be a stochastic kernel which agrees with $q_k(dx_{k+1}|y_k)$ for r_k almost every x_k. Define r_{k+1} by specifying it on measurable rectangles to be

$$r_{k+1}(\underline{X}_1\underline{X}_2\cdots\underline{X}_{k+1}) = \int_{\underline{X}_1\underline{X}_2\cdots\underline{X}_k} \bar{q}_k(\underline{X}_{k+1}|x_1, x_2, \ldots, x_k)\, dr_k$$

$$\forall \underline{X}_1 \in \mathscr{B}_{X_1}, \ldots, \underline{X}_{k+1} \in \mathscr{B}_{X_{k+1}}.$$

Proceed as in the case of $n = 2$ to prove the proposition for $n = k + 1$. (See also the proof of Proposition 7.28.)

The existence of $r \in P(Y)$ such that the marginal of r on X_n is r_n, $n = 2$, $3, \ldots$, is proved exactly as in Proposition 7.28.   Q.E.D.

In the course of proving Proposition 7.45, we have also established the following fact.

Proposition 7.46   Let X and Y be Borel spaces and let $f:XY \to R^*$ be universally measurable. Let $q(dy|x)$ be a universally measurable stochastic kernel on Y given X. Then the mapping $\lambda:X \to R^*$ defined by

$$\lambda(x) = \int f(x, y)q(dy|x)$$

is universally measurable.

Corollary 7.46.1   Let X be a Borel space and let $f:X \to R^*$ be universally measurable. Then the function $\theta_f:P(X) \to R^*$ defined by

$$\theta_f(p) = \int f \, dp$$

is universally measurable.

Proof   Define a universally measurable stochastic kernel on X given $P(X)$ by $q(dx|p) = p(dx)$. Apply Proposition 7.46.   Q.E.D.

As mentioned previously, the functions obtained by infimizing bivariate, extended real-valued, Borel-measurable functions over one of their variables have analytic lower level sets. We give these functions a name.

Definition 7.21   Let X be a Borel space, $D \subset X$, and $f:D \to R^*$. If D is analytic and the set $\{x \in D | f(x) < c\}$ is analytic for every $c \in R$, then f is said to be *lower semianalytic*.

It is apparent from the definition that a lower semianalytic function is analytically measurable. We state some characterizations and basic properties of lower semianalytic functions as a lemma.

Lemma 7.30   (1)   Let X be a Borel space, D an analytic subset of X, and $f:D \to X$. The following statements are equivalent.

(a)   The function f is lower semianalytic, i.e., the set

$$\{x \in D | f(x) < c\} \tag{101}$$

is analytic for every $c \in R$.

(b)   The set (101) is analytic for every $c \in R^*$.

(c)   The set

$$\{x \in D | f(x) \le c\} \tag{102}$$

is analytic for every $c \in R$.

(d)   The set (102) is analytic for every $c \in R^*$.

(2)   Let X be a Borel space, D an analytic subset of X, and $f_n : D \to R^*$, $n = 1, 2, \ldots$, a sequence of lower semianalytic functions. Then the functions $\inf_n f_n$, $\sup_n f_n$, $\liminf_{n \to \infty} f_n$, and $\limsup_{n \to \infty} f_n$ are lower semianalytic. In particular, if $f_n \to f$, then f is lower semianalytic.

(3)   Let X and Y be Borel spaces, $g : X \to Y$, and $f : g(X) \to R^*$. If g is Borel-measurable and f is lower semianalytic, then $f \circ g$ is lower semianalytic.

(4)   Let X be a Borel space, D an analytic subset of X, and $f, g : D \to R^*$. If f and g are lower semianalytic, then $f + g$ is lower semianalytic. If, in addition, g is Borel-measurable and $g \geq 0$ or if $f \geq 0$ and $g \geq 0$, then fg is lower semianalytic, where we define $0 \cdot \infty = \infty \cdot 0 = 0(-\infty) = (-\infty)0 = 0$.

Proof   (1)   We show (b) $\Rightarrow$ (a) $\Rightarrow$ (d) $\Rightarrow$ (c) $\Rightarrow$ (b). It is clear that (b) $\Rightarrow$ (a). If (a) holds, then

$$\{x \in D \,|\, f(x) \leq \infty\} = D$$

is analytic by definition, while the sets

$$\{x \in D \,|\, f(x) \leq -\infty\} = \bigcap_{n=1}^{\infty} \{x \in D \,|\, f(x) < -n\},$$

$$\{x \in D \,|\, f(x) \leq c\} = \bigcap_{n=1}^{\infty} \{x \in D \,|\, f(x) < c + (1/n)\}, \qquad c \in R,$$

are analytic by Corollary 7.35.2. Therefore (a) $\Rightarrow$ (d). It is clear that (d) $\Rightarrow$ (c). If (c) holds, then the sets

$$\{x \in D \,|\, f(x) < -\infty\} = \varnothing,$$

$$\{x \in D \,|\, f(x) < \infty\} = \bigcup_{n=1}^{\infty} \{x \in D \,|\, f(x) \leq n\},$$

$$\{x \in D \,|\, f(x) < c\} = \bigcup_{n=1}^{\infty} \{x \in D \,|\, f(x) \leq c - (1/n)\}, \qquad c \in R,$$

are analytic by Corollary 7.35.2. Therefore (c) $\Rightarrow$ (b).

(2)   For $c \in R$,

$$\{x \in D \,|\, \inf_n f_n(x) < c\} = \bigcup_{n=1}^{\infty} \{x \in D \,|\, f_n(x) < c\},$$

$$\{x \in D \,|\, \sup_n f_n(x) \leq c\} = \bigcap_{n=1}^{\infty} \{x \in D \,|\, f_n(x) \leq c\},$$

so $\inf_n f_n$ and $\sup_n f_n$ are lower semianalytic by Corollary 7.35.2 and part (1). Then

$$\liminf_{n \to \infty} f_n = \sup_{n \geq 1} \inf_{k \geq n} f_k$$

and

$$\limsup_{n \to \infty} f_n = \inf_{n \geq 1} \sup_{k \geq n} f_k$$

are lower semianalytic as well.

(3)   The domain $g(X)$ of f is analytic by Proposition 7.40. For $c \in R$,

$$\{x \in X | (f \circ g)(x) < c\} = g^{-1}(\{y \in g(X) | f(y) < c\})$$

is analytic by the same proposition.

(4)   For $c \in R$,

$$\{x \in D | f(x) + g(x) < c\} = \bigcup_{r \in Q} [\{x \in D | f(x) < r\} \cap \{x \in D | g(x) < c - r\}],$$

and this is true even if $f(x) + g(x) = \infty - \infty = \infty$ for some $x \in D$. From Corollary 7.35.2 it follows that $f + g$ is lower semianalytic whenever f and g are. Now suppose g is Borel-measurable and $g \geq 0$. For $c > 0$, we have

$$\{x \in D | f(x)g(x) < c\} = \{x \in D | f(x) \leq 0\} \cup \{x \in D | g(x) \leq 0\}$$

$$\cup \left[\bigcup_{r \in Q, r > 0} \{x \in D | f(x) < r, g(x) < c/r\} \right],$$

while if $c \leq 0$, we have

$$\{x \in D | f(x)g(x) < c\} = \bigcup_{r \in Q, r < 0} \{x \in D | f(x) < r, g(x) > c/r\}.$$

In both cases, the set $\{x \in D | f(x)g(x) < c\}$ is analytic by Corollary 7.35.2. Suppose f and g are both lower semianalytic and nonnegative. For $c > 0$, the set $\{x \in D | f(x)g(x) < c\}$ is analytic as before, and for $c \leq 0$, this set is empty. It follows that fg is lower semianalytic under either set of assumptions on f and g.   Q.E.D.

Note in connection with Lemma 7.30(3) that the composition of a Borel-measurable function with a lower semianalytic function can be guaranteed to be lower semianalytic only when the composition is in the order specified. To see this, let X be a Borel space and $A \subset X$ be an analytic set whose complement is not analytic (see Appendix B). Define $f(x) = -\chi_A(x)$, which is lower semianalytic, because $\{x \in X | f(x) < c\}$ is either $\varnothing$, A, or X, depending on the value of c. Let $g: R^* \to R^*$ be given by $g(c) = -c$. Then $\chi_A = g \circ f$, and this function is not lower semianalytic, since $\{x \in X | \chi_A(x) < \frac{1}{2}\} = A^c$. This also provides us with an example of an analytically measurable function which is not lower semianalytic.

Proposition 7.47   Let X and Y be Borel spaces, let D be an analytic subset of XY, and let $f: D \to R^*$ be lower semianalytic. Then the function

$f^*:\mathrm{proj}_X(D) \to R^*$ defined by

$$f^*(x) = \inf_{y \in D_x} f(x, y) \qquad (103)$$

is lower semianalytic. Conversely, if $f^*: X \to R^*$ is a given lower semianalytic function and Y is an uncountable Borel space, then there exists a Borel-measurable function $f: XY \to R^*$ which satisfies (103) with $D = XY$.

Proof For the first part of the theorem, observe that if $f: D \to R^*$ is lower semianalytic and $c \in R$, the set

$$\left\{ x \in \mathrm{proj}_X(D) \Big| \inf_{y \in D_x} f(x, y) < c \right\} = \mathrm{proj}_X(\{(x, y) \in D | f(x, y) < c\})$$

is analytic by Proposition 7.39.

For the converse part of the theorem, let $f^*: X \to R^*$ be lower semi-analytic and let Y be an uncountable Borel space. For $r \in Q$, let $A(r) = \{x \in X | f^*(x) < r\}$. Then $A(r)$ is analytic and, by Proposition 7.39, there exists $B(r) \in \mathscr{B}_{XY}$ such that $A(r) = \mathrm{proj}_X[B(r)]$. Define $G(r) = \bigcup_{s \in Q, s \leq r} B(s)$ and $f: XY \to R^*$ by

$$f(x, y) = \inf\{r \in Q | (x, y) \in G(r)\} = \inf_{r \in Q} \psi_r(x, y),$$

where $\psi_r(x, y) = r$ if $(x, y) \in G(r)$ and $\psi_r(x, y) = \infty$ otherwise. Then f is Borel-measurable. Let g be defined by $g(x) = \inf_{y \in Y} f(x, y)$. We show that $f^*(x) = g(x)$ for every $x \in X$.

If $f^*(x) < c$ for some $c \in R$, then there exists $r \in Q$ for which $f^*(x) < r < c$, and so $x \in A(r)$. There exists $y \in Y$ such that $(x, y) \in G(r)$, and, consequently, $f(x, y) \leq r$ and $g(x) \leq r < c$. Therefore $g(x)$ cannot be greater than $f^*(x)$.

If $g(x) < c$ for some $c \in R$, then there exists $r \in Q$ and $y \in Y$ for which $g(x) < r < c$ and $(x, y) \in G(r)$. Thus for some $s \in Q$, $s \leq r$, we have $(x, y) \in B(s)$ and $x \in A(s)$. This implies $f^*(x) < s \leq r < c$, which shows that $f^*(x)$ cannot be greater than $g(x)$.   Q.E.D.

Proposition 7.48 Let X and Y be Borel spaces, $f: XY \to R^*$ lower semi-analytic, and $q(dy|x)$ a Borel-measurable stochastic kernel on Y given X. Then the function $\lambda: X \to R^*$ defined by

$$\lambda(x) = \int f(x, y) q(dy|x)$$

is lower semianalytic.

Proof Suppose $f \geq 0$. Let $f_n(x, y) = \min\{n, f(x, y)\}$. Then each f_n is lower semianalytic and $f_n \uparrow f$. The set

$$E_n = \{(x, y, b) \in XYR | f_n(x, y) \leq b \leq n\}$$

$$= \bigcap_{k=1}^{\infty} \bigcup_{r \in Q} \{(x, y, b) \in XYR | f_n(x, y) < r, r \leq b + (1/k) \leq n + (1/k)\}$$

is analytic in XYR by Corollary 7.35.2 and Proposition 7.38. Let μ be Lebesgue measure on R, $p \in P(XY)$, and $p\mu$ the product measure on XYR. By Fubini's theorem,

$$(p\mu)(E_n) = \int_{XY} \int_R \chi_{E_n} \, d\mu \, dp = \int_{XY} [n - f_n(x, y)] \, dp$$

$$= n - \int_{XY} f_n(x, y) \, dp.$$

For $c \in R$ we have, by the monotone convergence theorem,

$$\left\{ p \in P(XY) \Big| \int f(x, y) \, dp \leq c \right\} = \bigcap_{n=1}^{\infty} \left\{ p \in P(XY) \Big| \int_{XY} f_n(x, y) \, dp \leq c \right\}$$

$$= \bigcap_{n=1}^{\infty} \{ p \in P(XY) | (p\mu)(E_n) \geq n - c \}.$$

Hence, by Proposition 7.43 and the fact that the mapping $p \to p\mu$ is continuous (Lemma 7.12), the function $\theta_f : P(XY) \to R^*$ defined by $\theta_f(p) = \int f(x, y) \, dp$ is lower semianalytic. We have

$$\lambda(x) = \theta_f[q(dy|x)p_x].$$

Since the mapping $x \to q(dy|x)$ is Borel-measurable from X to $P(Y)$ and the mappings $x \to p_x$ and $[q(dy|x), p_x] \to q(dy|x)p_x$ are continuous from X to $P(X)$ and $P(X)P(Y)$ to $P(XY)$, respectively (Corollary 7.21.1 and Lemma 7.12), it follows from Lemma 7.30(3) that λ is lower semianalytic.

Suppose $f \leq 0$. Let $f_n(x, y) = \max\{-n, f(x, y)\}$. Then each f_n is lower semianalytic and $f_n \downarrow f$. The sets $E_n = \{(x, y, b) \in XYR | f_n(x, y) \leq b \leq 0\}$ are analytic and

$$(p\mu)(E_n) = \int_{XY} \int_R \chi_{E_n} \, d\mu \, dp = - \int_{XY} f_n(x, y) \, dp.$$

For $c \in R$,

$$\left\{ p \in P(XY) \Big| \int f(x, y) \, dp < c \right\} = \bigcup_{n=1}^{\infty} \left\{ p \in P(XY) \Big| \int_{XY} f_n(x, y) \, dp < c \right\}$$

$$= \bigcup_{n=1}^{\infty} \{ p \in P(XY) | (p\mu)(E_n) > -c \}.$$

Proceed as before.

In the general case,

$$\int f(x, y) q(dy|x) = \int f^+(x, y) q(dy|x) - \int f^-(x, y) q(dy|x).$$

The functions f^+ and $-f^-$ are lower semianalytic, so by the preceding arguments each of the summands on the right is lower semianalytic. The result follows from Lemma 7.30(4).   Q.E.D.

Corollary 7.48.1   Let X be a Borel space and let $f : X \to R^*$ be lower semianalytic. Then the function $\theta_f : P(X) \to R^*$ defined by

$$\theta_f(p) = \int f \, dp$$

is lower semianalytic.

Proof   Define a Borel-measurable stochastic kernel on X given $P(X)$ by $q(dx|p) = p(dx)$. Apply Proposition 7.48.   Q.E.D.

As an aid in proving the selection theorem for lower semianalytic functions, we give a result concerning selection in an analytic subset of a product of Borel spaces. The reader will notice a strong resemblance between this result and Lemma 7.21, which was instrumental in proving the selection theorem for upper semicontinuous functions.

Proposition 7.49   (Jankov–von Neumann theorem)   Let X and Y be Borel spaces and A an analytic subset of XY. There exists an analytically measurable function $\varphi : \mathrm{proj}_X(A) \to Y$ such that $\mathrm{Gr}(\varphi) \subset A$.

Proof   (See Fig. 7.2.) Let $f : \mathscr{N} \to XY$ be continuous such that $A = f(\mathscr{N})$. Let $g = \mathrm{proj}_X \circ f$. Then $g : \mathscr{N} \to X$ is continuous from $\mathscr{N}$ onto $\mathrm{proj}_X(A)$. For $x \in \mathrm{proj}_X(A)$, $g^{-1}(\{x\})$ is a closed nonempty subset of $\mathscr{N}$. Let $\zeta_1(x)$ be the smallest integer which is the first component of an element $z_1 \in g^{-1}(\{x\})$. Let $\zeta_2(x)$ be the smallest integer which is the second component of an element $z_2 \in g^{-1}(\{x\})$ whose first component is $\zeta_1(x)$. In general, let $\zeta_k(x)$ be the smallest integer which is the kth component of an element $z_k \in g^{-1}(\{x\})$ whose first $(k-1)$st components are $\zeta_1(x), \ldots, \zeta_{k-1}(x)$. Let $\psi(x) = (\zeta_1(x), \zeta_2(x), \ldots)$. Since $z_k \to \psi(x)$, we have

$$\psi(x) \in g^{-1}(\{x\}). \tag{104}$$

Define $\varphi : \mathrm{proj}_X(A) \to Y$ by $\varphi = \mathrm{proj}_Y \circ f \circ \psi$, so that $\mathrm{Gr}(\varphi) \subset A$.

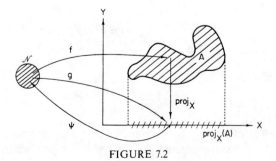

FIGURE 7.2

We show that φ is analytically measurable. As in the proof of Proposition 7.42, for $(\sigma_1, \ldots, \sigma_k) \in \Sigma$ let

$$N(\sigma_1, \ldots, \sigma_k) = \{(\zeta_1, \zeta_2, \ldots) \in \mathcal{N} \mid \zeta_1 = \sigma_1, \ldots, \zeta_k = \sigma_k\},$$
$$M(\sigma_1, \ldots, \sigma_k) = \{(\zeta_1, \zeta_2, \ldots) \in \mathcal{N} \mid \zeta_1 \leq \sigma_1, \ldots, \zeta_k \leq \sigma_k\}.$$

We first show that ψ is analytically measurable, i.e., $\psi^{-1}(\mathcal{B}_{\mathcal{N}}) \subset \mathcal{A}_X$. Since $\{N(s) \mid s \in \Sigma\}$ is a base for the topology on $\mathcal{N}$, by the remark following Definition 7.6, we have $\sigma(\{N(s) \mid s \in \Sigma\}) = \mathcal{B}_{\mathcal{N}}$. Then

$$\psi^{-1}(\mathcal{B}_{\mathcal{N}}) = \psi^{-1}[\sigma(\{N(s) \mid s \in \Sigma\})] = \sigma[\psi^{-1}(\{N(s) \mid s \in \Sigma\})],$$

and it suffices to prove

$$\psi^{-1}[N(s)] \in \mathcal{A}_X \qquad \forall s \in \Sigma. \tag{105}$$

We claim that for $s = (\sigma_1, \sigma_2, \ldots, \sigma_k) \in \Sigma$

$$\psi^{-1}[N(s)] = g[M(s)] - \bigcup_{j=1}^{k} g[M(\sigma_1, \ldots, \sigma_{j-1}, \sigma_j - 1)], \tag{106}$$

where $M(\sigma_1, \ldots, \sigma_{j-1}, \sigma_j - 1) = \varnothing$ if $\sigma_j - 1 = 0$. We show this by proving that $\psi^{-1}[N(s)]$ is a subset of the set on the right-hand side of (106) and vice versa.

Suppose $x \in \psi^{-1}[N(s)]$. Let $\psi(x) = (\zeta_1(x), \zeta_2(x), \ldots)$. Then

$$\psi(x) \in N(s) \subset M(s), \tag{107}$$

so (104) implies

$$x = g[\psi(x)] \in g[M(s)]. \tag{108}$$

Relation (107) also implies $\zeta_1(x) = \sigma_1, \ldots, \zeta_k(x) = \sigma_k$. By the construction of ψ, we have that σ_1 is the smallest integer which is the first component of an element of $g^{-1}(\{x\})$, and for $j = 2, \ldots, k$, σ_j is the smallest integer which is the jth component of an element of $g^{-1}(\{x\})$ whose first $(j-1)$ components are $\sigma_1, \ldots, \sigma_{j-1}$. In other words,

$$g^{-1}(\{x\}) \cap M(\sigma_1, \ldots, \sigma_{j-1}, \sigma_j - 1) = \varnothing, \qquad j = 1, \ldots, k.$$

It follows that

$$x \notin \bigcup_{j=1}^{k} g[M(\sigma_1, \ldots, \sigma_{j-1}, \sigma_j - 1)]. \tag{109}$$

Relations (108) and (109) imply

$$\psi^{-1}[N(s)] \subset g[M(s)] - \bigcup_{j=1}^{k} g[M(\sigma_1, \ldots, \sigma_{j-1}, \sigma_j - 1)]. \tag{110}$$

To prove the reverse set containment, suppose

$$x \in g[M(s)] - \bigcup_{j=1}^{k} g[M(\sigma_1, \ldots, \sigma_{j-1}, \sigma_j - 1)]. \tag{111}$$

Since $x \in g[M(s)]$, there must exist $y = (\eta_1, \eta_2, \ldots) \in g^{-1}(\{x\})$ such that

$$\eta_1 \le \sigma_1, \ldots, \eta_k \le \sigma_k. \tag{112}$$

Clearly, $x \in \text{proj}_X(A) = g(\mathcal{N})$, so $\psi(x)$ is defined. Let $\psi(x) = (\zeta_1(x), \zeta_2(x), \ldots)$. By (104), we have $g[\psi(x)] = x$, so (111) implies

$$\psi(x) \notin M(\sigma_1, \ldots, \sigma_{j-1}, \sigma_j - 1), \qquad j = 1, 2, \ldots, k.$$

Since $\psi(x) \notin M(\sigma_1 - 1)$, we know that $\zeta_1(x) \ge \sigma_1$. But $\zeta_1(x)$ is the smallest integer which is the first component of an element of $g^{-1}(\{x\})$, so (112) implies $\zeta_1(x) \le \eta_1 \le \sigma_1$. Therefore $\zeta_1(x) = \sigma_1$. Similarly, since $\psi(x) \notin M(\zeta_1(x), \sigma_2 - 1)$, we have $\zeta_2(x) \ge \sigma_2$, Again from (112) we see that $\zeta_2(x) \le \eta_2 \le \sigma_2$, so $\zeta_2(x) = \sigma_2$. Continuing in this manner, we show that $\psi(x) \in N(s)$, i.e., $x \in \psi^{-1}[N(s)]$ and

$$\psi^{-1}[N(s)] \supset g[M(s)] - \bigcup_{j=1}^{k} g[M(\sigma_1, \ldots, \sigma_{j-1}, \sigma_j - 1)]. \tag{113}$$

Relations (110) and (113) imply (106).

We note now that $M(t)$ is open in $\mathcal{N}$ for every $t \in \Sigma$, so $g[M(t)]$ is analytic by Proposition 7.40. Relation (105) now follows from (106), so ψ is analytically measurable.

By the definition of φ and the Borel-measurability of f and proj_Y, we have

$$\varphi^{-1}(\mathcal{B}_Y) = \psi^{-1}(f^{-1}[\text{proj}_Y^{-1}(\mathcal{B}_Y)]) \subset \psi^{-1}(f^{-1}[\mathcal{B}_{XY}]) \subset \psi^{-1}(\mathcal{B}_{\mathcal{N}}).$$

We have just proved $\psi^{-1}(\mathcal{B}_{\mathcal{N}}) \subset \mathcal{A}_X$, and the analytic measurability of φ follows.   Q.E.D.

This brings us to the selection theorem for lower semianalytic functions.

Proposition 7.50   Let X and Y be Borel spaces, $D \subset XY$ an analytic set, and $f : D \to R^*$ a lower semianalytic function. Define $f^* : \text{proj}_X(D) \to R^*$ by

$$f^*(x) = \inf_{y \in D_x} f(x, y). \tag{114}$$

(a)   For every $\varepsilon > 0$, there exists an analytically measurable function $\varphi : \text{proj}_X(D) \to Y$ such that $\text{Gr}(\varphi) \subset D$ and for all $x \in \text{proj}_X(D)$,

$$f[x, \varphi(x)] \le \begin{cases} f^*(x) + \varepsilon & \text{if } f^*(x) > -\infty, \\ -1/\varepsilon & \text{if } f^*(x) = -\infty. \end{cases}$$

(b)   The set

$$I = \{x \in \text{proj}_X(D) | \text{for some } y_x \in D_x, f(x, y_x) = f^*(x)\}$$

is universally measurable, and for every $\varepsilon > 0$ there exists a universally measurable function $\varphi : \text{proj}_X(D) \to Y$ such that $\text{Gr}(\varphi) \subset D$ and for all $x \in \text{proj}_X(D)$

$$f[x, \varphi(x)] = f^*(x) \qquad \text{if} \quad x \in I, \tag{115}$$

$$f[x, \varphi(x)] \leq \begin{cases} f^*(x) + \varepsilon & \text{if} \quad x \notin I, \quad f^*(x) > -\infty, \\ -1/\varepsilon & \text{if} \quad x \notin I, \quad f^*(x) = -\infty. \end{cases} \tag{116}$$

Proof   (a)   (Cf. proof of Proposition 7.34 and Fig. 7.1.) The function f^* is lower semianalytic by Proposition 7.47. For $k = 0, \pm 1, \pm 2, \ldots,$ define

$$A(k) = \{(x, y) \in D | f(x, y) < k\varepsilon\},$$
$$B(k) = \{x \in \text{proj}_X(D) | (k - 1)\varepsilon \leq f^*(x) < k\varepsilon\},$$
$$B(-\infty) = \{x \in \text{proj}_X(D) | f^*(x) = -\infty\}.$$
$$B(\infty) = \{x \in \text{proj}_X(D) | f^*(x) = \infty\}.$$

The sets $A(k)$, $k = 0, \pm 1, \pm 2, \ldots,$ and $B(-\infty)$ are analytic, while the sets $B(k)$, $k = 0, \pm 1, \pm 2, \ldots,$ and $B(\infty)$ are analytically measurable. By the Jankov–von Neumann theorem (Proposition 7.49) there exists, for each $k = 0, \pm 1, \pm 2, \ldots,$ an analytically measurable $\varphi_k : \text{proj}_X[A(k)] \to Y$ with $(x, \varphi_k(x)) \in A(k)$ for all $x \in \text{proj}_X[A(k)]$ and an analytically measurable $\overline{\varphi} : \text{proj}_X(D) \to Y$ such that $(x, \overline{\varphi}(x)) \in D$ for all $x \in \text{proj}_X(D)$. Let k^* be an integer such that $k^* \leq -1/\varepsilon^2$. Define $\varphi : \text{proj}_X(D) \to Y$ by

$$\varphi(x) = \begin{cases} \varphi_k(x) & \text{if} \quad x \in B(k), \quad k = 0, \pm 1, \pm 2, \ldots, \\ \overline{\varphi}(x) & \text{if} \quad x \in B(\infty), \\ \varphi_{k^*}(x) & \text{if} \quad x \in B(-\infty). \end{cases}$$

Since $B(k) \subset \text{proj}_X[A(k)]$ and $B(-\infty) \subset \text{proj}_X[A(k)]$ for all k, this definition is possible. It is clear that φ is analytically measurable and $\text{Gr}(\varphi) \subset D$. If $x \in B(k)$, then $(x, \varphi_k(x)) \in A(k)$ and we have

$$f[x, \varphi(x)] = f[x, \varphi_k(x)] < k\varepsilon \leq f^*(x) + \varepsilon.$$

If $x \in B(\infty)$, then $f(x, y) = \infty$ for all $y \in D_x$ and $f[x, \varphi(x)] = \infty = f^*(x)$. If $x \in B(-\infty)$, we have

$$f[x, \varphi(x)] = f[x, \varphi_{k^*}(x)] < k^*\varepsilon \leq -1/\varepsilon.$$

Hence φ has the required properties.

(b)   Consider the set $E \subset XYR^*$ defined by

$$E = \{(x, y, b) | (x, y) \in D, f(x, y) \le b\}.$$

Since

$$E = \bigcap_{k=1}^{\infty} \bigcup_{r \in Q^*} \{(x, y, b) | (x, y) \in D, \, f(x, y) \le r, \, r \le b + (1/k)\},$$

it follows from Corollary 7.35.2 and Proposition 7.38 that E is analytic in XYR^*, and hence the set

$$A = \text{proj}_{XR^*}(E)$$

is analytic in XR^*. The mapping $T: \text{proj}_X(D) \to XR^*$ defined by

$$T(x) = (x, f^*(x))$$

is analytically measurable, and

$$I = \{x | (x, f^*(x)) \in A\} = T^{-1}(A).$$

Hence I is universally measurable by Corollary 7.44.2.

Since E is analytic, there is, by the Jankov–von Neumann Theorem, an analytically measurable $\rho: A \to Y$ such that $(x, \rho(x, b), b) \in E$ for every $(x, b) \in A$. Define $\psi: I \to Y$ by

$$\psi(x) = \rho(x, f^*(x)) = (\rho \circ T)(x) \qquad \forall x \in I.$$

Then ψ is universally measurable by Corollary 7.44.2, and by construction $f[x, \psi(x)] \le f^*(x)$ for $x \in I$. Hence

$$f[x, \psi(x)] = f^*(x) \qquad \forall x \in I. \tag{117}$$

By part (a) there exists an analytically measurable $\psi_\varepsilon: \text{proj}_X(D) \to Y$ such that

$$f[x, \psi_\varepsilon(x)] \le \begin{cases} f^*(x) + \varepsilon & \text{if} \quad f^*(x) > -\infty, \\ -1/\varepsilon & \text{if} \quad f^*(x) = -\infty. \end{cases} \tag{118}$$

Define $\varphi: \text{proj}_X(D) \to Y$ by

$$\varphi(x) = \begin{cases} \psi(x) & \text{if} \quad x \in I, \\ \psi_\varepsilon(x) & \text{if} \quad x \in \text{proj}_X(D) - I. \end{cases}$$

Then φ is universally measurable and, by (117) and (118), it has the required properties.   Q.E.D.

Since the composition of analytically measurable functions can fail to be analytically measurable (Appendix B), the selector obtained in the proof of Proposition 7.50(b) can fail to be analytically measurable. The composition of universally measurable functions is universally measurable, and so we obtained a selector which is universally measurable. However, there is a σ-algebra, which we call the *limit σ-algebra*, lying between $\mathscr{A}_X$ and $\mathscr{U}_X$ such that the composition of limit measurable functions is again limit-measurable. We discuss this σ-algebra in Appendix B and state a strengthened version of Proposition 7.50 in Section 11.1.

Chapter 8

The Finite Horizon Borel Model

In Chapters 8–10 we will treat a model very similar to that of Section 2.3.2. An applications-oriented treatment of that model can be found in "Dynamic Programming and Stochastic Control" by Bertsekas [B4], hereafter referred to as DPSC. The model of Section 2.3.2 and DPSC has a countable disturbance space and arbitrary state and control spaces, whereas the model treated here will have Borel state, control, and disturbance spaces.

8.1 The Model

Definition 8.1 A *finite horizon stochastic optimal control model* is a nine-tuple $(S, C, U, W, p, f, \alpha, g, N)$ as described here. The letters x and u are used to denote elements of S and C, respectively.

    S *State space.* A nonempty Borel space.
    C *Control space.* A nonempty Borel space.
    U *Control constraint.* A function from S to the set of nonempty subsets of C. The set

$$\Gamma = \{(x, u) | x \in S, u \in U(x)\} \tag{1}$$

is assumed to be analytic in SC.

188

W   Disturbance space. A nonempty Borel space.

$p(dw|x, u)$ *Disturbance kernel.* A Borel-measurable stochastic kernel on W given SC.

f   System function. A Borel-measurable function from SCW to S.

α *Discount factor.* A positive real number.

g   One-stage cost function. A lower semianalytic function from Γ to R^*.

N   Horizon. A positive integer.

We envision a system moving from state x_k to state x_{k+1} via the system equation

$$x_{k+1} = f(x_k, u_k, w_k), \qquad k = 0, 1, \ldots, N - 2,$$

and incurring cost at each stage of $g(x_k, u_k)$. The disturbances w_k are random objects with probability distributions $p(dw_k|x_k, u_k)$. The goal is to choose u_k dependent on the history $(x_0, u_0, \ldots, x_{k-1}, u_{k-1}, x_k)$ so as to minimize

$$E\left\{ \sum_{k=0}^{N-1} \alpha^k g(x_k, u_k) \right\}. \tag{2}$$

The meaning of this statement will be made precise shortly. We have the constraint that when x_k is the kth state, the kth control u_k must be chosen to lie in $U(x_k)$.

In the models in Section 2.3.2 and DPSC, the one-stage cost g is also a function of the disturbance, i.e., has the form $g(x, u, w)$. If this is the case, then $g(x, u, w)$ can be replaced by

$$\bar{g}(x, u) = \int g(x, u, w) p(dw|x, u).$$

If $g(x, u, w)$ is lower semianalytic, so is $\bar{g}(x, u)$ (Proposition 7.48). If $p(dw|x, u)$ is continuous and $g(x, u, w)$ is lower semicontinuous and bounded below or upper semicontinuous and bounded above, then $\bar{g}(x, u)$ is lower semicontinuous and bounded below or upper semicontinuous and bounded above, respectively (Proposition 7.31). Since these are the three cases we deal with, there is no loss of generality in considering a one-stage cost function which is independent of the disturbance.

The model posed in Definition 8.1 is stationary, i.e., the data does not vary from stage to stage. A reduction of the nonstationary model to this form is discussed in Section 10.1.

A notational device which simplifies the presentation is the *state transition stochastic kernel* on S given SC defined by

$$t(B|x, u) = p(\{w | f(x, u, w) \in B\}|x, u) = p(f^{-1}(B)_{(x, u)}|x, u). \tag{3}$$

Thus $t(B|x, u)$ is the probability that the $(k + 1)$st state is in B given that the kth state is x and the kth control is u. Proposition 7.26 and Corollary 7.26.1 imply that $t(dx'|x, u)$ is Borel-measurable.

Definition 8.2 A *policy* for the model of Definition 8.1 is a sequence $\pi = (\mu_0, \mu_1, \ldots, \mu_{N-1})$ such that, for each k, $\mu_k(du_k|x_0, u_0, \ldots, u_{k-1}, x_k)$ is a universally measurable stochastic kernel on C given $SC \cdots CS$ satisfying

$$\mu_k(U(x_k)|x_0, u_0, \ldots, u_{k-1}, x_k) = 1$$

for every $(x_0, u_0, \ldots, u_{k-1}, x_k)$. If, for each k, μ_k is parameterized only by (x_0, x_k), π is a *semi-Markov policy*. If μ_k is parameterized only by x_k, π is a *Markov policy*. If, for each k and $(x_0, u_0, \ldots, u_{k-1}, x_k)$, $\mu_k(du_k|x_0, u_0, \ldots, u_{k-1}, x_k)$ assigns mass one to some point in C, π is *nonrandomized*. In this case, by a slight abuse of notation, π can be considered to be a sequence of universally measurable (Corollary 7.44.3) mappings $\mu_k : SC \cdots CS \to C$ such that

$$\mu_k(x_0, u_0, \ldots, u_{k-1}, x_k) \in U(x_k)$$

for every $(x_0, u_0, \ldots, u_{k-1}, x_k)$. If $\mathscr{F}$ is a type of σ-algebra on Borel spaces and all the stochastic kernel components of a policy are $\mathscr{F}$-measurable, we say the *policy is $\mathscr{F}$-measurable*. (For example, $\mathscr{F}$ could represent the Borel σ-algebras or the analytic σ-algebras.)

We denote by Π' the set of all policies for the model of Definition 8.1 and by Π the set of all Markov policies. We will show that in many cases it is not necessary to go outside Π to find the "best" available policy. In most cases, this "best" policy can be taken to be nonrandomized. Since Γ is analytic, the Jankov–von Neumann theorem (Proposition 7.49) guarantees that there exists at least one nonrandomized Markov policy, so Π and Π' are nonempty.

If $\pi = (\mu_0, \mu_1, \ldots, \mu_{N-1})$ is a nonrandomized Markov policy, then π is a finite horizon version of a policy in the sense of Section 2.1. The notion of policy as set forth in Definition 8.2 is wider than the concept of Section 2.1 in that randomized non-Markov policies are permitted. It is narrower in that universal measurability is required.

We are now in a position to make precise expression (2). In this and subsequent discussions, we often index the state and control spaces for clarity. However, except in Chapter 10 when the nonstationary model is treated, we will always understand S_k to be a copy of S and C_k to be a copy of C. Suppose $p \in P(S)$ and $\pi = (\mu_0, \mu_1, \ldots, \mu_{N-1})$ is a policy for the model of Definition 8.1. By Proposition 7.45, there is a unique probability measure $r_N(\pi, p)$ on $S_0 C_0 \cdots S_{N-1} C_{N-1}$ such that for any universally measurable function $h : S_0 C_0 \cdots S_{N-1} C_{N-1} \to R^*$ which satisfies either $\int h^+ \, dr_N(\pi, p) < \infty$

or $\int h^- \, dr_N(\pi, p) < \infty$, we have

$$
\int h \, dr_N(\pi, p) = \int_{S_0} \int_{C_0} \int_{S_1} \cdots \int_{S_{N-1}} \int_{C_{N-1}} h(x_0, u_0, \ldots, x_{N-1}, u_{N-1})
$$
$$
\times \mu_{N-1}(du_{N-1}|x_0, u_0, \ldots, u_{N-2}, x_{N-1})
$$
$$
\times t(dx_{N-1}|x_{N-2}, u_{N-2}) \cdots t(dx_1|x_0, u_0)\mu_0(du_0|x_0)p(dx_0), \quad (4)
$$

where $t(dx'|x, u)$ is the Borel-measurable stochastic kernel defined by (3). Furthermore we have from (4) that $\int h dr_N(\pi, p_x)$ is a universally measurable function of x (Proposition 7.46), and if h and π are Borel-measurable, then $\int h dr_N(\pi, p_x)$ is a Borel-measurable function of x (Proposition 7.29).

Definition 8.3   Suppose $\pi = (\mu_0, \mu_1, \ldots, \mu_{N-1})$ is a policy for the model of Definition 8.1. For $K \leq N$, the *K-stage cost corresponding to* π at $x \in S$ is

$$
J_{K,\pi}(x) = \int \left[\sum_{k=0}^{K-1} \alpha^k g(x_k, u_k) \right] dr_N(\pi, p_x), \tag{5}
$$

where, for each $\pi \in \Pi'$ and $p \in P(S)$, $r_N(\pi, p)$ is the unique probability measure satisfying (4). The *K-stage optimal cost* at x is

$$
J_K^*(x) = \inf_{\pi \in \Pi'} J_{K,\pi}(x). \tag{6}
$$

If $\varepsilon > 0$, the policy π is *K-stage ε-optimal at x* provided

$$
J_{K,\pi}(x) \leq \begin{cases} J_K^*(x) + \varepsilon & \text{if} \quad J_K^*(x) > -\infty, \\ -1/\varepsilon & \text{if} \quad J_K^*(x) = -\infty. \end{cases}
$$

If $J_{K,\pi}(x) = J_K^*(x)$, then π is *K-stage optimal at x*. If π is K-stage ε-optimal or K-stage optimal at every $x \in S$, it is said to be *K-stage ε-optimal* or *K-stage optimal*, respectively. If $\{\varepsilon_n\}$ is a sequence of positive numbers with $\varepsilon_n \downarrow 0$, a sequence of policies $\{\pi_n\}$ exhibits $\{\varepsilon_n\}$-*dominated convergence to K-stage optimality* provided

$$
\lim_{n \to \infty} J_{K,\pi_n} = J_K^*,
$$

and for $n = 2, 3, \ldots$

$$
J_{K,\pi_n}(x) \leq \begin{cases} J_K^*(x) + \varepsilon_n & \text{if} \quad J_K^*(x) > -\infty, \\ J_{K,\pi_{n-1}}(x) + \varepsilon_n & \text{if} \quad J_K^*(x) = -\infty. \end{cases}
$$

If $K = N$, we suppress the qualifier "K-stage" in the preceding terms.

Note that J_K^* is independent of the horizon N as long as $K \leq N$. Note also that $J_{K,\pi}(x)$ is universally measurable in x. If π is a Borel-measurable policy and g is Borel-measurable, then $J_{K,\pi}(x)$ is Borel-measurable in x.

For $\pi = (\mu_0, \mu_1, \ldots, \mu_{N-1}) \in \Pi'$ and $p \in P(S)$, let $q_k(\pi, p)$ be the marginal of $r_N(\pi, p)$ on $S_k C_k$. If we take $h = \chi_{S_0 \cdots C_{k-1} \underline{S}_k \underline{C}_k S_{k+1} \cdots C_{N-1}}$ in (4), we obtain

$$
\begin{aligned}
q_k(\pi, p)(\underline{S}_k \underline{C}_k) &= \int_{S_0} \int_{C_0} \int_{S_1} \cdots \int_{C_{k-1}} \int_{\underline{S}_k} \mu_k(\underline{C}_k | x_0, u_0, \ldots, u_{k-1}, x_k) \\
&\quad \times t(dx_k | x_{k-1}, u_{k-1}) \mu_{k-1}(du_{k-1} | x_0, u_0, \ldots, u_{k-2}, x_{k-1}) \cdots \\
&\quad \times t(dx_1 | x_0, u_0) \mu_0(du_0 | x_0) p(dx_0) \\
&= \int_{S_0 C_0 \ldots S_{k-1} C_{k-1}} \int_{\underline{S}_k} \mu_k(\underline{C}_k | x_0, u_0, \ldots, u_{k-1}, x_k) \\
&\quad \times t(dx_k | x_{k-1}, u_{k-1}) \, dr_{k-1}(\pi, p) \qquad \forall \underline{S}_k \in \mathscr{B}_S, \underline{C}_k \in \mathscr{B}_C. \quad (7)
\end{aligned}
$$

From (1) and (7), we see that $q_k(\pi, p)(\Gamma) = 1$. If π is Markov, (7) becomes

$$
q_k(\pi, p)(\underline{S}_k \underline{C}_k) = \int_{S_{k-1} C_{k-1}} \int_{\underline{S}_k} \mu_k(\underline{C}_k | x_k) t(dx_k | x_{k-1}, u_{k-1}) \, dq_{k-1}(\pi, p)
$$
$$
\forall \underline{S}_k \in \mathscr{B}_S, \quad \underline{C}_k \in \mathscr{B}_C. \quad (8)
$$

If either

$$
\int_{S_k C_k} g^- \, dq_k(\pi, p_x) < \infty \qquad \forall \pi \in \Pi', \quad x \in S, \quad k = 0, \ldots, N-1, \quad (\mathrm{F}^+)
$$

or

$$
\int_{S_k C_k} g^+ \, dq_k(\pi, p_x) < \infty \qquad \forall \pi \in \Pi', \quad x \in S, \quad k = 0, \ldots, N-1, \quad (\mathrm{F}^-)
$$

then Lemma 7.11(b) implies that for every $\pi \in \Pi'$ and $x \in S$

$$
J_{K, \pi}(x) = \sum_{k=0}^{K-1} \alpha^k \int_{S_k C_k} g \, dq_k(\pi, p_x), \qquad K = 1, \ldots, N. \quad (9)
$$

If (F^+) [respectively (F^-)] appears preceding the statement of a proposition, then (F^+) [respectively (F^-)] is understood to be a part of the hypotheses of the proposition. If both (F^+) and (F^-) appear, then the proposition is valid when either (F^+) or (F^-) is included among the hypotheses.

If $\pi' \in \Pi'$ is a given policy, there may not exist a Markov policy which does at least as well as π' for every $x \in S$, i.e., a policy $\pi \in \Pi$ for which

$$
J_{N, \pi}(x) \le J_{N, \pi'}(x) \quad (10)
$$

for every $x \in S$. However, if x is held fixed, then a Markov policy π can be found for which (10) holds.

Proposition 8.1 $(\mathrm{F}^+)(\mathrm{F}^-)$ If $x \in S$ and $\pi' \in \Pi'$, then there is a Markov policy π such that

$$
J_{K, \pi}(x) = J_{K, \pi'}(x), \qquad K = 1, \ldots, N. \quad (11)
$$

Proof Let $\pi' = (\mu_0', \mu_1', \ldots, \mu_{N-1}')$ be a policy and let $x \in S$ be given. For $k = 0, 1, \ldots, N - 1$, let $\mu_k(du_k | x_k)$ be the Borel-measurable stochastic kernel obtained by decomposing $q_k(\pi', p_x)$ (Corollary 7.27.2), i.e.,

$$q_k(\pi', p_x)(\underline{S_k}\underline{C_k}) = \int_{\underline{S_k}} \mu_k(\underline{C_k} | x_k) p_k(\pi', p_x)(dx_k) \qquad \forall \underline{S_k} \in \mathcal{B}_S, \quad \underline{C_k} \in \mathcal{B}_C, \quad (12)$$

where $p_k(\pi', p_x)$ is the marginal of $q_k(\pi', p_k)$ on S_k. From (12) we see that

$$1 = q_k(\pi', p_x)(\Gamma) = \int_{S_k} \mu_k(U(x_k) | x_k) p_k(\pi', p_x)(dx_k),$$

so we must have $\mu_k(U(x_k) | x_k) = 1$ for $p_k(\pi', p_x)$ almost every x_k. By altering $\mu_k(\cdot | x_k)$ on a set of $p_k(\pi', p_x)$ measure zero if necessary, we may assume that (12) holds and $\pi = (\mu_0, \mu_1, \ldots, \mu_{N-1})$ is a policy as set forth in Definition 8.2. In light of (9), (11) will follow if we show that $q_k(\pi', p_x) = q_k(\pi, p_x)$ for $k = 0, 1, \ldots, N - 1$. For this, it suffices to show that, for $k = 0, 1, \ldots, N - 1$,

$$q_k(\pi', p_x)(\underline{S_k}\underline{C_k}) = q_k(\pi, p_x)(\underline{S_k}\underline{C_k}) \qquad \forall \underline{S_k} \in \mathcal{B}_S, \quad \underline{C_k} \in \mathcal{B}_C. \quad (13)$$

We prove (13) by induction. For $k = 0$, $\underline{S_0} \in \mathcal{B}_S$ and $\underline{C_0} \in \mathcal{B}_C$, we have, from (12),

$$q_0(\pi', p_x)(\underline{S_0}\underline{C_0}) = \int_{\underline{S_0}} \mu_0(\underline{C_0} | x_0) p_x(dx_0) = q_0(\pi, p_x)(\underline{S_0}\underline{C_0}).$$

If $q_k(\pi', p_x) = q_k(\pi, p_x)$, then for $\underline{S_{k+1}} \in \mathcal{B}_S$, $\underline{C_{k+1}} \in \mathcal{B}_C$, we have, from (12),

$$q_{k+1}(\pi', p_x)(\underline{S_{k+1}}\underline{C_{k+1}}) = \int_{\underline{S_{k+1}}} \mu_{k+1}(\underline{C_{k+1}} | x_{k+1}) p_{k+1}(\pi', p_x)(dx_{k+1}). \quad (14)$$

From (7) we see that

$$p_{k+1}(\pi', p_x)(\underline{S_{k+1}}) = \int_{S_k C_k} t(\underline{S_{k+1}} | x_k, u_k) \, dq_k(\pi', p_x),$$

so if $h: S_{k+1} \to [0, \infty]$ is a Borel-measurable indicator function, then

$$\int_{S_{k+1}} h(x_{k+1}) \, dp_{k+1}(\pi', p_x)(dx_{k+1}) = \int_{S_k C_k} \int_{S_{k+1}} h(x_{k+1}) t(dx_{k+1} | x_k, u_k)$$
$$\times dq_k(\pi', p_x). \quad (15)$$

Then (15) holds for Borel-measurable simple functions, and finally, for all Borel-measurable functions $h: S_{k+1} \to [0, \infty]$. Letting $h(x_{k+1})$ in (15) be $\mu_{k+1}(\underline{C_{k+1}} | x_{k+1})$, we obtain from (14), the induction hypothesis, and (8)

$$q_{k+1}(\pi', p_x)(\underline{S_{k+1}}\underline{C_{k+1}}) = \int_{S_k C_k} \int_{S_{k+1}} \mu_{k+1}(\underline{C_{k+1}} | x_{k+1}) t(dx_{k+1} | x_k, u_k)$$
$$\times dq_k(\pi', p_x)$$
$$= \int_{S_k C_k} \int_{S_{k+1}} \mu_{k+1}(\underline{C_{k+1}} | x_{k+1}) t(dx_{k+1} | x_k, u_k) \, dq_k(\pi, p_x)$$
$$= q_{k+1}(\pi, p_x)(\underline{S_{k+1}}\underline{C_{k+1}}),$$

which proves (13) for $k + 1$. Q.E.D.

Corollary 8.1.1 $(F^+)(F^-)$ For $K = 1, 2, \ldots, N$, we have

$$J_K^*(x) = \inf_{\pi \in \Pi} J_{K,\pi}(x) \qquad \forall x \in S,$$

where Π is the set of all Markov policies.

Corollary 8.1.1 shows that the admission of non-Markov policies to our discussion has not resulted in a reduction of the optimal cost function. The advantage of allowing non-Markov policies is that an ε-optimal nonrandomized policy can then be guaranteed to exist (Proposition 8.3), whereas one may not exist within the class of Markov policies (Example 2).

8.2   The Dynamic Programming Algorithm—Existence of Optimal and ε-Optimal Policies

Let $U(C|S)$ denote the set of universally measurable stochastic kernels μ on C given S which satisfy $\mu(U(x)|x) = 1$ for every $x \in S$. Thus the set of Markov policies is $\Pi = U(C|S)U(C|S) \cdots U(C|S)$, where there are N factors.

Definition 8.4   Let $J : S \to R^*$ be universally measurable and $\mu \in U(C|S)$. The *operator* T_μ mapping J into $T_\mu(J) : S \to R^*$ is defined by

$$T_\mu(J)(x) = \int_C \left[g(x, u) + \alpha \int_S J(x') t(dx'|x, u) \right] \mu(du|x)$$

for every $x \in S$.

The operator T_μ can also be written in terms of the system function f and the disturbance kernel $p(dw|x, u)$ as [cf. (3)]

$$T_\mu(J)(x) = \int_C \left[g(x, u) + \alpha \int_W J[f(x, u, w)] p(dw|x, u) \right] \mu(du|x).$$

By Proposition 7.46, $T_\mu(J)$ is universally measurable. We show that under (F^+) or (F^-), the cost corresponding to a policy $\pi = (\mu_0, \ldots, \mu_{N-1})$ can be defined in terms of the composition of operators $T_{\mu_0} T_{\mu_1} \cdots T_{\mu_{N-1}}$.

Lemma 8.1 $(F^+)(F^-)$ Let $\pi = (\mu_0, \mu_1, \ldots, \mu_{N-1})$ be a Markov policy and let $J_0 : S \to R^*$ be identically zero. Then for $K = 1, 2, \ldots, N$ we have

$$J_{K,\pi} = (T_{\mu_0} \cdots T_{\mu_{K-1}})(J_0), \qquad (16)$$

where $T_{\mu_0} \cdots T_{\mu_{K-1}}$ denotes the composition of $T_{\mu_0}, \ldots, T_{\mu_{K-1}}$.

Proof   We proceed by induction. For $x \in S$,

$$J_{1,\pi}(x) = \int g \, dq_0(\pi, p_x)$$

$$= \int_{C_0} g(x, u_0) \mu_0(du_0|x) = T_{\mu_0}(J_0)(x).$$

Suppose the lemma holds for $K - 1$. Let $\bar{\pi} = (\mu_1, \mu_2, \ldots, \mu_{N-1}, \mu)$, where μ is some element of $U(C|S)$. Then for any $x \in S$, the (F^+) or (F^-) assumption along with Lemma 7.11(b) implies that (5) can be rewritten as

$$J_{K,\pi}(x) = \int_{C_0} g(x, u_0)\mu_0(du_0|x) + \alpha \int_{C_0} \int_{S_1} \int_{C_1} \cdots \int_{C_{K-1}} \left[\sum_{k=1}^{K-1} \alpha^{k-1} g(x_k, u_k) \right]$$

$$\times \mu_{K-1}(du_{K-1}|x_{K-1})t(dx_{K-1}|x_{K-2}, u_{K-2}) \cdots$$

$$\times \mu_1(du_1|x_1)t(dx_1|x, u_0)\mu_0(du_0|x)$$

$$= \int_{C_0} g(x, u_0)\mu_0(du_0|x) + \alpha \int_{C_0} \int_{S_1} J_{K-1,\bar{\pi}}(x_1)t(dx_1|x, u_0)\mu_0(du_0|x). \quad (17)$$

Under (F^-), $\int_{C_0} g^+(x, u_0)\mu_0(du_0|x) < \infty$ and

$$\int_{C_0} \int_{S_1} [J_{K-1,\bar{\pi}}(x_1)t(dx_1|x, u_0)]^+ \mu_0(du_0|x) < \infty,$$

while under (F^+) a similar condition holds, so Lemma 7.11(b) and the induction hypothesis can be applied to the right-hand side of (17) to conclude

$$J_{K,\pi}(x) = \int_{C_0} \left[g(x, u_0) + \alpha \int_{S_1} (T_{\mu_1} \cdots T_{\mu_{K-1}})(J_0)(x_1)t(dx_1|x, u_0) \right] \mu_0(du_0|x)$$

$$= (T_{\mu_0} T_{\mu_1} \cdots T_{\mu_{K-1}})(J_0)(x). \qquad \text{Q.E.D.}$$

Definition 8.5 Let $J: S \to R^*$ be universally measurable. The *operator* T mapping J into $T(J): S \to R^*$ is defined by

$$T(J)(x) = \inf_{u \in U(x)} \left\{ g(x, u) + \alpha \int_S J(x')t(dx'|x, u) \right\}$$

for every $x \in S$.

Similarly as for T_μ, the operator T may be written in terms of f and $p(dw|x, u)$ as

$$T(J)(x) = \inf_{u \in U(x)} \left\{ g(x, u) + \alpha \int_W J[f(x, u, w)]p(dw|x, u) \right\}.$$

If μ is nonrandomized, the operators T_μ and T of Definitions 8.4 and 8.5 are, except for measurability restrictions on J and μ, special cases of those defined in Section 2.1. In the present case, the mapping H of Section 2.1 is

$$H(x, u, J) = g(x, u) + \alpha \int_S J(x')t(dx'|x, u)$$

$$= g(x, u) + \alpha \int_W J[f(x, u, w)]p(dw|x, u).$$

We will state and prove versions of Assumptions F.1 and F.3 of Section 3.1 for this function H. Assumption F.2 is clearly true. Furthermore, if $\mu \in U(C|S)$,

$J_1, J_2: S \to R^*$ are universally measurable, and $J_1 \le J_2$, then $T_\mu(J_1) \le T_\mu(J_2)$ and $T(J_1) \le T(J_2)$. If $r \in (0, \infty)$, then $T_\mu(J_1 + r) = T_\mu(J_1) + \alpha r$ and $T(J_1 + r) = T(J_1) + \alpha r$. We will make frequent use of these properties. The reader should not be led to believe, however, that the model of this chapter is a special case of the model of Chapters 2 and 3. The earlier model does not admit measurability restrictions on policies.

By Lemma 7.30(4) and Propositions 7.47 and 7.48, $T(J)$ is lower semi-analytic whenever J is. The composition of T with itself k times is denoted by T^k, i.e., $T^k(J) = T[T^{k-1}(J)]$, where $T^0(J) = J$. We show in Proposition 8.2 that under (F^+) or (F^-) the optimal cost can be defined in terms of T^N. Three preparatory lemmas are required.

Lemma 8.2   Let $J: S \to R^*$ be lower semianalytic. Then for $\varepsilon > 0$, there exists $\mu \in U(C|S)$ such that

$$T_\mu(J)(x) \le T(J)(x) + \varepsilon \qquad \forall x \in S,$$

where $T(J)(x) + \varepsilon$ may be $-\infty$.

Proof   By Proposition 7.50, there are universally measurable selectors $\mu_m: S \to C$ such that for $m = 1, 2, \ldots$ and $x \in S$, we have $\mu_m(x) \in U(x)$ and

$$T_{\mu_m}(J)(x) \le \begin{cases} T(J)(x) + \varepsilon & \text{if} \quad T(J)(x) > -\infty, \\ -2^m & \text{if} \quad T(J)(x) = -\infty. \end{cases}$$

Let $\mu(du|x)$ assign mass one to $\mu_1(x)$ if $T(J)(x) > -\infty$ and assign mass $1/2^m$ to $\mu_m(x)$, $m = 1, 2, \ldots$, if $T(J)(x) = -\infty$.

For each $\underline{C} \in \mathscr{B}_C$,

$$\mu(\underline{C}|x) = \begin{cases} \chi_{\underline{C}}[\mu_1(x)] & \text{if} \quad T(J)(x) > -\infty, \\ \displaystyle\sum_{m=1}^{\infty} (1/2^m)\chi_{\underline{C}}[\mu_m(x)] & \text{if} \quad T(J)(x) = -\infty, \end{cases}$$

is a universally measurable function of x, and therefore μ is a universally measurable stochastic kernel [Lemma 7.28(a),(b)]. This μ has the desired properties.   Q.E.D.

Lemma 8.3 (F^+) If $J_0: S \to R^*$ is identically zero, then $T^K(J_0)(x) > -\infty$ for every $x \in S$, $K = 1, \ldots, N$, where T^K denotes the composition of T with itself K times.

Proof   Suppose for some $K \le N$ and $\bar{x} \in S$ that

$$T^j(J_0)(x) > -\infty, \qquad j = 0, \ldots, K-1,$$

for every $x \in S$, and

$$T^K(J_0)(\bar{x}) = -\infty.$$

By Proposition 7.50, there are universally measurable selectors $\mu_j: S \to C$, $j = 1, \ldots, K - 1$, such that $\mu_j(x) \in U(x)$ and

$$(T_{\mu_{K-j}} T^{j-1})(J_0)(x) \le T^j(J_0)(x) + 1, \qquad j = 1, \ldots, K - 1,$$

for every $x \in S$. Then

$$
\begin{aligned}
(T_{\mu_1} \cdots T_{\mu_{K-1}})(J_0) &\le (T_{\mu_1} \cdots T_{\mu_{K-2}})[T(J_0) + 1] \\
&\le (T_{\mu_1} \cdots T_{\mu_{K-3}})[T^2(J_0) + 1 + \alpha] \\
&\le T^{K-1}(J_0) + 1 + \alpha + \cdots + \alpha^{K-2},
\end{aligned}
$$

where the last inequality is obtained by repeating the process used to obtain the first two inequalities. By Lemma 8.2, there is a stochastic kernel $\mu_0 \in U(C|S)$ such that

$$(T_{\mu_0} T^{K-1})(J_0)(\bar{x}) = -\infty.$$

Then

$$(T_{\mu_0} T_{\mu_1} \cdots T_{\mu_{K-1}})(J_0)(\bar{x}) \le T_{\mu_0}[T^{K-1}(J_0) + 1 + \alpha + \cdots + \alpha^{K-2}](\bar{x}) = -\infty.$$

Choose any $\mu \in U(C|S)$ and let $\pi = (\mu_0, \ldots, \mu_{K-1}, \mu, \ldots, \mu)$, so that $\pi \in \Pi$. By Lemma 8.1,

$$\sum_{k=0}^{K-1} \alpha^k \int g \, dq_k(\pi, p_{\bar{x}}) = J_{K, \pi}(\bar{x}) = (T_{\mu_0} \cdots T_{\mu_{K-1}})(J_0)(\bar{x}) = -\infty,$$

so for some $k \le K - 1$, $\int g^- \, dq_k(\pi, p_{\bar{x}}) = \infty$. This contradicts the (F^+) assumption.   Q.E.D.

Lemma 8.4   Let $\{J_k\}$ be a sequence of extended real-valued, universally measurable functions on S and let μ be an element of $U(C|S)$.

(a) If $T_\mu(J_1)(x) < \infty$ for every $x \in S$ and $J_k \downarrow J$, then $T_\mu(J_k) \downarrow T_\mu(J)$.
(b) If $T_\mu(J_1^-)(x) < \infty$ for every $x \in S, g \ge 0$, and $J_k \uparrow J$, then $T_\mu(J_k) \uparrow T_\mu(J)$.
(c) If $\{J_k\}$ is uniformly bounded, g is bounded, and $J_k \to J$, then $T_\mu(J_k) \to T_\mu(J)$.

Proof   Assume first that $T_\mu(J_1) < \infty$ and $J_k \downarrow J$. Fix x. Since

$$\int [g(x, u) + \alpha \int J_1(x') t(dx'|x, u)] \mu(du|x) < \infty,$$

we have

$$g(x, u) + \alpha \int J_1(x') t(dx'|x, u) < \infty$$

for $\mu(du|x)$ almost all u. By the monotone convergence theorem [Lemma 7.11(f)],

$$g(x, u) + \alpha \int J_k(x') t(dx'|x, u) \downarrow g(x, u) + \alpha \int J(x') t(dx'|x, u)$$

for $\mu(du|x)$ almost all u. Apply the monotone convergence theorem again to conclude $T_\mu(J_k)(x) \downarrow T_\mu(J)(x)$.

If $T_\mu(J_1^-) < \infty$, $g \geq 0$, and $J_k \uparrow J$, the same type of argument applies. If $\{J_k\}$ is uniformly bounded, g bounded, and $J_k \to J$, a similar argument using the bounded convergence theorem applies.   Q.E.D.

The dynamic programming algorithm over a finite horizon is executed by beginning with the identically zero function on S and applying the operator T successively N times. The next theorem says that this procedure generates the optimal cost function. In Proposition 8.3, we show how ε-optimal policies can also be obtained from this algorithm.

Proposition 8.2 $(F^+)(F^-)$ Let J_0 be the identically zero function on S. Then

$$J_K^* = T^K(J_0), \qquad K = 1, \ldots, N. \tag{18}$$

Proof   It suffices to prove (18) for $K = N$, since the horizon N can be chosen to be any positive integer. For any $\pi = (\mu_0, \ldots, \mu_{N-1}) \in \Pi$ and $K \leq N$, we have

$$J_{K,\pi} = (T_{\mu_0} \cdots T_{\mu_{K-1}})(J_0) \geq (T_{\mu_0} \cdots T_{\mu_{K-2}} T)(J_0) \geq T^K(J_0), \tag{19}$$

where the last inequality is obtained by repeating the process used to obtain the first inequality. Infimizing over $\pi \in \Pi$ when $K = N$ and using Corollary 8.1.1, we obtain

$$J_N^* \geq T^N(J_0). \tag{20}$$

If (F^+) holds, then, by Lemma 8.3, $T^k(J_0) > -\infty$, $k = 1, \ldots, N$. For $\varepsilon > 0$, there are universally measurable selectors $\hat{\mu}_k : S \to C$, $k = 0, \ldots, N-1$, with $\hat{\mu}_k(x) \in U(x)$ and

$$-\infty < T_{\hat{\mu}_{N-k}}[T^{k-1}(J_0)](x)$$
$$\leq T^k(J_0)(x) + \varepsilon/(1 + \alpha + \alpha^2 + \cdots + \alpha^{N-1}), \qquad k = 1, \ldots, N,$$

for every $x \in S$ (Proposition 7.50). Then

$$(T_{\hat{\mu}_0} T_{\hat{\mu}_1} \cdots T_{\hat{\mu}_{N-1}})(J_0) \leq (T_{\hat{\mu}_0} T_{\hat{\mu}_1} \cdots T_{\hat{\mu}_{N-2}})[T(J_0)$$
$$+ \varepsilon/(1 + \alpha + \alpha^2 + \cdots + \alpha^{N-1})]$$
$$\leq (T_{\hat{\mu}_0} T_{\hat{\mu}_1} \cdots T_{\hat{\mu}_{N-3}})[T^2(J_0)$$
$$+ \varepsilon(1 + \alpha)/(1 + \alpha + \alpha^2 + \cdots + \alpha^{N-1})]$$
$$\leq T^N(J_0) + \varepsilon, \tag{21}$$

where the last inequality is obtained by repeating the process used to obtain the first two inequalities. It follows that

$$J_N^* \leq T^N(J_0). \tag{22}$$

Combining (20) and (22), we see that the proposition holds under the (F^+) assumption.

If (F^-) holds, then $J_{K,\pi}(x) < \infty$ for every $x \in S$, $\pi \in \Pi$, $K = 1, \dots, N$. Use Proposition 7.50 to choose nonrandomized policies $\pi^i = (\mu_0^i, \dots, \mu_{N-1}^i) \in \Pi$ such that

$$(T_{\mu_k^i} T^{N-k-1})(J_0) \downarrow T^{N-k}(J_0), \qquad k = 0, \dots, N-1,$$

as $i \to \infty$. By (19) and Lemma 8.4(a),

$$
\begin{aligned}
J_N^* &\leq \inf_{(i_0, \dots, i_{N-1})} (T_{\mu_0^{i_0}} \cdots T_{\mu_{N-1}^{i_{N-1}}})(J_0) \\
&= \inf_{i_0} \cdots \inf_{i_{N-1}} (T_{\mu_0^{i_0}} \cdots T_{\mu_{N-1}^{i_{N-1}}})(J_0) \\
&= \inf_{i_0} \cdots \inf_{i_{N-2}} (T_{\mu_0^{i_0}} \cdots T_{\mu_{N-2}^{i_{N-2}}}) \left[\inf_{i_{N-1}} T_{\mu_{N-1}^{i_{N-1}}}(J_0) \right] \\
&= \inf_{i_0} \cdots \inf_{i_{N-2}} (T_{\mu_0^{i_0}} \cdots T_{\mu_{N-2}^{i_{N-2}}} T)(J_0) \\
&= T^N(J_0),
\end{aligned}
\tag{23}
$$

where the last equality is obtained by repeating the process used to obtain the previous equality. Combining (20) and (23), we see that the proposition holds under the (F^-) assumption.   Q.E.D.

When the state, control, and disturbance spaces are countable, the model of Definition 8.1 falls within the framework of Part I. Consider such a model, and, as in Part I, let M be the set of mappings $\mu: S \to C$ for which $\mu(x) \in U(x)$ for every $x \in S$. In Section 3.2, it was often assumed that for every $x \in S$ and $\mu_j \in M$, $j = 0, \dots, K-1$, we have

$$(T_{\mu_0} \cdots T_{\mu_{K-1}})(J_0)(x) < \infty, \qquad K = 1, \dots, N, \tag{24}$$

or else for every $x \in S$

$$\inf_{\mu_j \in M, \, 0 \leq j \leq K-1} (T_{\mu_0} \cdots T_{\mu_{K-1}})(J_0)(x) > -\infty, \qquad K = 1, \dots, N. \tag{25}$$

Under the (F^+) assumption, Lemma 8.3 implies that

$$-\infty < T^K(J_0) \leq \inf_{\mu_j \in M, \, 0 \leq j \leq K-1} (T_{\mu_0} \cdots T_{\mu_{K-1}})(J_0),$$

so (25) is satisfied. Under (F^-), we have from Lemma 8.1 that

$$(T_{\mu_0} \cdots T_{\mu_{K-1}})(J_0) = J_{K,\pi} < \infty,$$

where $\pi = (\mu_0, \dots, \mu_{K-1})$, so (24) holds. The primary reason for introducing the stronger (F^+) and (F^-) assumptions is to enable us to prove Lemma 8.1. If one chooses instead to take (16) as the definition of $J_{K,\pi}$ (as is done in

Section 3.1), then (24) or (25) suffices to prove Proposition 8.2 along the lines of the proof of Proposition 3.1 of Part I.

Proposition 8.2 implies the following property of the optimal cost function.

Corollary 8.2.1 $(F^+)(F^-)$ For $K = 1, 2, \ldots, N$, the function J_K^* is lower semianalytic.

Proof As observed following Definition 8.5, $T(J)$ is lower semianalytic whenever J is. Since $J_K^* = T^K(J_0)$ and $J_0 \equiv 0$ is lower semianalytic, the result follows.   Q.E.D.

We give an example to show that even when $\Gamma = SC$ and the one-stage cost $g: SC \to R^*$ is Borel-measurable, J_1^* can fail to be Borel-measurable.

EXAMPLE 1 Let A be an analytic subset of $[0, 1]$ which is not Borel-measurable (Appendix B). By Proposition 7.39, there is a closed set $F \subset [0,1] \mathcal{N}$ such that $A = \text{proj}_{[0,1]}(F)$. Let $S = [0,1]$, $C = \mathcal{N}$, $\Gamma = SC$, and $g = \chi_{F^c}$. Then

$$J_1^*(x) = \inf_{u \in C} g(x, u) = \chi_{A^c}(x) \qquad \forall x \in S,$$

which is a lower semianalytic but not Borel-measurable function. We could also choose $C = [0,1]$, $\Gamma = SC$, B a G_δ-subset of the unit square SC, and $g = \chi_{B^c}$. This is because $\mathcal{N}$ and $\mathcal{N}_0$, the space of irrational numbers in $[0,1]$, are homeomorphic (Proposition 7.5). But

$$\mathcal{N}_0 = \bigcap_{r \in Q}([0,1] - \{r\})$$

is a G_δ-subset of $[0, 1]$, so there is a homeomorphism $\varphi: \mathcal{N} \to [0,1]$ such that $\varphi(\mathcal{N})$ is a G_δ-subset of $[0, 1]$. Let $\Phi: [0,1] \mathcal{N} \to [0,1][0,1]$ be the homeomorphism defined by

$$\Phi(x, z) = (x, \varphi(z)).$$

Then $\Phi([0,1] \mathcal{N}) = [0,1]\varphi(\mathcal{N})$ is a G_δ-subset of $SC = [0,1][0,1]$, and since F is a G_δ-set in $[0,1]\mathcal{N}$, $B = \Phi(F)$ is a G_δ-subset of SC which satisfies $\text{proj}_S(B) = A$. If $g = \chi_{B^c}$, then again $J_1^* = \chi_{A^c}$.

We now use Proposition 8.2 to establish existence of ε-optimal policies.

Proposition 8.3

(F^+) For each $\varepsilon > 0$, there exists a nonrandomized Markov ε-optimal policy.

(F^-) For each $\varepsilon > 0$, there exists a nonrandomized semi-Markov ε-optimal policy and a (randomized) Markov ε-optimal policy.

Proof If (F^+) holds, then the policy $(\hat{\mu}_0, \ldots, \hat{\mu}_{N-1})$ constructed in the proof of Proposition 8.2 is ε-optimal, nonrandomized, and Markov.

Assume (F^-) holds. We show first the existence of an ε-optimal, non-randomized, semi-Markov policy. Let $\pi^i = (\mu_0^i, \ldots, \mu_{N-1}^i)$ be as in the proof of Proposition 8.2. Then

$$J_N^* = T^N(J_0) = \inf_{(i_0, \ldots, i_{N-1})} (T_{\mu_0^{i_0}} \cdots T_{\mu_{N-1}^{i_{N-1}}})(J_0)$$

$$= \inf_{(i_0, \ldots, i_{N-1})} J_{N, \pi^{(i_0, \ldots, i_{N-1})}},$$

where $\pi^{(i_0, \ldots, i_{N-1})} = (\mu_0^{i_0}, \ldots, \mu_{N-1}^{i_{N-1}})$. Choose $\varepsilon > 0$ and define

$$\theta(x) = \begin{cases} J_N^*(x) + \varepsilon & \text{if } J_N^*(x) > -\infty, \\ -1/\varepsilon & \text{if } J_N^*(x) = -\infty. \end{cases}$$

Order linearly the countable set $\{\pi^{(i_0, \ldots, i_{N-1})} | i_0, \ldots, i_{N-1}$ are positive integers$\}$ and define $\pi(x)$ to be the first $\pi^{(i_0, \ldots, i_{N-1})}$ such that

$$J_{N, \pi^{(i_0, \ldots, i_{N-1})}}(x) \le \theta(x).$$

Let the components of $\pi(x)$ be

$$(\mu_0^x(x_0), \mu_1^x(x_1), \ldots, \mu_{N-1}^x(x_{N-1})).$$

The set $\{x | \pi(x) = \pi^{(i_0, \ldots, i_{N-1})}\}$ is universally measurable for each $(i_0, \ldots, i_{N-1})$, so

$$(\mu_0(x_0), \mu_1(x_0, x_1), \ldots, \mu_{N-1}(x_0, x_{N-1})),$$

where $\mu_0(x_0) = \mu_0^{x_0}(x_0)$ and $\mu_k(x_0, x_k) = \mu_k^{x_0}(x_k)$, $k = 1, \ldots, N-1$, is an ε-optimal nonrandomized semi-Markov policy.

We now show the existence of an ε-optimal (randomized) Markov policy. By Lemma 8.2, there exist $\mu_{N-k} \in U(C|S)$ such that for $k = 1, \ldots, N$

$$(T_{\mu_{N-k}} T^{k-1})(J_0) \le T^k(J_0) + \varepsilon/(1 + \alpha + \alpha^2 + \cdots + \alpha^{N-1}).$$

Proceed as in (21).   Q.E.D.

If the (F^-) assumption holds and $\varepsilon > 0$, it may not be possible to find a nonrandomized Markov ε-optimal policy, as the following example demonstrates.

EXAMPLE 2   Let $S = \{0, 1, 2, \ldots\}$, $C = \{1, 2, \ldots\}$, $W = \{w_1, w_2\}$, $\Gamma = SC$, $N = 2$ and define

$$g(x, u) = \begin{cases} -u & \text{if } x = 1, \\ 0 & \text{if } x \ne 1, \end{cases}$$

$$f(x, u, w) = \begin{cases} 0 & \text{if } x = 0 \text{ or } x = 1 \text{ or } w = w_1, \\ 1 & \text{if } x \ne 0, x \ne 1, \text{ and } w = w_2, \end{cases}$$

$$p(\{w_1\}|x, u) = 1 - 1/x \qquad \text{if } x \ne 0, \quad x \ne 1,$$
$$p(\{w_2\}|x, u) = 1/x \qquad\qquad \text{if } x \ne 0, \quad x \ne 1.$$

The (F^-) assumption is satisfied. Let $\pi = (\mu_0, \mu_1)$ be a nonrandomized Markov policy. If the initial state x_0 is neither zero nor one, then regardless of the policy employed, $x_1 = 0$ with probability $1 - (1/x_0)$, and $x_1 = 1$ with probability $1/x_0$. Once the system reaches zero, it remains there at no further cost. If the system reaches one, it moves to $x_2 = 0$ at a cost of $-\mu_1(1)$. Thus $J_{N,\pi}(x_0) = -\mu_1(1)/x_0$ if $x_0 \neq 0$, $x_0 \neq 1$, and $J_N^*(x_0) = -\infty$ if $x_0 \neq 0$, $x_0 \neq 1$. For any $\varepsilon > 0$, π cannot be ε-optimal.

In Example 2, it is possible to find a sequence of nonrandomized Markov policies $\{\pi_n\}$ such that $J_{N,\pi_n} \downarrow J_N^*$. This example motivates the idea of policies exhibiting $\{\varepsilon_n\}$-dominated convergence to optimality (Definition 8.3) and Proposition 8.4, which we prove with the aid of the next lemma.

Lemma 8.5   Let $\{J_k\}$ be a sequence of universally measurable functions from S to R^* and μ a universally measurable function from S to C whose graph lies in Γ. Suppose for some sequence $\{\varepsilon_k\}$ of positive numbers with $\sum_{k=1}^{\infty} \varepsilon_k < \infty$, we have, for every $x \in S$,

$$\int J_1^+(x')t(dx'|x, \mu(x)) < \infty, \qquad \lim_{k \to \infty} J_k(x) = J(x)$$

and for $k = 2, 3, \ldots$

$$J(x) \leq J_k(x) \leq J(x) + \varepsilon_k \qquad \text{if} \quad J(x) > -\infty,$$
$$J_k(x) \leq J_{k-1}(x) + \varepsilon_k \qquad \text{if} \quad J(x) = -\infty.$$

Then

$$\lim_{k \to \infty} T_\mu(J_k) = T_\mu(J). \tag{26}$$

Proof   Since $J \leq J_k$ for every k, it is clear that

$$T_\mu(J) \leq \liminf_{k \to \infty} T_\mu(J_k). \tag{27}$$

For $x \in S$,

$$\limsup_{k \to \infty} T_\mu(J_k)(x) \leq g[x, \mu(x)] + \alpha \limsup_{k \to \infty} \int_{\{x'|J(x') > -\infty\}} J_k(x')t(dx'|x, \mu(x))$$

$$+ \alpha \limsup_{k \to \infty} \int_{\{x'|J(x') = -\infty\}} J_k(x')t(dx'|x, \mu(x)).$$

Now

$$\limsup_{k \to \infty} \int_{\{x'|J(x') > -\infty\}} J_k(x')t(dx'|x, \mu(x))$$

$$\leq \limsup_{k \to \infty} \left[\int_{\{x'|J(x') > -\infty\}} J(x')t(dx'|x, \mu(x)) + \varepsilon_k \right]$$

$$= \int_{\{x'|J(x') > -\infty\}} J(x')t(dx'|x, \mu(x)).$$

If $J(x') = -\infty$, then

$$J_k(x') + \sum_{n=k+1}^{\infty} \varepsilon_n \downarrow J(x'),$$

and since $\int [J_1^+(x') + \sum_{n=2}^{\infty} \varepsilon_n] t(dx'|x, \mu(x)) < \infty$, we have

$$\limsup_{k \to \infty} \int_{\{x'|J(x') = -\infty\}} J_k(x') t(dx'|x, \mu(x))$$

$$\leq \lim_{k \to \infty} \int_{\{x'|J(x') = -\infty\}} \left[J_k(x') + \sum_{n=k+1}^{\infty} \varepsilon_n \right] t(dx'|x, \mu(x))$$

$$= \int_{\{x'|J(x') = -\infty\}} J(x') t(dx'|x, \mu(x)).$$

If follows that

$$\limsup_{k \to \infty} T_\mu(J_k) \leq T_\mu(J). \tag{28}$$

Combine (27) and (28) to conclude (26).   Q.E.D.

Proposition 8.4 (F^-) Let $\{\varepsilon_n\}$ be a sequence of positive numbers with $\varepsilon_n \downarrow 0$. There exists a sequence of nonrandomized Markov policies $\{\pi_n\}$ exhibiting $\{\varepsilon_n\}$-dominated convergence to optimality. In particular, if $J_N^*(x) > -\infty$ for all $x \in S$, then for every $\varepsilon > 0$ there exists an ε-optimal nonrandomized Markov policy.

Proof For $N = 1$, by Proposition 7.50 there exists a sequence of nonrandomized Markov policies $\pi_n = (\mu_0^n)$ such that for all n

$$T_{\mu_0^n}(J_0)(x) \leq \begin{cases} T(J_0)(x) + \varepsilon_n & \text{if } T(J_0)(x) > -\infty, \\ -1/\varepsilon_n & \text{if } T(J_0)(x) = -\infty. \end{cases}$$

We may assume without loss of generality that

$$T_{\mu_0^n}(J_0) \leq T_{\mu_0^{n-1}}(J_0).$$

Therefore $\{\pi_n\}$ exhibits $\{\varepsilon_n\}$-dominated convergence to one-stage optimality.

Suppose the result holds for $N - 1$. Let $\pi_n = (\mu_1^n, \ldots, \mu_{N-1}^n)$ be a sequence of $(N-1)$-stage nonrandomized Markov policies exhibiting $\{\varepsilon_n/2\alpha\}$-dominated convergence to $(N-1)$-stage optimality, i.e.,

$$\lim_{n \to \infty} J_{N-1, \pi_n} = J_{N-1}^*,$$

$$J_{N-1, \pi_n}(x) \leq \begin{cases} J_{N-1}^*(x) + (\varepsilon_n/2\alpha) & \text{if } J_{N-1}^*(x) > -\infty, \tag{29} \\ J_{N-1, \pi_{n-1}}(x) + (\varepsilon_n/2\alpha) & \text{if } J_{N-1}^*(x) = -\infty. \tag{30} \end{cases}$$

We assume without loss of generality that $\sum_{n=1}^{\infty} \varepsilon_n < \infty$. By Proposition 7.50, there exists a sequence $\{\mu^n\}$ of universally measurable functions from S to

C whose graphs lie in Γ such that

$$T_{\mu^n}(J^*_{N-1})(x) \le \begin{cases} J^*_N(x) + (\varepsilon_n/2) & \text{if } J^*_N(x) > -\infty, \\ -2/\varepsilon_n & \text{if } J^*_N(x) = -\infty. \end{cases} \tag{31}$$

We may assume without loss of generality that

$$T_{\mu^n}(J^*_{N-1}) \le T_{\mu^{n-1}}(J^*_{N-1}), \qquad n = 2, 3, \dots. \tag{32}$$

By Proposition 7.48, the set

$$\begin{aligned} A(J^*_{N-1}) &= \{(x,u) \in \Gamma \,|\, t(\{x'|J^*_{N-1}(x') = -\infty\}|x,u) > 0\} \\ &= \left\{ (x,u) \in \Gamma \,\middle|\, \int -\chi_{\{x'|J^*_{N-1}(x') = -\infty\}}(x')t(dx'|x,u) < 0 \right\} \end{aligned} \tag{33}$$

is analytic in SC; and the Jankov-von Neumann theorem (Proposition 7.49) implies the existence of a universally measurable $\mu:\text{proj}_S[A(J^*_{N-1})] \to C$ whose graph lies in $A(J^*_{N-1})$. Define

$$\hat{\mu}^n(x) = \begin{cases} \mu(x) & \text{if } x \in \text{proj}_S[A(J^*_{N-1})], \\ \mu^n(x) & \text{otherwise.} \end{cases}$$

Then $\hat{\pi}_n = (\hat{\mu}^n, \pi_n)$ is an N-stage nonrandomized Markov policy which will be shown to exhibit $\{\varepsilon_n\}$-dominated convergence to optimality.

For $x \in \text{proj}_S[A(J^*_{N-1})]$, we have, from Lemma 8.5 and the choice of μ,

$$\begin{aligned} \limsup_{n \to \infty} J_{N, \hat{\pi}_n}(x) &= \limsup_{n \to \infty} T_\mu(J_{N-1, \pi_n})(x) \\ &= T_\mu(J^*_{N-1})(x) = -\infty. \end{aligned}$$

For $x \notin \text{proj}_S[A(J^*_{N-1})]$, we have $t(\{x'|J^*_{N-1}(x') = -\infty\}|x,u) = 0$ for every $u \in U(x)$, so by (29)

$$J_{N, \hat{\pi}_n}(x) = T_{\mu^n}(J_{N-1, \pi_n})(x) \le T_{\mu^n}(J^*_{N-1})(x) + \varepsilon_n/2, \tag{34}$$

and

$$\limsup_{n \to \infty} J_{N, \hat{\pi}_n}(x) \le \limsup_{n \to \infty} T_{\mu^n}(J^*_{N-1})(x) \le J^*_N(x)$$

by (31). It follows that

$$\lim_{n \to \infty} J_{N, \hat{\pi}_n} = J^*_N. \tag{35}$$

Suppose for fixed $x \in S$ we have $J^*_N(x) > -\infty$. Then $x \notin \text{proj}_S[A(J^*_{N-1})]$, and we have, from (31) and (34),

$$J_{N, \hat{\pi}_n}(x) \le T_{\mu^n}(J^*_{N-1})(x) + \varepsilon_n/2 \le J^*_N(x) + \varepsilon_n. \tag{36}$$

Suppose now that $J_N^*(x) = -\infty$. If $x \notin \text{proj}_S[A(J_{N-1}^*)]$, then (32) and (34) imply, for $n \geq 2$,

$$J_{N,\hat{\pi}_n}(x) \leq T_{\mu^n}(J_{N-1}^*)(x) + \varepsilon_n/2$$
$$\leq T_{\mu^{n-1}}(J_{N-1}^*)(x) + \varepsilon_n/2$$
$$\leq T_{\mu^{n-1}}(J_{N-1,\pi_{n-1}})(x) + \varepsilon_n/2$$
$$\leq J_{N,\hat{\pi}_{n-1}}(x) + \varepsilon_n/2,$$

while if $x \in \text{proj}_S[A(J_{N-1}^*)]$, we have, from (29) and (30),

$$J_{N,\hat{\pi}_n}(x) = T_\mu(J_{N-1,\pi_n})(x)$$
$$\leq T_\mu(J_{N-1,\pi_{n-1}})(x) + \varepsilon_n/2$$
$$= J_{N,\hat{\pi}_{n-1}}(x) + \varepsilon_n/2.$$

In either case,

$$J_{N,\hat{\pi}_n}(x) \leq J_{N,\hat{\pi}_{n-1}}(x) + \varepsilon_n. \tag{37}$$

From (35)–(37) we see that $\{\hat{\pi}_n\}$ exhibits $\{\varepsilon_n\}$-dominated convergence to optimality.   Q.E.D.

We conclude our discussion of the ramifications of Proposition 8.2 with a technical result needed for the development in Chapter 10.

Lemma 8.6 $(F^+)(F^-)$ For every $p \in P(S)$,

$$\int J_N^*(x)p(dx) = \inf_{\pi \in \Pi} \int J_{N,\pi}(x)p(dx).$$

Proof   For $p \in P(S)$ and $\pi \in \Pi$,

$$\int J_N^*(x)p(dx) \leq \int J_{N,\pi}(x)p(dx),$$

which implies

$$\int J_N^*(x)p(dx) \leq \inf_{\pi \in \Pi} \int J_{N,\pi}(x)p(dx). \tag{38}$$

Choose $\varepsilon > 0$ and let $\hat{\pi} \in \Pi$ be ε-optimal. If $p(\{x \mid J_N^*(x) = -\infty\}) = 0$, then

$$\int J_{N,\hat{\pi}}(x)p(dx) \leq \int J_N^*(x)p(dx) + \varepsilon,$$

and it follows that

$$\inf_{\pi \in \Pi} \int J_{N,\pi}(x)p(dx) \leq \int J_N^*(x)p(dx). \tag{39}$$

If $p(\{x|J_N^*(x) = -\infty\}) > 0$, then

$$\int J_{N,\tilde{\pi}}(x)p(dx) \le -p(\{x|J_N^*(x) = -\infty\})/\varepsilon$$

$$+ \int_{\{x|J_N^*(x) > -\infty\}} J_N^*(x)p(dx) + \varepsilon. \qquad (40)$$

If $\int_{\{x|J_N^*(x) > -\infty\}} J_N^*(x)p(dx) = \infty$, then $\int J_N^*(x)p(dx) = \infty$ and (39) follows. Otherwise, the right-hand side of (40) can be made arbitrarily small by letting ε approach zero, so $\inf_{\pi \in \Pi} \int J_{N,\pi}(x)p(dx) = -\infty$ and (39) is again valid. The lemma follows from (38) and (39).   Q.E.D.

We now consider the question of constructing an optimal policy, if this is at all possible. When the dynamic programming algorithm can be used to construct an optimal policy, this policy usually satisfies a condition stronger than mere optimality. This condition is given in the next definition.

Definition 8.6   Let $\pi = (\mu_0, \ldots, \mu_{N-1})$ be a Markov policy and $\pi^{N-k} = (\mu_k, \ldots, \mu_{N-1}), k = 0, \ldots, N - 1$. The policy π is *uniformly N-stage optimal* if

$$J_{N-k,\pi^{N-k}} = J_{N-k}^*, \qquad k = 0, \ldots, N - 1.$$

Lemma 8.7 $(F^+)(F^-)$ The policy $\pi = (\mu_0, \ldots, \mu_{N-1}) \in \Pi$ is uniformly N-stage optimal if and only if

$$(T_{\mu_k} T^{N-k-1})(J_0) = T^{N-k}(J_0), \qquad k = 0, \ldots, N - 1.$$

Proof   If $\pi = (\mu_0, \ldots, \mu_{N-1})$ is uniformly N-stage optimal, then

$$T^{N-k}(J_0) = J_{N-k}^* = J_{N-k, \pi^{N-k}} = T_{\mu_k}(J_{N-k-1, \pi^{N-k-1}})$$
$$= T_{\mu_k}(J_{N-k-1}^*) = (T_{\mu_k} T^{N-k-1})(J_0), \qquad k = 0, \ldots, N - 1,$$

where $J_{0,\pi^0} = J_0^* \equiv 0$. If $(T_{\mu_k} T^{N-k-1})(J_0) = T^{N-k}(J_0), k = 0, \ldots, N - 1$, then for all k

$$J_{N-k}^* = T^{N-k}(J_0) = (T_{\mu_k} T^{N-k-1})(J_0)$$
$$= (T_{\mu_k} T_{\mu_{k+1}} T^{N-k-2})(J_0)$$
$$= (T_{\mu_k} \cdots T_{\mu_{N-1}})(J_0)$$
$$= J_{N-k, \pi^{N-k}},$$

where the next to last equality is obtained by continuing the process used to obtain the previous equalities.   Q.E.D.

Lemma 8.7 is the analog for the Borel model of Proposition 3.3 for the model of Part I. Because (F^+) or (F^-) is a required assumption in Lemma 8.1, one of them is also required in Lemma 8.7, as the following example shows. If we take (16) as the definition of $J_{k,\pi}$, then Lemma 8.7, Proposition 8.5, and

Corollaries 8.5.1 and 8.5.2 hold without the (F^+) and (F^-) assumptions. The proofs are similar to those of Section 3.2.

EXAMPLE 3   Let $S = \{s, t\} \cup \{(k, j) | k = 1, 2, \ldots; \ j = 1, 2\}$, $C = \{a, b\}$, $U(s) = \{a, b\}$, $U(t) = U(k, j) = \{b\}$, $k = 1, 2, \ldots, j = 1, 2$, $W = S$, and $\alpha = 1$. Let the disturbance kernel be given by $p(s|s, a) = 1$,

$$p[(k, 1)|s, b] = p[(k, 2)|(l, 1), b] = k^{-2} \left(\sum_{n=1}^{\infty} \frac{1}{n^2} \right)^{-1}, \qquad k, l = 1, 2, \ldots,$$

$p[t|(k, 2), b] = 1$, $k = 1, 2, \ldots$, and $p(t|t, b) = 1$. Let the system function be $f(x, u, w) = w$. Thus if the system begins at state s, we can hold it at s or allow it to move to some state $(l, 1)$, from which it subsequently moves to some $(k, 2)$ and then to t. Having reached t, the system remains there. The relevant costs are $g(s, a) = g(s, b) = g(t, b) = 0$, $g[(k, 1), b] = k$, $g[(k, 2), b] = -k$, $k = 1, 2, \ldots$. Let $\pi = (\mu_0, \mu_1, \mu_2)$ be a policy with $\mu_0(s) = b$, $\mu_1(s) = \mu_2(s) = a$. Then

$$T(J_0)(x_2) = T_{\mu_2}(J_0)(x_2) \qquad = \begin{cases} 0 & \text{if } x_2 = s, \\ k & \text{if } x_2 = (k, 1), \\ -k & \text{if } x_2 = (k, 2), \\ 0 & \text{if } x_2 = t, \end{cases}$$

$$T^2(J_0)(x_1) = (T_{\mu_1} T_{\mu_2})(J_0)(x_1) \qquad = \begin{cases} 0 & \text{if } x_1 = s, \\ -\infty & \text{if } x_1 = (k, 1), \\ -k & \text{if } x_1 = (k, 2), \\ 0 & \text{if } x_1 = t, \end{cases}$$

$$T^3(J_0)(x_0) = (T_{\mu_0} T_{\mu_1} T_{\mu_2})(J_0)(x_0) = \begin{cases} -\infty & \text{if } x_0 = s, \\ -\infty & \text{if } x_0 = (k, 1), \\ -k & \text{if } x_0 = (k, 2), \\ 0 & \text{if } x_0 = t. \end{cases}$$

However, $J_{\pi, 3}(s) = \infty > J_{\bar{\pi}, 3}(s) = 0$, where $\bar{\pi} = (\bar{\mu}_0, \bar{\mu}_1, \bar{\mu}_2)$ and $\bar{\mu}_0(s) = \bar{\mu}_1(s) = \bar{\mu}_2(s) = a$, so π is not optimal and $T^3(J_0) \neq J_3^*$. It is easily verified that $\bar{\pi}$ is a uniformly three-stage optimal policy, so Corollary 3.3.1(b) also fails to hold for the Borel model of this chapter. Here both assumptions (F^+) and (F^-) are violated.

Proposition 8.5 $(F^+)(F^-)$ If the infimum in

$$\inf_{u \in U(x)} \left\{ g(x, u) + \alpha \int J_k^*(x') t(dx'|x, u) \right\}, \qquad k = 0, \ldots, N - 1, \qquad (41)$$

is achieved for each $x \in S$, where J_0^* is identically zero, then a uniformly N-stage optimal (and hence optimal) nonrandomized Markov policy exists. This policy is generated by the dynamic programming algorithm, i.e., by measurably selecting for each x a control u which achieves the infimum.

Proof   Let $\pi = (\mu_0, \ldots, \mu_{N-1})$, where $\mu_{N-k-1} : S \to C$ achieves the infimum in (41) and satisfies $\mu_{N-k-1}(x) \in U(x)$ for every $x \in S, k = 0, \ldots, N-1$ (Proposition 7.50). Apply Lemma 8.7.   Q.E.D.

Corollary 8.5.1 $(F^+)(F^-)$ If $U(x)$ is a finite set for each $x \in S$, then a uniformly N-stage optimal nonrandomized Markov policy exists.

Corollary 8.5.2 $(F^+)(F^-)$ If for each $x \in S, \lambda \in R$, and $k = 0, \ldots, N-1$, the set

$$U_k(x, \lambda) = \left\{ u \in U(x) \,\middle|\, g(x, u) + \alpha \int J_k^*(x')t(dx'|x, u) \le \lambda \right\}$$

is compact, then there exists a uniformly N-stage optimal nonrandomized Markov policy.

Proof   Apply Lemma 3.1 to Proposition 8.5.   Q.E.D.

8.3   The Semicontinuous Models

Along the lines of our development of lower and upper semicontinuous functions in Section 7.5, we can consider lower and upper semicontinuous decision models. Our models will be designed to take advantage of the possibility for Borel-measurable selection (Propositions 7.33 and 7.34), and in the case of lower semicontinuity, the attainment of the infimum in (55) of Chapter 7. We discuss the lower semicontinuous model first.

Definition 8.7   The *lower semicontinuous, finite horizon, stochastic, optimal control model* is a nine-tuple $(S, C, U, W, p, f, \alpha, g, N)$ as given in Definition 8.1 which has the following additional properties:

(a)   The control space C is compact.
(b)   The set Γ defined by (1) has the form $\Gamma = \bigcup_{j=1}^{\infty} \Gamma^j$, where $\Gamma^1 \subset \Gamma^2 \subset \cdots$, each Γ^j is a closed subset of SC, and

$$\lim_{j \to \infty} \inf_{(x, u) \in \Gamma^j - \Gamma^{j-1}} g(x, u) = \infty.^\dagger$$

(c)   The disturbance kernel $p(dw|x, u)$ is continuous on Γ.

---

† By convention, the infimum over the empty set is ∞, so this condition is satisfied if the Γ^j are all identical for j larger than some index k.

(d) The system function f is continuous on ΓW.

(e) The one-stage cost function g is lower semicontinuous and bounded below on Γ.

Conditions (c) and (d) of Definition 8.7 and Proposition 7.30 imply that $t(dx'|x, u)$ defined by (3) is continuous on Γ, since for any $h \in C(S)$ we have

$$\int h(x')t(dx'|x, u) = \int h[f(x, u, w)]p(dw|x, u).$$

Condition (e) implies that the (F^+) assumption holds.

Proposition 8.6 Consider the lower semicontinuous finite horizon model of Definition 8.7. For $k = 1, 2, \ldots, N$, the k-stage optimal cost function J_k^* is lower semicontinuous and bounded below, and $J_k^* = T^k(J_0)$. Furthermore, a Borel-measurable, uniformly N-stage optimal, nonrandomized Markov policy exists.

Proof Suppose $J: S \to R^*$ is lower semicontinuous and bounded below, and define $K: \Gamma \to R^*$ by

$$K(x, u) = g(x, u) + \alpha \int J(x')t(dx'|x, u). \tag{42}$$

Extend K to all of SC by defining

$$\hat{K}(x, u) = \begin{cases} K(x, u) & \text{if } (x, u) \in \Gamma, \\ \infty & \text{if } (x, u) \notin \Gamma. \end{cases}$$

By Proposition 7.31(a) and the remarks following Lemma 7.13, the function K is lower semicontinuous on Γ. For $c \in R$, the set $\{(x, u) \in SC | \hat{K}(x, u) \le c\}$ must be contained in some Γ^k by Definition 7.8(b), so the set

$$\{(x, u) \in SC | \hat{K}(x, u) \le c\} = \{(x, u) \in \Gamma^k | K(x, u) \le c\}$$

is closed in Γ^k and thus closed in SC as well. It follows that $\hat{K}(x, u)$ is lower semicontinous and bounded below on SC and, by Proposition 7.32, the function

$$T(J)(x) = \inf_{u \in C} \hat{K}(x, u) \tag{43}$$

is as well. In fact, Proposition 7.33 states that the infimum in (43) is achieved for every $x \in S$, and there exists a Borel-measurable $\varphi: S \to C$ such that

$$T(J)(x) = \hat{K}[x, \varphi(x)] \qquad \forall x \in S.$$

For $j = 1, 2, \ldots$, let $\varphi_j: \text{proj}_S(\Gamma^j) \to C$ be a Borel-measurable function with graph in Γ^j. (Set $D = \Gamma^j$ in Proposition 7.33 to establish the existence of such a function.) Define $\mu: S \to C$ so that $\mu(x) = \varphi(x)$ if $T(J)(x) < \infty$, $\mu(x) = \varphi_1(x)$ if $T(J)(x) = \infty$ and $x \in \text{proj}_S(\Gamma^1)$; and for $j = 2, 3, \ldots$, define $\mu(x) = \varphi_j(x)$ if

$T(J)(x) = \infty$ and $x \in \text{proj}_S(\Gamma^j) - \text{proj}_S(\Gamma^{j-1})$. Then μ is Borel-measurable, $\mu(x) \in U(x)$ for every $x \in S$, and $T_\mu(J) = T(J)$.

Since $J_0 \equiv 0$ is lower semicontinuous and bounded below, the above argument shows that $J_k^* = T^k(J_0)$ has these properties also, and furthermore, for each $k = 0, \ldots, N - 1$, there exists a Borel-measurable $\mu_k : S \to C$ such that $\mu_k(x) \in U(x)$ for every $x \in S$ and $(T_{\mu_k} T^{N-k-1})(J_0) = T^{N-k}(J_0)$. The proposition follows from Lemma 8.7.   Q.E.D.

We note that although condition (a) of Definition 8.7 requires the compactness of C, the conclusion of Proposition 8.6 still holds if C is not compact but can be homeomorphically embedded in a compact space $\hat{C}$ in such a way that the image of $\Gamma^j, j = 1, 2, \ldots$, is closed in $S\hat{C}$. That is to say, the conclusion holds if there is a compact space $\hat{C}$ and a homeomorphism $\varphi : C \to \hat{C}$ such that for $j = 1, 2, \ldots$, $\Phi(\Gamma^j)$ is closed in $S\hat{C}$, where

$$\Phi(x, u) = (x, \varphi(u)).$$

The continuity of f and $p(dw|x, u)$ and the lower semicontinuity of g are unaffected by this embedding. In particular, if Γ^j is compact for each j, we can take $\hat{C} = \mathcal{H}$ and use Urysohn's theorem (Proposition 7.2) and the fact that the continuous image of a compact set is compact to accomplish this transformation. We state this last result as a corollary.

Corollary 8.6.1   The conclusions of Proposition 8.6 hold if instead of assuming that C is compact and each Γ^j is closed in Definition 8.7, we assume that each Γ^j is compact.

Definition 8.8   The *upper semicontinuous, finite horizon, stochastic, optimal control model* is a nine-tuple $(S, C, U, W, p, f, \alpha, g, N)$ as given in Definition 8.1 which has the following additional properties:

(a)   The set Γ defined by (1) is open in SC.
(b)   The disturbance kernel $p(dw|x, u)$ is continuous on Γ.
(c)   The system function f is continuous on ΓW
(d)   The one-stage cost g is upper semicontinuous and bounded above on Γ.

As in the lower semicontinuous model, the stochastic kernel $t(dx'|x, u)$ is continuous in the upper semicontinuous model. In the upper semicontinuous model, the (F^-) assumption holds. If $J : S \to R^*$ is upper semicontinuous and bounded above, then $K : \Gamma \to R^*$ defined by (42) is upper semicontinuous and bounded above. By Proposition 7.34, the function

$$T(J)(x) = \inf_{u \in U(x)} K(x, u)$$

is upper semicontinuous, and for every $\varepsilon > 0$ there exists a Borel-measurable

$\mu: S \to C$ such that $\mu(x) \in U(x)$ for every $x \in S$, and

$$T_\mu(J)(x) \leq \begin{cases} T(J)(x) + \varepsilon & \text{if} \quad T(J)(x) > -\infty \\ -1/\varepsilon & \text{if} \quad T(J)(x) = -\infty. \end{cases}$$

Since $J_0 \equiv 0$ is upper semicontinuous and bounded above, so is $J_k^* = T^k(J_0)$, $k = 1, 2, \ldots, N$. The following proposition is obtained by using these facts to parallel the proof of the (F$^-$) part of Proposition 8.3.

Proposition 8.7   Consider the upper semicontinuous finite horizon model of Definition 8.8. For $k = 1, 2, \ldots, N$, the k-stage optimal cost function J_k^* is upper semicontinuous and bounded above, and $J_k^* = T^k(J_0)$. For each $\varepsilon > 0$, there exists a Borel-measurable, nonrandomized, semi-Markov, ε-optimal policy and a Borel-measurable, (randomized) Markov, ε-optimal policy.

Actually, it is not necessary that S and C be Borel spaces for Proposition 8.7 to hold. Assuming only that S and C are separable metrizable spaces, one can use the results on upper semicontinuity of Section 7.5 and the other assumptions of the upper semicontinuous model to prove the conclusion of Proposition 8.7.

It is not possible to parallel the proof of Proposition 8.4 to show for the upper semicontinuous model that given a sequence of positive numbers $\{\varepsilon_n\}$ with $\varepsilon_n \downarrow 0$, a sequence of Borel-measurable, nonrandomized, Markov policies exhibiting $\{\varepsilon_n\}$-dominated convergence to optimality exists. The set $A(J_{N-1}^*)$ defined by (33) may not be open, so the proof breaks down when one is restricted to Borel-measurable policies.

We conclude this section by pointing out one important case when the disturbance kernel $p(dw|x, u)$ is continuous. If W is n-dimensional Euclidean space and the distribution of w is given by a density $d(w|x, u)$ which is jointly continuous in (x, u) for fixed w, then $p(dw|x, u)$ is continuous. To see this, let G be an open set in W and let $(x_n, u_n) \to (x, u)$ in SC. Then

$$\liminf_{k \to \infty} p(G|x_k, u_k) = \liminf_{k \to \infty} \int_G d(w|x_k, u_k)\, dw$$

$$\geq \int_G d(w|x, u)\, dw = p(G|x, u)$$

by Fatou's lemma. The continuity of $p(dw|x, u)$ follows from Proposition 7.21.[†]

[†] Note that by the same argument,

$$\liminf_{k \to \infty} p(G^c|x_k, u_k) \geq p(G^c|x, u),$$

so $p(G|x_k, u_k) \to p(G|x, u)$. Under this condition, the assumption that the system function is continuous in the state (Definitions 8.7(d) and 8.8(c)) can be weakened. See [H3] and [S5].

In fact, it is not necessary that d be continuous in (x, u) for each w, but only that $(x_n, u_n) \to (x, u)$ imply $d(w|x_n, u_n) \to d(w|x, u)$ for Lebesgue almost all w. For example, if $W = R$, the exponential density

$$d(w|x, u) = \begin{cases} \exp[-(w - m(x, u))] & \text{if} \quad w \geq m(x, u), \\ 0 & \text{if} \quad w < m(x, u), \end{cases}$$

where $m: SC \to R$ is continuous, has this property, but need not be continuous in (x, u) for any $w \in R$.

Chapter 9

The Infinite Horizon Borel Models

A first approach to the analysis of the infinite horizon decision model is to treat it as the limit of the finite horizon model as the horizon tends to infinity. In the case (N) of a nonpositive cost per stage and the case (D) of bounded cost per stage and discount factor less than one, this procedure has merit. However, in the case (P) of nonnegative cost per stage, the finite horizon optimal cost functions can fail to converge to the infinite horizon optimal cost function (Example 1 in this chapter), and this failure to converge can occur in such a way that each finite horizon optimal cost function is Borel-measurable, while the infinite horizon optimal cost function is not (Example 2). We thus must develop an independent line of analysis for the infinite horizon model. Our strategy is to define two models, a stochastic one and its deterministic equivalent. There are no measurability restrictions on policies in the deterministic model, and the theory of Part I or of Bertsekas [B4], hereafter abbreviated DPSC, can be applied to it directly. We then transfer this theory to the stochastic model. Sections 9.1–9.3 set up the two models and establish their relationship. Sections 9.4–9.6 analyze the stochastic model via its deterministic counterpart.

9.1 The Stochastic Model

Definition 9.1 An *infinite horizon stochastic optimal control model*, denoted by (SM), is an eight-tuple $(S, C, U, W, p, f, \alpha, g)$ as described in

Definition 8.1. We consider three cases, where Γ is defined by (1) of Chapter 8:

(P) $0 \le g(x, u)$ for every $(x, u) \in \Gamma$.
(N) $g(x, u) \le 0$ for every $(x, u) \in \Gamma$.
(D) $0 < \alpha < 1$, and for some $b \in R$, $-b \le g(x, u) \le b$ for every $(x, u) \in \Gamma$.

Thus we are really treating three models: (P), (N), and (D). If a result is applicable to one of these models, the corresponding symbol will appear. The assumptions (P), (N), and (D) replace the (F$^+$) and (F$^-$) conditions of Chapter 8.

Definition 9.2   A *policy* for (SM) is a sequence $\pi = (\mu_0, \mu_1, \ldots)$ such that for each k, $\mu_k(du_k | x_0, u_0, \ldots, u_{k-1}, x_k)$ is a universally measurable stochastic kernel on C given $SC \cdots CS$ satisfying

$$\mu_k(U(x_k) | x_0, u_0, \ldots, u_{k-1}, x_k) = 1$$

for every $(x_0, u_0, \ldots, u_{k-1}, x_k)$. The concepts of *semi-Markov, Markov, non-randomized*, and *$\mathscr{F}$-measurable policies* are the same as in Definition 8.2. *We denote by Π' the set of all policies for (SM) and by Π the set of all Markov policies. If π is a Markov policy of the form $\pi = (\mu, \mu, \ldots)$, it is said to be stationary.*

As in Chapter 8, we often index S and C for clarity, understanding S_k to be a copy of S and C_k to be a copy of C. Suppose $p \in P(S)$ and $\pi = (\mu_0, \mu_1, \ldots)$ is a policy for (SM). By Proposition 7.45, there is a sequence of unique probability measures $r_N(\pi, p)$ on $S_0 C_0 \cdots S_{N-1} C_{N-1}$, $N = 1, 2, \ldots$, such that for any N and any universally measurable function $h : S_0 C_0 \cdots S_{N-1} C_{N-1} \to R^*$ which satisfies either $\int h^+ \, dr_N(\pi, p) < \infty$ or $\int h^- \, dr_N(\pi, p) < \infty$, (4) of Chapter 8 is satisfied. Furthermore, there exists a unique probability measure $r(\pi, p)$ on $S_0 C_0 S_1 C_1 \cdots$ such that for each N the marginal of $r(\pi, p)$ on $S_0 C_0 \cdots S_{N-1} C_{N-1}$ is $r_N(\pi, p)$. With $r_N(\pi, p)$ and $r(\pi, p)$ determined in this manner, we are ready to define the cost corresponding to a policy.

Definition 9.3   Suppose π is a policy for (SM). The (infinite horizon) *cost corresponding to π at $x \in S$ is*

$$J_\pi(x) = \int \left[\sum_{k=0}^{\infty} \alpha^k g(x_k, u_k) \right] dr(\pi, p_x)$$

$$= \sum_{k=0}^{\infty} \alpha^k \int g(x_k, u_k) \, dr_k(\pi, p_x).^\dagger \qquad (1)$$

---

† The interchange of integration and summation is justified by appeal to the monotone convergence theorem under (P) and (N), and the bounded convergence theorem under (D).

If $\pi = (\mu, \mu, \ldots)$ is stationary, we sometimes write J_μ in place of J_π. The (infinite horizon) *optimal cost* at x is

$$J^*(x) = \inf_{\pi \in \Pi'} J_\pi(x). \tag{2}$$

If $\varepsilon > 0$, the policy π is ε-*optimal at* x provided

$$J_\pi(x) \leq \begin{cases} J^*(x) + \varepsilon & \text{if } J^*(x) > -\infty, \\ -1/\varepsilon & \text{if } J^*(x) = -\infty. \end{cases}$$

If $J_\pi(x) = J^*(x)$, then π is *optimal at* x. If π is ε-optimal or optimal at every $x \in S$, it is said to be ε-*optimal* or *optimal*, respectively.

It is easy to see, using Propositions 7.45 and 7.46, that, for any policy π, $J_\pi(x)$ is universally measurable in x. In fact, if $\pi = (\mu_0, \mu_1, \ldots)$ and $\pi^k = (\mu_0, \ldots, \mu_{k-1})$, then $J_{k,\pi^k}(x)$ defined by (5) of Chapter 8 is universally measurable in x and

$$\lim_{k \to \infty} J_{k,\pi^k}(x) = J_\pi(x) \qquad \forall x \in S. \tag{3}$$

If π is Markov, then (3) can be rewritten in terms of the operators T_{μ_k} of Definition 8.4 as

$$\lim_{k \to \infty} (T_{\mu_0} \cdots T_{\mu_{k-1}})(J_0)(x) = J_\pi(x) \qquad \forall x \in S, \tag{4}$$

which is the infinite horizon analog of Lemma 8.1. If π is a Borel-measurable policy and g is Borel-measurable, then $J_\pi(x)$ is Borel-measurable in x (Proposition 7.29).

It may occur under (P), however, that $\lim_{k \to \infty} J_k^*(x) \neq J^*(x)$, where $J_k^*(x)$ is the optimal k-stage cost defined by (6) of Chapter 8. We offer an example of this.

EXAMPLE 1   Let $S = \{0, 1, 2, \ldots\}$, $C = \{1, 2, \ldots\}$, $U(x) = C$ for every $x \in S$, $\alpha = 1$,

$$f(x, u) = \begin{cases} u & \text{if } x = 0, \\ x - 1 & \text{if } x \neq 0, \end{cases} \qquad g(x, u) = \begin{cases} 1 & \text{if } x = 1, \\ 0 & \text{if } x \neq 1. \end{cases}$$

The problem is deterministic, so the choice of W and $p(dw|x, u)$ is irrelevant. Beginning at $x_0 = 0$, the system moves to some positive integer u_0 at no cost. It then successively moves to $u_0 - 1, u_0 - 2, \ldots$, until it returns to zero and the process begins again. The only transition which incurs a nonzero cost is the transition from one to zero. If the horizon k is finite and u_0 is chosen larger than k, then no cost is incurred before termination, so $J_k^*(0) = 0$. Over the infinite horizon, the transition from one to zero will be made infinitely often, regardless of the policy employed, so $J^*(0) = \infty$.

For $\pi = (\mu_0, \mu_1, \ldots) \in \Pi'$ and $p \in P(S)$, let $q_k(\pi, p)$ be the marginal of $r(\pi, p)$ on $S_k C_k$, $k = 0, 1, \ldots$. Then (7) of Chapter 8 holds, and if π is Markov, (8) holds as well. Furthermore, from (1) we have

$$J_\pi(x) = \sum_{k=0}^{\infty} \alpha^k \int_{S_k C_k} g \, dq_k(\pi, p_x) \qquad \forall x \in S, \tag{5}$$

which is the infinite horizon analog of (9) of Chapter 8. Using these facts to parallel the proof of Proposition 8.1, we obtain the following infinite horizon version.

Proposition 9.1 (P)(N)(D)   If $x \in S$ and $\pi' \in \Pi'$, then there is a Markov policy π such that

$$J_\pi(x) = J_{\pi'}(x).$$

Corollary 9.1.1 (P)(N)(D)   We have

$$J^*(x) = \inf_{\pi \in \Pi} J_\pi(x) \qquad \forall x \in S,$$

where Π is the set of all Markov policies for (SM).

9.2   The Deterministic Model

Definition 9.4   Let $(S, C, U, W, p, f, \alpha, g)$ be an infinite horizon stochastic optimal control model as given by Definition 9.1. The corresponding *infinite horizon deterministic optimal control model*, denoted by (DM), consists of the following:

$P(S)$ *State space.*
$P(SC)$ *Control space.*
$\bar{U}$ *Control constraint.* A function from $P(S)$ to the set of nonempty subsets of $P(SC)$ defined for each $p \in P(S)$ by

$$\bar{U}(p) = \{q \in P(SC) | q(\Gamma) = 1 \text{ and the marginal of } q \text{ on } S \text{ is } p\}, \tag{6}$$

where Γ is given by (1) of Chapter 8.
$\bar{f}$ *System function.* The function from $P(SC)$ to $P(S)$ defined by

$$\bar{f}(q)(\underline{S}) = \int_{SC} t(\underline{S} | x, u) q(d(x, u)) \qquad \forall \underline{S} \in \mathcal{B}_S, \tag{7}$$

where $t(dx' | x, u)$ is given by (3) of Chapter 8.
α *Discount factor.*
$\bar{g}$ *One-stage cost function.* The function from $P(SC)$ to R^* given by

$$\bar{g}(q) = \int_{SC} g(x, u) q(d(x, u)). \tag{8}$$

The model (DM) inherits considerable regularity from (SM). Its state and control spaces $P(S)$ and $P(SC)$ are Borel spaces (Corollary 7.25.1). The system function $\bar{f}$ is Borel-measurable (Proposition 7.26 and Corollary 7.29.1), and the one-stage cost function $\bar{g}$ is lower semianalytic (Corollary 7.48.1). Furthermore, under assumption (P) in (SM), we have $\bar{g} \geq 0$, while under (N), $\bar{g} \leq 0$, and under (D), $0 < \alpha < 1$ and $-b \leq \bar{g} \leq b$.

Definition 9.5 A *policy* for (DM) is a sequence of mappings $\bar{\pi} = (\bar{\mu}_0, \bar{\mu}_1, \ldots)$ such that for each k, $\bar{\mu}_k : P(S) \to P(SC)$ and $\bar{\mu}_k(p) \in \bar{U}(p)$ for every $p \in P(S)$. The set of all policies in (DM) will be denoted by $\bar{\Pi}$. We place no measurability requirements on these mappings. A policy $\bar{\pi}$ of the form $\bar{\pi} = (\bar{\mu}, \bar{\mu}, \ldots)$ is said to be *stationary*.

Definition 9.6 Given $p_0 \in P(S)$ and a policy $\bar{\pi} = (\bar{\mu}_0, \bar{\mu}_1, \ldots)$ for (DM), the *cost corresponding to* $\bar{\pi}$ *at* p_0 is

$$\bar{J}_{\bar{\pi}}(p_0) = \sum_{k=0}^{\infty} \alpha^k \bar{g}(q_k), \tag{9}$$

where the control sequence $\{q_k\}$ is generated recursively by means of the equation

$$q_k = \bar{\mu}_k(p_k), \qquad k = 0, 1, \ldots, \tag{10}$$

and the system equation

$$p_{k+1} = \bar{f}(q_k), \qquad k = 0, 1, \ldots. \tag{11}$$

If $\bar{\pi} = (\bar{\mu}, \bar{\mu}, \ldots)$ is stationary, we write $\bar{J}_{\bar{\mu}}$ in place of $\bar{J}_{\bar{\pi}}$. The *optimal cost* at p_0 is

$$\bar{J}^*(p_0) = \inf_{\bar{\pi} \in \bar{\Pi}} \bar{J}_{\bar{\pi}}(p_0).$$

The concepts of *ε-optimal* and *optimal policies* for (DM) are the same as those given in Definition 9.3 for (SM).

Definition 9.7 A sequence $(p_0, q_0, q_1, \ldots) \in P(S)P(SC)P(SC) \cdots$ is *admissible in* (DM) if $q_0 \in \bar{U}(p_0)$ and $q_{k+1} \in \bar{U}[\bar{f}(q_k)]$, $k = 0, 1, \ldots$. The set of all admissible sequences will be denoted by Δ.

The admissible sequences are just the sequences of controls $q_0, q_1, \ldots$ together with the initial state p_0 which can be generated by some policy for (DM) via (10) and (11). Except for p_0, the measures p_k are not included in the sequence, but can be recovered as the marginals of the measures q_k on S [cf. (6)].

Definition 9.8 Let $\bar{J} : P(S) \to R^*$ be given and let $\bar{\mu} : P(S) \to P(SC)$ be such that $\bar{\mu}(p) \in \bar{U}(p)$ for every $p \in P(S)$. The *operator* $\bar{T}_{\bar{\mu}}$ mapping $\bar{J}$ into

$\overline{T}_{\overline{\mu}}(\overline{J}):P(S) \rightarrow R^*$ is defined by

$$\overline{T}_{\overline{\mu}}(\overline{J})(p) = \overline{g}[\overline{\mu}(p)] + \alpha\overline{J}[\overline{f}(\overline{\mu}(p))] \qquad \forall p \in P(S).$$

The *operator* $\overline{T}$ mapping $\overline{J}$ into $\overline{T}(\overline{J}):P(S) \rightarrow R^*$ is defined by

$$\overline{T}(\overline{J})(p) = \inf_{q \in \overline{U}(p)} \{\overline{g}(q) + \alpha\overline{J}[\overline{f}(q)]\} \qquad \forall p \in P(S).$$

Because (DM) is deterministic, it can be studied using results from Part I, Chapters 4 and 5 or from DPSC. This is because there is no need to place measurability restrictions on policies in a deterministic model. The operators $\overline{T}_{\mu}$ and $\overline{T}$ of Definition 9.8 are special cases of those defined in Section 2.1. In the present case, we take $H(p, q, \overline{J})$ to be

$$H(p, q, \overline{J}) = \overline{g}(q) + \alpha\overline{J}[\overline{f}(q)].$$

The monotonicity assumption of Section 2.1 is satisfied by this choice of H. The cost corresponding to a policy $\overline{\pi} = (\overline{\mu}_0, \overline{\mu}_1, \ldots)$ as given by (9) is easily seen to be of the form (cf. Section 2.2)

$$\overline{J}_{\overline{\pi}} = \lim_{N \to \infty} (\overline{T}_{\overline{\mu}_0} \cdots \overline{T}_{\overline{\mu}_{N-1}})(\overline{J}_0),$$

where $\overline{J}_0(p) = 0$ for every $p \in P(S)$. It is a straightforward matter to verify that under (D) the contraction assumption of Section 4.1 is satisfied when $\overline{B}$ is taken to be the set of bounded real-valued functions on $P(S)$, m is taken to be one, and $\rho = \alpha$. Under (P), Assumptions I, I.1, and I.2 of Section 5.1 are satisfied, while under (N), Assumptions D, D.1, and D.2 of the same section are in force.

9.3   Relations between the Models

Definition 9.9   Let $\pi = (\mu_0, \mu_1, \ldots) \in \Pi$ be a Markov policy for (SM) and $\overline{\pi} = (\overline{\mu}_0, \overline{\mu}_1, \ldots) \in \overline{\Pi}$ a policy for (DM). Let $p_0 \in P(S)$ be given. If for all k

$$\int_{\underline{S}} \mu_k(\underline{C}|x)p_k(dx) = \overline{\mu}_k(p_k)(\underline{SC}) \qquad \forall \underline{S} \in \mathscr{B}_S, \quad \underline{C} \in \mathscr{B}_C, \qquad (12)$$

where p_k is generated from p_0 by $\overline{\pi}$ via (10) and (11), then π and $\overline{\pi}$ are said to *correspond at* p_0. If π and $\overline{\pi}$ correspond at every $p \in P(S)$, then π and $\overline{\pi}$ are said to *correspond*.

If π and $\overline{\pi}$ correspond at p_0, then the sequence of measures $[q_0(\pi, p_0),$ $q_1(\pi, p_0), \ldots]$ generated from p_0 by π via (8) of Chapter 8 is the same as the sequence $(q_0, q_1, \ldots)$ generated from p_0 by $\overline{\pi}$ via (10) and (11). If π and $\overline{\pi}$ correspond, then they generate the same sequence $(q_0, q_1, \ldots)$ for any initial p_0.

Proposition 9.2 (P)(N)(D)   Given a Markov policy $\pi \in \Pi$, there is a corresponding $\bar{\pi} \in \bar{\Pi}$. If $\bar{\pi} \in \bar{\Pi}$ and $p_0 \in P(S)$ are given, then there is a Markov policy $\pi \in \Pi$ corresponding to $\bar{\pi}$ at p_0.

Proof If $\pi = (\mu_0, \mu_1, \ldots) \in \Pi$ is given, then for each k and any $p_k \in P(S)$, there is a unique probability measure on SC, which we denote by $\bar{\mu}_k(p_k)$, satisfying (12) (Proposition 7.45). Furthermore,

$$\bar{\mu}_k(p_k)(\Gamma) = \int_S \mu_k(U(x)|x) p_k(dx) = 1, \qquad (13)$$

so $\bar{\pi} = (\bar{\mu}_0, \bar{\mu}_1, \ldots)$ is in $\bar{\Pi}$ and corresponds to π. If $\bar{\pi} = (\bar{\mu}_0, \bar{\mu}_1, \ldots) \in \bar{\Pi}$ and $p_0 \in P(S)$ are given, let $(p_0, p_1, p_2, \ldots)$ be generated from p_0 by $\bar{\pi}$ via (10) and (11). For each k, choose a Borel-measurable stochastic kernel $\mu_k(du|x)$ which satisfies (12) for this particular p_k (Corollary 7.27.2). Then (13) holds, so

$$\mu_k(U(x)|x) = 1 \qquad (14)$$

for p_k almost every x. Altering $\mu_k(du|x)$ on a set of p_k-measure zero if necessary, we may assume that (14) holds for every $x \in S$ and (12) is satisfied. Then $\pi = (\mu_0, \mu_1, \ldots) \in \Pi$ corresponds to $\bar{\pi}$ at p_0.   Q.E.D.

Proposition 9.3 (P)(N)(D)   Let $p \in P(S)$, $\pi \in \Pi$, and $\bar{\pi} \in \bar{\Pi}$ be given. If π and $\bar{\pi}$ correspond at p, then

$$\bar{J}_{\bar{\pi}}(p) = \int J_\pi(x) p(dx).$$

Proof We have from (7) of Chapter 8, (5), (8), (9), and the monotone or bounded convergence theorems

$$\int J_\pi(x) p(dx) = \int_S \left[\sum_{k=0}^{\infty} \alpha^k \int_{S_k C_k} g \, dq_k(\pi, p_x) \right] p(dx)$$

$$= \sum_{k=0}^{\infty} \alpha^k \int_S \int_{S_k C_k} g \, dq_k(\pi, p_x) p(dx)$$

$$= \sum_{k=0}^{\infty} \alpha^k \int_{S_k C_k} g \, dq_k(\pi, p)$$

$$= \sum_{k=0}^{\infty} \alpha^k \bar{g}[q_k(\pi, p)]$$

$$= \bar{J}_{\bar{\pi}}(p).   \text{Q.E.D.}$$

Corollary 9.3.1 (P)(N)(D)   Let $x \in S$, $\pi \in \Pi$, and $\bar{\pi} \in \bar{\Pi}$ be given. If π and $\bar{\pi}$ correspond at p_x, then

$$\bar{J}_{\bar{\pi}}(p_x) = J_\pi(x).$$

Corollary 9.3.2   (P)(N)(D)   For every $x \in S$,

$$\bar{J}^*(p_x) = J^*(x).$$

Proof   Corollaries 9.1.1, 9.3.1, and Proposition 9.2 imply that, for every $x \in S$,

$$\bar{J}^*(p_x) = \inf_{\bar{\pi} \in \bar{\Pi}} \bar{J}_{\bar{\pi}}(p_x) = \inf_{\pi \in \Pi} J_\pi(x) = J^*(x). \qquad \text{Q.E.D.}$$

Corollary 9.3.2 shows that J^* and $\bar{J}^*$ are related, but in a rather weak way that involves $\bar{J}^*$ only on $\bar{S} = \{p_x \in P(S) | x \in S\}$. In Proposition 9.5 we strengthen this relationship, but in order to state that proposition we must show a measurability property of J^*. This is the subject of Proposition 9.4, which we prove with the aid of the following lemma.

Lemma 9.1   The set Δ of admissible sequences in (DM) is an analytic subset of $P(S)P(SC)P(SC) \cdots$.

Proof   The set Δ is equal to $A_0 \cap [\bigcap_{k=0}^\infty B_k]$, where

$$A_0 = \{(p_0, q_0, q_1, \ldots) | q_0 \in \bar{U}(p_0)\},$$
$$B_k = \{(p_0, q_0, q_1, \ldots) | q_{k+1} \in \bar{U}[\bar{f}(q_k)]\}.$$

By Corollary 7.35.2, it suffices to show that A_0 and B_k, $k = 0, 1, \ldots$, are analytic. Using the result of Proposition 7.38, this will follow if we show that

$$A = \{(p_1, q_1) \in P(S)P(SC) | q_1 \in \bar{U}(p_1)\},$$
$$B = \{(q_0, q_1) \in P(SC)P(SC) | q_1 \in \bar{U}[\bar{f}(q_0)]\}$$

are analytic. Let $P(\Gamma) = \{q \in P(SC) | q(\Gamma) = 1\}$, where Γ is given by (1) of Chapter 8. Then $P(\Gamma)$ is analytic (Proposition 7.43). Equation (6) implies that A is the intersection of the analytic set $P(S)P(\Gamma)$ (Proposition 7.38) with the graph of the function $\sigma: P(SC) \to P(S)$ which maps q into its marginal on S. It is easily verified that σ is continuous (Proposition 7.21(a) and (b)), so $\text{Gr}(\sigma)$ is Borel (Corollary 7.14.1). Therefore, A is analytic. The set B is the inverse image of A under the Borel-measurable mapping $(q_0, q_1) \to [\bar{f}(q_0), q_1]$, so is also analytic (Proposition 7.40).   Q.E.D.

Proposition 9.4   (P)(N)(D)   The function $\bar{J}^*: P(S) \to R^*$ is lower semi-analytic.

Proof   Define $G: \Delta \to R^*$ by

$$G(p_0, q_0, q_1, \ldots) = \sum_{k=0}^\infty \alpha^k \bar{g}(q_k), \qquad (15)$$

where Δ is the set of admissible sequences (Definition 9.7). Then G is lower semianalytic by Lemma 7.30(2), (4) and Lemma 9.1. By the definition of $\bar{J}^*$ and Δ, we have

$$\bar{J}^*(p_0) = \inf_{(q_0, q_1, \ldots) \in \Delta_{p_0}} G(p_0, q_0, q_1, \ldots) \qquad \forall p_0 \in P(S), \tag{16}$$

so $\bar{J}^*$ is lower semianalytic by Proposition 7.47. Q.E.D.

Corollary 9.4.1 (P)(N)(D) The function $J^*: S \to R^*$ is lower semi-analytic.

Proof By Corollary 9.3.2,

$$J^*(x) = \bar{J}^*[\delta(x)] \qquad \forall x \in S,$$

where $\delta(x) = p_x$ is the homeomorphism defined in Corollary 7.21.1. Apply Lemma 7.30(3) and Proposition 9.4 to conclude that J^* is lower semianalytic. Q.E.D.

Lemma 9.2 Given $p \in P(S)$ and $\varepsilon > 0$, there exists a policy $\bar{\pi}$ for (DM) such that

(P)(D) $\qquad \bar{J}_{\bar{\pi}}(p) \leq \int J^*(x)p(dx) + \varepsilon,$

(N) $\qquad \bar{J}_{\bar{\pi}}(p) \leq \begin{cases} \int J^*(x)p(dx) + \varepsilon & \text{if } \int J^*(x)p(dx) > -\infty, \\ -1/\varepsilon & \text{if } \int J^*(x)p(dx) = -\infty. \end{cases}$

Proof As a consequence of Corollary 9.4.1, $\int J^*(x)p(dx)$ is well defined. Let $p \in P(S)$ and $\varepsilon > 0$ be given. Let $G: \Delta \to R^*$ be defined by (15). Proposition 7.50 guarantees that under (P) and (D) there exists a universally measurable selector $\varphi: P(S) \to P(SC)P(SC) \cdots$ such that $(p, \varphi(p)) \in \Delta$ for every $p \in P(S)$ and

$$G[p, \varphi(p)] \leq \bar{J}^*(p) + \varepsilon \qquad \forall p \in P(S).$$

Let $\sigma: S \to P(SC)P(SC) \cdots$ be defined by $\sigma(x) = \varphi(p_x)$. Then σ is universally measurable (Proposition 7.44) and

$$G[p_x, \sigma(x)] \leq J^*(x) + \varepsilon \qquad \forall x \in S. \tag{17}$$

Under (N), there exists a universally measurable $\sigma: S \to P(SC)P(SC) \cdots$ such that for every $x \in S$, $(p_x, \sigma(x)) \in \Delta$ and

$$G[p_x, \sigma(x)] \leq \begin{cases} J^*(x) + \varepsilon & \text{if } J^*(x) > -\infty, \\ -(1 + \varepsilon^2)/\varepsilon p_\infty(J^*) & \text{if } J^*(x) = -\infty, \end{cases} \tag{18}$$

where $p_\infty(J^*) = p(\{x | J^*(x) = -\infty\})$ if $p(\{x | J^*(x) = -\infty\}) > 0$ and $p_\infty(J^*) = 1$ otherwise.

Denote $\sigma(x) = [q_0(d(x_0, u_0)|x), q_1(d(x_1, u_1)|x), \ldots]$. Each $q_k(d(x_k, u_k)|x)$ is a universally measurable stochastic kernel on $S_k C_k$ given S. Furthermore,

$$q_0(d(x_0, u_0)|x) \in \bar{U}(p_x) \qquad \forall x \in S,$$

and, for $k = 0, 1, \ldots,$

$$q_{k+1}(d(x_{k+1}, u_{k+1})|x) \in \bar{U}(\bar{f}[q_k(d(x_k, u_k)|x)]) \qquad \forall x \in S.$$

For $k = 0, 1, \ldots,$ define $\bar{q}_k \in P(SC)$ by

$$\bar{q}_k(B) = \int q_k(B|x)p(dx) \qquad \forall B \in \mathscr{B}_{SC}.$$

Then $\bar{q}_k(\Gamma) = 1, k = 0, 1, \ldots.$ We show that $(p, \bar{q}_0, \bar{q}_1, \ldots) \in \Delta$. Since the marginal of $q_0(d(x_0, u_0)|x)$ on S_0 is p_x, we have

$$\bar{q}_0(\underline{S}_0 C_0) = \int_S q_0(\underline{S}_0 C_0|x)p(dx) = \int_S \chi_{\underline{S}_0}(x)p(dx) = p(\underline{S}_0) \qquad \forall \underline{S}_0 \in \mathscr{B}_S,$$

so $\bar{q}_0 \in \bar{U}(p)$. For $k = 0, 1, \ldots,$ we have

$$\bar{q}_{k+1}(\underline{S}_{k+1} C_{k+1}) = \int_S q_{k+1}(\underline{S}_{k+1} C_{k+1}|x)p(dx)$$

$$= \int_S \int_{S_k C_k} t(\underline{S}_{k+1}|x_k, u_k)q_k(d(x_k, u_k)|x)p(dx)$$

$$= \int_{S_k C_k} t(\underline{S}_{k+1}|x_k, u_k)\bar{q}_k(d(x_k, u_k)) \qquad \forall \underline{S}_{k+1} \in \mathscr{B}_S.$$

Therefore $\bar{q}_{k+1} \in \bar{U}[\bar{f}(\bar{q}_k)]$ and $(p, \bar{q}_0, \bar{q}_1, \ldots) \in \Delta$.

Let $\bar{\pi}$ be any policy for (DM) which generates the admissible sequence $(p, \bar{q}_0, \bar{q}_1, \ldots)$. Then under (P) and (D), we have from (17) and the monotone or bounded convergence theorem

$$\bar{J}_{\bar{\pi}}(p) = G(p, \bar{q}_0, \bar{q}_1, \ldots)$$

$$= \sum_{k=0}^{\infty} \alpha^k \int_{S_k C_k} g(x_k, u_k)\bar{q}_k(d(x_k, u_k))$$

$$= \sum_{k=0}^{\infty} \alpha^k \int_S \int_{S_k C_k} g(x_k, u_k)q_k(d(x_k, u_k)|x)p(dx)$$

$$= \int_S \left[\sum_{k=0}^{\infty} \alpha^k \int_{S_k C_k} g(x_k, u_k)q_k(d(x_k, u_k)|x) \right] p(dx)$$

$$= \int_S G[p_x, \sigma(x)]p(dx)$$

$$\leq \int_S J^*(x)p(dx) + \varepsilon.$$

Under (N), we have from the monotone convergence theorem

$$\bar{J}_{\bar{\pi}}(p) = \int_S G[p_x, \sigma(x)]p(dx). \tag{19}$$

If $p(\{x|J^*(x) = -\infty\}) = 0$, (18) and (19) imply

$$\bar{J}_{\bar{\pi}}(p) \leq \int J^*(x)p(dx) + \varepsilon,$$

where both sides may be $-\infty$. If $p(\{x|J^*(x) = -\infty\}) > 0$, then $\int J^*(x) p(dx) = -\infty$ and we have, from (18) and (19),

$$\bar{J}_{\bar{\pi}}(p) \leq \int_{\{x|J^*(x) > -\alpha\}} [J^*(x) + \varepsilon]p(dx) - (1 + \varepsilon^2)/\varepsilon$$

$$\leq \varepsilon - (1 + \varepsilon^2)/\varepsilon = -1/\varepsilon. \qquad \text{Q.E.D.}$$

Proposition 9.5 (P)(N)(D) For every $p \in P(S)$,

$$\bar{J}^*(p) = \int J^*(x)p(dx).$$

Proof Lemma 9.2 shows that

$$\bar{J}^*(p) \leq \int J^*(x)p(dx) \qquad \forall p \in P(S).$$

For the reverse inequality, let p be in $P(S)$ and let $\bar{\pi}$ be a policy for (DM). There exists a policy $\pi \in \Pi$ corresponding to $\bar{\pi}$ at p (Proposition 9.2), and, by Proposition 9.3,

$$\bar{J}_{\bar{\pi}}(p) = \int J_{\pi}(x)p(dx) \geq \int J^*(x)p(dx).$$

By taking the infimum of the left-hand side over $\bar{\pi} \in \bar{\Pi}$, we obtain the desired result. Q.E.D.

Propositions 9.3 and 9.5 are the key relationships between (SM) and (DM). As an example of their implications, consider the following corollary.

Corollary 9.5.1 (P)(N)(D) Suppose $\pi \in \Pi$ and $\bar{\pi} \in \bar{\Pi}$ are corresponding policies for (SM) and (DM). Then π is optimal if and only if $\bar{\pi}$ is optimal.

Proof If π is optimal, then

$$\bar{J}_{\bar{\pi}}(p) = \int J_{\pi}(x)p(dx) = \int J^*(x)p(dx) = \bar{J}^*(p) \qquad \forall p \in P(S).$$

If $\bar{\pi}$ is optimal, then

$$J_{\pi}(x) = \bar{J}_{\bar{\pi}}(p_x) = \bar{J}^*(p_x) = J^*(x) \qquad \forall x \in S. \qquad \text{Q.E.D.}$$

The next corollary is a technical result needed for Chapter 10.

Corollary 9.5.2   (P)(N)(D)   For every $p \in P(S)$,

$$\int J^*(x)p(dx) = \inf_{\pi \in \Pi} \int J_\pi(x)p(dx).$$

Proof   By Propositions 9.2 and 9.3,

$$\bar{J}^*(p) = \inf_{\pi \in \Pi} \int J_\pi(x)p(dx) \qquad \forall p \in P(S).$$

Apply Proposition 9.5.   Q.E.D.

We now explore the connections between the operators T_μ and $\bar{T}_{\bar{\mu}}$ and the operators T and $\bar{T}$. The first proposition is a direct consequence of the definitions. We leave the verification to the reader.

Proposition 9.6   (P)(N)(D)   Let $J : S \to R^*$ be universally measurable and satisfy $J \geq 0$; $J \leq 0$, or $-c \leq J \leq c$, $c < \infty$, according as (P), (N), or (D) is in force. Let $\bar{J} : P(S) \to R^*$ be defined by

$$\bar{J}(p) = \int J(x)p(dx) \qquad \forall p \in P(S),$$

and suppose $\bar{\mu} : P(S) \to P(SC)$ is of the form

$$\bar{\mu}(p)(\underline{SC}) = \int_{\underline{S}} \mu(\underline{C}|x)p(dx) \qquad \forall \underline{S} \in \mathscr{B}_S, \quad \underline{C} \in \mathscr{B}_C$$

for some $\mu \in U(C|S)^\dagger$. Then $\bar{\mu}(p) \in \bar{U}(p)$ for every $p \in P(S)$, and

$$\bar{T}_{\bar{\mu}}(\bar{J})(p) = \int T_\mu(J)(x)p(dx) \qquad \forall p \in P(S).$$

Proposition 9.7   (P)(N)(D)   Let $J : S \to R^*$ be lower semianalytic and satisfy $J \geq 0$, $J \leq 0$, or $-c \leq J \leq c$, $c < \infty$, according as (P), (N), or (D) is in force. Let $\bar{J} : P(S) \to R^*$ be defined by

$$\bar{J}(p) = \int J(x)p(dx) \qquad \forall p \in P(S). \tag{20}$$

Then

$$\bar{T}(\bar{J})(p) = \int T(J)(x)p(dx) \qquad \forall p \in P(S).$$

Proof   For $p \in P(S)$ and $q \in \bar{U}(p)$ we have

$$\bar{g}(q) + \alpha \bar{J}[\bar{f}(q)] = \int_{SC}[g(x, u) + \alpha \int_S J(x')t(dx'|x, u)]q(d(x, u))$$

$$\geq \int_S T(J)(x)p(dx),$$

---

† The set $U(C|S)$, defined in Section 8.2, is the collection of universally measurable stochastic kernels μ on C given S which satisfy $\mu(U(x)|x) = 1$ for every $x \in S$.

which implies

$$\bar{T}(\bar{J})(p) \geq \int T(J)(x)p(dx).$$

Given $\varepsilon > 0$, Lemma 8.2 implies that there exists $\mu \in U(C|S)$ such that

$$\int_C [g(x,u) + \alpha \int_S J(x')t(dx'|x,u)]\mu(du|x) \leq T(J)(x) + \varepsilon.$$

Let $q \in \bar{U}(p)$ be such that

$$q(\underline{SC}) = \int_{\underline{S}} \mu(\underline{C}|x)p(dx) \qquad \forall \underline{S} \in \mathscr{B}_S, \underline{C} \in \mathscr{B}_C.$$

Then

$$\bar{T}(\bar{J})(p) \leq \int_{SC} \left[g(x,u) + \alpha \int_S J(x')t(dx'|x,u) \right] q(d(x,u))$$

$$= \int_S \int_C \left[g(x,u) + \alpha \int_S J(x')t(dx'|x,u) \right] \mu(du|x)p(dx)$$

$$\leq \int T(J)(x)p(dx) + \varepsilon,$$

where $\int T(J)(x)p(dx) + \varepsilon$ may be $-\infty$. Therefore,

$$\bar{T}(\bar{J})(p) \leq \int T(J)(x)p(dx). \qquad \text{Q.E.D.}$$

9.4   The Optimality Equation—Characterization of Optimal Policies

As noted following Definition 9.8, the model (DM) is a special case of that considered in Part I and DPSC[†]. This allows us to easily obtain many results for both (SM) and (DM). A prime example of this is the next proposition.

Proposition 9.8   (P)(N)(D)   We have

$$\bar{J}^* = \bar{T}(\bar{J}^*), \tag{21}$$

$$J^* = T(J^*). \tag{22}$$

Proof   The optimality equation (21) for (DM) follows from Propositions 4.2(a), 5.2, and 5.3 or from DPSC, Chapter 6, Proposition 2 and Chapter 7,

---

[†] Whereas we allow $\bar{g}$ to be extended real-valued, in Chapter 7 of DPSC the one-stage cost function is assumed to be real-valued. This more restrictive assumption is not essential to any of the results we quote from DPSC.

Proposition 1. We have then, for any $x \in S$,

$$J^*(x) = \bar{J}^*(p_x) = \bar{T}(\bar{J}^*)(p_x) = T(J^*)(x)$$

by Propositions 9.5 and 9.7, so (22) holds as well.   Q.E.D.

Proposition 9.9 (P)(N)(D) If $\bar{\pi} = (\bar{\mu}, \bar{\mu}, \ldots)$ is a stationary policy for (DM), then $\bar{J}_{\bar{\mu}} = \bar{T}_{\bar{\mu}}(\bar{J}_{\bar{\mu}})$. If $\pi = (\mu, \mu, \ldots)$ is a stationary policy for (SM), then $J_\mu = T_\mu(J_\mu)$.

Proof For (DM) this result follows from Proposition 4.2(b), Corollary 5.2.1, and Corollary 5.3.2 or from DPSC, Chapter 6, Corollary 2.1 and Chapter 7, Corollary 1.1. Let $\pi = (\mu, \mu, \ldots)$ be a stationary policy for (SM) and let $\bar{\pi} = (\bar{\mu}, \bar{\mu}, \ldots)$ be a policy for (DM) corresponding to π. Then for each $x \in S$,

$$J_\mu(x) = \bar{J}_{\bar{\mu}}(p_x) = \bar{T}_{\bar{\mu}}(\bar{J}_{\bar{\mu}})(p_x) = T_\mu(J_\mu)(x)$$

by Propositions 9.3 and 9.6.   Q.E.D.

Note that Proposition 9.9 for (SM) cannot be deduced from Proposition 9.8 by considering a modified (SM) with control constraint of the form

$$U_\mu(x) = \{\mu(x)\}   \forall x \in S, \tag{23}$$

as was done in the proof of Corollary 5.2.1. Even if μ is nonrandomized so that (23) makes sense, the set

$$\Gamma_\mu = \{(x, u) | x \in S, u \in U_\mu(x)\}$$

may not be analytic, so U_μ is not an acceptable control constraint.

The optimality equations are necessary conditions for the optimal cost functions, but except in case (D) they are by no means sufficient. We have the following partial sufficiency results.

Proposition 9.10

(P) If $\bar{J} : P(S) \to [0, \infty]$ and $\bar{J} \geq \bar{T}(\bar{J})$, then $\bar{J} \geq \bar{J}^*$.
     If $J : S \to [0, \infty]$ is lower semianalytic and $J \geq T(J)$, then $J \geq J^*$.
(N) If $\bar{J} : P(S) \to [-\infty, 0]$ and $\bar{J} \leq \bar{T}(\bar{J})$, then $\bar{J} \leq \bar{J}^*$.
     If $J : S \to [-\infty, 0]$ is lower semianalytic and $J \leq T(J)$, then $J \leq J^*$.
(D) If $\bar{J} : P(S) \to [-c, c]$, $c < \infty$, and $\bar{J} = \bar{T}(\bar{J})$, then $\bar{J} = \bar{J}^*$.
     If $J : S \to [-c, c]$, $c < \infty$, is lower semianalytic and $J = T(J)$, then $J = J^*$.

Proof We consider first the statements for (DM). The result under (P) follows from Proposition 5.2, the result under (N) from Proposition 5.3, and the result under (D) from Proposition 4.2(a). These results for (DM) follow from Proposition 2 and trivial modifications of the proof of Proposition 9 of DPSC, Chapter 6.

We now establish the (SM) part of the proposition under (P). Cases (N) and (D) are handled in the same manner. Given a lower semianalytic function $J:S \to [0, \infty]$ satisfying $J \geq T(J)$, define $\bar{J}:P(S) \to [0, \infty]$ by (20). Then

$$\bar{J}(p) = \int J(x)p(dx) \geq \int T(J)(x)p(dx) = \bar{T}(\bar{J})(p) \qquad \forall p \in P(S)$$

by Proposition 9.7. By the result for (DM), $\bar{J} \geq \bar{J}^*$. In particular,

$$J(x) = \bar{J}(p_x) \geq \bar{J}^*(p_x) = J^*(x) \qquad \forall x \in S. \qquad \text{Q.E.D.}$$

Proposition 9.11 Let $\bar{\pi} = (\bar{\mu}, \bar{\mu}, \ldots)$ and $\pi = (\mu, \mu, \ldots)$ be stationary policies in (DM) and (SM), respectively.

(P) If $\bar{J}:P(S) \to [0, \infty]$ and $\bar{J} \geq \bar{T}_{\bar{\mu}}(\bar{J})$, then $\bar{J} \geq \bar{J}_{\bar{\mu}}$.
     If $J:S \to [0, \infty]$ is universally measurable and $J \geq T_\mu(J)$, then $J \geq J_\mu$.

(N) If $\bar{J}:P(S) \to [-\infty, 0]$ and $\bar{J} \leq \bar{T}_{\bar{\mu}}(\bar{J})$, then $\bar{J} \leq \bar{J}_{\bar{\mu}}$.
     If $J:S \to [-\infty, 0]$ is universally measurable and $J \leq T_\mu(J)$, then $J \leq J_\mu$.

(D) If $\bar{J}:P(S) \to [-c, c]$, $c < \infty$, and $\bar{J} = \bar{T}_{\bar{\mu}}(\bar{J})$, then $\bar{J} = \bar{J}_{\bar{\mu}}$.
     If $J:S \to [-c, c]$, $c < \infty$, is universally measurable and $J - T_\mu(J)$, then $J = J_\mu$.

Proof The (DM) results follow from Proposition 4.2(b) and Corollaries 5.2.1 and 5.3.2 or from DPSC, Corollary 2.1 and trivial modifications of Corollary 9.1 of Chapter 6. The (SM) results follow from the (DM) results and Proposition 9.6 in a manner similar to the proof of Proposition 9.10. Q.E.D.

Proposition 9.11 implies that under (P), J_μ is the smallest nonnegative universally measurable solution to the functional equation

$$J = T_\mu(J).$$

Under (D), J_μ is the only bounded universally measurable solution to this equation. This provides us with a simple necessary and sufficient condition for a stationary policy to be optimal under (P) and (D).

Proposition 9.12 (P)(D) Let $\bar{\pi} = (\bar{\mu}, \bar{\mu}, \ldots)$ and $\pi = (\mu, \mu, \ldots)$ be stationary policies in (DM) and (SM), respectively. The policy $\bar{\pi}$ is optimal if and only if $\bar{J}^* = \bar{T}_{\bar{\mu}}(\bar{J}^*)$. The policy π is optimal if and only if $J^* = T_\mu(J^*)$.

Proof If $\bar{\pi}$ is optimal, then $\bar{J}_{\bar{\mu}} = \bar{J}^*$. By Proposition 9.9, $\bar{J}^* = \bar{T}_{\bar{\mu}}(\bar{J}^*)$. Conversely, if $\bar{J}^* = \bar{T}_{\bar{\mu}}(\bar{J}^*)$, then, by Proposition 9.11, $\bar{J}^* \geq \bar{J}_{\bar{\mu}}$ and $\bar{\pi}$ is optimal. The proof for (SM) follows from the (SM) parts of the same propositions.   Q.E.D.

Corollary 9.12.1   (P)(D)   There is an optimal nonrandomized stationary policy for (SM) if and only if for each $x \in S$ the infimum in

$$\inf_{u \in U(x)} \left\{ g(x, u) + \alpha \int J^*(x')t(dx'|x, u) \right\} \tag{24}$$

is achieved.

Proof   If the infimum in (24) is achieved for every $x \in S$, then by Proposition 7.50 there is a universally measurable selector $\mu: S \to C$ whose graph lies in Γ and for which

$$g[x, \mu(x)] + \alpha \int J^*(x')t(dx'|x, \mu(x))$$

$$= \inf_{u \in U(x)} \left\{ g(x, u) + \alpha \int J^*(x')t(dx'|x, u) \right\} \qquad \forall x \in S.$$

Then by Proposition 9.8

$$T_\mu(J^*) = T(J^*) = J^*,$$

so $\pi = (\mu, \mu, \ldots)$ is optimal by Proposition 9.12.

If $\pi = (\mu, \mu, \ldots)$ is an optimal nonrandomized stationary policy for (SM), then by Propositions 9.8 and 9.9

$$T_\mu(J^*) = T_\mu(J_\mu) = J_\mu = J^* = T(J^*),$$

so $\mu(x)$ achieves the infimum in (24) for every $x \in S$.   Q.E.D.

In Proposition 9.19, we show that under (P) or (D), the existence of any optimal policy at all implies the existence of an optimal policy that is nonrandomized and stationary. This means that Corollary 9.12.1 actually gives a necessary and sufficient condition for the existence of an optimal policy.

Under (N) we can use Proposition 9.10 to obtain a necessary and sufficient condition for a stationary policy to be optimal. This condition is not as useful as that of Proposition 9.12, however, since it cannot be used to construct a stationary optimal policy in the manner of Corollary 9.12.1.

Proposition 9.13   (N)(D)   Let $\bar{\pi} = (\bar{\mu}, \bar{\mu}, \ldots)$ and $\pi = (\mu, \mu, \ldots)$ be stationary policies in (DM) and (SM), respectively. The policy $\bar{\pi}$ is optimal if and only if $\bar{J}_{\bar{\mu}} = \bar{T}(\bar{J}_{\bar{\mu}})$. The policy π is optimal if and only if $J_\mu = T(J_\mu)$.

Proof   If $\bar{\pi}$ is optimal, then $\bar{J}_{\bar{\mu}} = \bar{J}^*$. By Proposition 9.8

$$\bar{J}_{\bar{\mu}} = \bar{J}^* = \bar{T}(\bar{J}^*) = \bar{T}(\bar{J}_{\bar{\mu}}).$$

Conversely, if $\bar{J}_{\bar{\mu}} = \bar{T}(\bar{J}_{\bar{\mu}})$, then Proposition 9.10 implies that $\bar{J}_{\bar{\mu}} \leq \bar{J}^*$ and $\bar{\pi}$ is optimal.

If π is optimal, $J_\mu = T(J_\mu)$ by the (SM) part of Proposition 9.8. The converse is more difficult, since the (SM) part of Proposition 9.10 cannot be invoked without knowing that J_μ is lower semianalytic. Let $\bar{\pi} = (\bar{\mu}, \bar{\mu}, \ldots)$ be a policy for (DM) corresponding to $\pi = (\mu, \mu, \ldots)$, so that $\bar{J}_{\bar{\mu}}(p) = \int J_\mu(x)p(dx)$ for every $p \in P(S)$. Then for fixed $p \in P(S)$ and $q \in \bar{U}(p)$,

$$\bar{g}(q) + \alpha \bar{J}_{\bar{\mu}}[\bar{f}(q)] = \int_{SC}\left[g(x,u) + \alpha \int_S J_\mu(x')t(dx'|x,u)\right]q(d(x,u))$$

$$\geq \int \inf_{u \in U(x)}\left\{ g(x,u) + \alpha \int_S J_\mu(x')t(dx'|x,u)\right\}p(dx),$$

provided the integrand

$$T(J_\mu)(x) = \inf_{u \in U(x)}\left\{ g(x,u) + \alpha \int_S J_\mu(x')t(dx'|x,u)\right\}$$

is universally measurable in x. But $T(J_\mu) = J_\mu$ by assumption, which is universally measurable, so

$$\bar{g}(q) + \alpha \bar{J}_{\bar{\mu}}[\bar{f}(q)] \geq \int J_\mu(x)p(dx) = \bar{J}_{\bar{\mu}}(p).$$

By taking the infimum of the left-hand side over $q \in \bar{U}(p)$ and using Proposition 9.9, we see that

$$\bar{T}(\bar{J}_{\bar{\mu}}) \geq \bar{J}_{\bar{\mu}} = \bar{T}_{\bar{\mu}}(\bar{J}_{\bar{\mu}}).$$

The reverse inequality always holds, and by the result already proved for (DM), $\bar{\pi}$ is optimal. The optimality of π follows from Corollary 9.5.1.   Q.E.D.

9.5   Convergence of the Dynamic Programming Algorithm— Existence of Stationary Optimal Policies

Definition 9.10   The *dynamic programming algorithm* is defined recursively for (DM) and (SM) by

$$\bar{J}_0(p) = 0 \qquad \forall p \in P(S),$$
$$\bar{J}_{k+1}(p) = \bar{T}(\bar{J}_k)(p) \qquad \forall p \in P(S), \quad k = 0,1,\ldots,$$
$$J_0(x) = 0 \qquad \forall x \in S,$$
$$J_{k+1}(x) = T(J_k)(x) \qquad \forall x \in S, \quad k = 0,1,\ldots.$$

We know from Proposition 8.2 that this algorithm generates the k-stage optimal cost functions J_k^*. For simplicity of notation, we suppress the $*$ here. At present we are concerned with the infinite horizon case and the possibility that J_k may converge to J^* as $k \to \infty$.

Under (P), $\bar{J}_0 \le \bar{J}_1$ and so $\bar{J}_1 = \bar{T}(\bar{J}_0) \le \bar{T}(\bar{J}_1) = \bar{J}_2$. Continuing, we see that $\bar{J}_k$ is an increasing sequence of functions, and so $\bar{J}_\infty = \lim_{k \to \infty} \bar{J}_k$ exists and takes values in $[0, +\infty]$. Under (N), $\bar{J}_k$ is a decreasing sequence of functions and $\bar{J}_\infty$ exists, taking values in $[-\infty, 0]$. Under (D), we have

$$\bar{J}_0 \le b + \bar{T}(\bar{J}_0),$$
$$0 \le b + \bar{T}(\bar{J}_0) \le b + \bar{T}[b + \bar{T}(\bar{J}_0)] = (1 + \alpha)b + \bar{T}^2(\bar{J}_0),$$
$$0 \le b + \bar{T}[b + \bar{T}(\bar{J}_0)] = (1 + \alpha)b + \bar{T}^2(\bar{J}_0) \le b + \bar{T}[(1 + \alpha)b + \bar{T}^2(\bar{J}_0)]$$
$$= (1 + \alpha + \alpha^2)b + \bar{T}^3(\bar{J}_0),$$

and, in general,

$$0 \le b \sum_{j=0}^{k-2} \alpha^j + \bar{T}^{k-1}(\bar{J}_0) \le b \sum_{j=0}^{k-1} \alpha^j + \bar{T}^k(\bar{J}_0).$$

As $k \to \infty$, we see that $b \sum_{j=0}^{k-1} \alpha^j + \bar{T}^k(\bar{J}_0)$ increases to a limit. But $b \sum_{j=0}^{\infty} \alpha^j = b/(1 - \alpha)$, so $\bar{J}_\infty = \lim_{k \to \infty} \bar{T}^k(\bar{J}_0)$ exists and satisfies

$$-b/(1 - \alpha) \le \bar{J}_\infty.$$

Similarly, we have

$$\bar{J}_\infty \le b/(1 - \alpha).$$

Now if $\bar{J}: P(S) \to [-c, c]$, $c < \infty$, then

$$\bar{J}_0 \le \bar{J} + c, \qquad \bar{T}(\bar{J}_0) \le \bar{T}(\bar{J} + c) = \alpha c + \bar{T}(\bar{J}),$$

and, in general,

$$\bar{T}^k(\bar{J}_0) \le \alpha^k c + \bar{T}^k(\bar{J}).$$

It follows that

$$\bar{J}_\infty \le \liminf_{k \to \infty} \bar{T}^k(\bar{J}),$$

and by a similar argument beginning with $\bar{J} - c \le \bar{J}_0$, we can show that $\limsup_{k \to \infty} \bar{T}^k(\bar{J}) \le \bar{J}_\infty$. This shows that under (D), if $\bar{J}$ is any bounded real-valued function on $P(S)$, then $\bar{J}_\infty = \lim_{k \to \infty} \bar{T}^k(\bar{J})$.

The same arguments can be used to establish the existence of $J_\infty = \lim_{k \to \infty} J_k$. Under (P), $J_\infty : S \to [0, +\infty]$; under (N), $J_\infty : S \to [-\infty, 0]$; and under (D), $J_\infty = \lim_{k \to \infty} T^k(J)$ takes values in $[-b/(1 - \alpha), b/(1 - \alpha)]$ where $J : S \to [-c, c]$, $c < \infty$, is lower semianalytic. Note that in every case, J_∞ is lower semianalytic by Lemma 7.30(2).

Lemma 9.3 (P)(N)(D)   For every $p \in P(S)$,

$$\bar{J}_k(p) = \int J_k(x)p(dx), \qquad k = 0, 1, \ldots, \quad k = \infty.$$

Proof   For $k = 0, 1, \ldots,$ the lemma follows from Proposition 9.7 by induction. When $k = \infty$, the lemma follows from the monotone convergence theorem under (P) and (N) and the bounded convergence theorem under (D).

$$\text{Q.E.D.}$$

Proposition 9.14   (N)(D)   We have

$$\bar{J}_{\infty} = \bar{J}^{*}, \tag{25}$$

$$J_{\infty} = J^{*}. \tag{26}$$

Indeed, under (D) the dynamic programming algorithm can be initiated from any $\bar{J}: P(S) \rightarrow [-c, c]$, $c < \infty$, or lower semianalytic $J: S \rightarrow [-c, c]$, $c < \infty$, and converges uniformly, i.e.,

$$\lim_{k \to \infty} \sup_{p \in P(S)} |\bar{T}^{k}(\bar{J})(p) - \bar{J}^{*}(p)| = 0, \tag{27}$$

$$\lim_{k \to \infty} \sup_{x \in S} |T^{k}(J)(x) - J^{*}(x)| = 0. \tag{28}$$

Proof   The result for (DM) follows from Proposition 4.2(c) and 5.7 or from DPSC, Chapter 6, Proposition 3 and Chapter 7, Proposition 4. By Lemma 9.3,

$$J_{k}(x) = \bar{J}_{k}(p_{x}) \qquad \forall x \in S, \quad k = 0, 1, \ldots, \quad k = \infty,$$

so (25) implies (26). Under (D), if a lower semianalytic function $J: S \rightarrow [-c, c]$, $c < \infty$, is given, then define $\bar{J}: P(S) \rightarrow [-c, c]$ by (20). Equation (28) now follows from (27) and Propositions 9.5 and 9.7.   Q.E.D.

Case (D) is the best suited for computational procedures. The machinery developed thus far can be applied to Proposition 4.6 or to DPSC, Chapter 6, Proposition 4, to show the validity for (SM) of the error bounds given there. We provide the theorem for (SM). The analogous result is of course true for (DM).

Proposition 9.15   (D)   Let $J: S \rightarrow [-c, c], c < \infty$, be lower semianalytic. Then for all $x \in S$ and $k = 0, 1, \ldots,$

$$T^{k}(J)(x) + b_{k} \le T^{k+1}(J)(x) + b_{k+1}$$
$$\le J^{*}(x) \le T^{k+1}(J)(x) + \bar{b}_{k+1} \le T^{k}(J)(x) + \bar{b}_{k}, \tag{29}$$

where

$$b_{k} = [\alpha/(1 - \alpha)] \inf_{x \in S}[T^{k}(J)(x) - T^{k-1}(J)(x)], \tag{30}$$

$$\bar{b}_{k} = [\alpha/(1 - \alpha)] \sup_{x \in S}[T^{k}(J)(x) - T^{k-1}(J)(x)]. \tag{31}$$

Proof   Given a lower semianalytic function $J:S \to [-c, c], c < \infty$, define $\bar{J}:P(S) \to [-c, c]$ by (20). By Proposition 9.7,

$$\bar{T}^k(\bar{J})(p) = \int T^k(J)(x)p(dx) \qquad \forall p \in P(S), \quad k = 0, 1, \dots.$$

Therefore

$$b_k = [\alpha/(1-\alpha)] \inf_{p \in P(S)} [\bar{T}^k(\bar{J})(p) - \bar{T}^{k-1}(\bar{J})(p)],$$

$$\bar{b}_k = [\alpha/(1-\alpha)] \sup_{p \in P(S)} [\bar{T}^k(\bar{J})(p) - \bar{T}^{k-1}(\bar{J})(p)],$$

where b_k and $\bar{b}_k$ are defined by (30) and (31). Taking $\alpha_1 = \alpha_2 = \alpha$ in Proposition 4.6 or using the proof of Proposition 4, Chapter 6 of DPSC, we obtain

$$\bar{T}^k(\bar{J})(p) + b_k \le \bar{T}^{k+1}(\bar{J})(p) + b_{k+1}$$
$$\le \bar{J}^*(p) \le \bar{T}^{k+1}(\bar{J})(p) + \bar{b}_{k+1} \le \bar{T}^k(\bar{J})(p) + \bar{b}_k.$$

Substituting $p = p_x$ in this equation, we obtain (29).   Q.E.D.

It is not possible to develop a policy iteration algorithm for (SM) along the lines of Proposition 4.8 or 4.9. One difficulty is this. If at the kth iteration we have constructed a policy $(\mu_k, \mu_k, \dots)$, where $\mu_k \in U(C|S)$, then J_{μ_k} is universally measurable but not necessarily lower semianalytic. We would like to find $\mu_{k+1} \in U(C|S)$ such that $T_{\mu_{k+1}}(J_{\mu_k}) \le T(J_{\mu_k}) + \varepsilon$, where $\varepsilon > 0$ is some prescribed small number, but Proposition 7.50 does not apply to this case.

We turn now to the question of convergence of the dynamic programming algorithm under (P). Without additional assumptions, we have only the following result.

Proposition 9.16   (P)   We have

$$\bar{J}_\infty \le \bar{J}^*, \tag{32}$$

$$J_\infty \le J^*. \tag{33}$$

Furthermore, the following statements are equivalent:

(a) $\bar{J}_\infty = \bar{T}(\bar{J}_\infty)$,
(b) $\bar{J}_\infty = \bar{J}^*$,
(c) $J_\infty = T(J_\infty)$,
(d) $J_\infty = J^*$.

Proof   It is clear that (32) holds and, by Proposition 9.10, implies the equivalence of (a) and (b). Lemma 9.3, Proposition 9.5, and (32) imply (33). Conditions (a) and (c) are equivalent by Lemma 9.3 and Proposition 9.7. Conditions (b) and (d) are equivalent by Lemma 9.3 and Proposition 9.5.
                                                                    Q.E.D.

In Example 1, we have $J_\infty(0) = 0$ and $J^*(0) = \infty$, so strict inequality in (32) and (33) is possible. We present now an example in which not only is J_∞ different from J^*, but J_∞ is Borel-measurable while J^* is not.

EXAMPLE 2   (Blackwell)   Let Σ be the set of finite sequences of positive integers and H the set of functions h from Σ into $\{0,1\}$. Then H can be regarded as the countable Cartesian product of copies of $\{0,1\}$ indexed by Σ. Let $\{0,1\}$ have the discrete topology and H the product topology, so H is a complete separable metrizable space (Proposition 7.4). A typical basic open set in H is $\{h \in H | h(s) = 1 \; \forall s \in \Sigma_1, h(s) = 0 \; \forall s \in \Sigma_2\}$, where Σ_1 and Σ_2 are finite subsets of Σ. Consider a Suslin scheme $R: \Sigma \to \mathscr{B}_H$ defined by

$$R(s) = \{h \in H | h(s) = 1\} \qquad \forall s \in \Sigma.$$

Then

$$N(R) = \{h \in H | \exists (\zeta_1, \zeta_2, \ldots) \in \mathscr{N} \text{ such that } h(\zeta_1, \zeta_2, \ldots, \zeta_n) = 1 \; \forall n\}$$

is an analytic subset of H (Proposition 7.36). We show with the aid of Appendix B that $N(R)$ is not Borel-measurable. Let Y be an uncountable Borel space and $Q: \Sigma \to \mathscr{B}_Y$ a Suslin scheme such that $N(Q)$ is not Borel-measurable (Proposition B.6). Define $\psi: Y \to H$ by

$$\psi(y)(s) = \begin{cases} 1 & \text{if } y \in Q(s), \\ 0 & \text{if } y \notin Q(s). \end{cases}$$

If Σ_1 and Σ_2 are finite subsets of Σ, then

$$\psi^{-1}(\{h \in H | h(s) = 1 \; \forall s \in \Sigma_1, h(s) = 0 \; \forall s \in \Sigma_2\})$$

$$= \left[\bigcap_{s \in \Sigma_1} Q(s) \right] \cap \left[\bigcap_{s \in \Sigma_2} (Y - Q(s)) \right]$$

is in $\mathscr{B}_Y$. The collection $\mathscr{E}$ of subsets E of H for which $\psi^{-1}(E) \in \mathscr{B}_Y$ is a σ-algebra containing a base for the topology on H, so, by the remark following Definition 7.6, $\mathscr{E}$ contains $\mathscr{B}_H$ and ψ is Borel-measurable. For each $s \in \Sigma$, we have $Q(s) = \psi^{-1}[R(s)]$, so

$$N(Q) = \bigcup_{z \in \mathscr{N}} \bigcap_{s < z} Q(s) = \bigcup_{z \in \mathscr{N}} \bigcap_{s < z} \psi^{-1}[R(s)]$$

$$= \psi^{-1} \left[\bigcup_{z \in \mathscr{N}} \bigcap_{s < z} R(s) \right] = \psi^{-1}[N(R)].$$

Since $N(Q)$ is not Borel-measurable, $N(R)$ is also not Borel-measurable.

Define the decision model by taking $S = H\Sigma^*$, where $\Sigma^* = \Sigma \cup \{0\}$, $C = \{1, 2, \ldots\}$, $U(x) = C$ for every $x \in S$, and

$$f([h, 0], u) = (h, u),$$
$$f([h, (\zeta_1, \zeta_2, \ldots, \zeta_n)], u) = [h, (\zeta_1, \zeta_2, \ldots, \zeta_n, u)].$$

The system transition is deterministic, so the choice of W and $p(dw|x, u)$ is irrelevant. Choose $\alpha = 1$ and

$$g([h, 0], u) = \begin{cases} 0 & \text{if } h(u) = 1, \\ 1 & \text{if } h(u) = 0, \end{cases}$$

$$g([h, (\zeta_1, \zeta_2, \ldots, \zeta_n)], u) = \begin{cases} 0 & \text{if } h(\zeta_1, \zeta_2, \ldots, \zeta_n, u) = 1, \\ 1 & \text{if } h(\zeta_1, \zeta_2, \ldots, \zeta_n, u) = 0. \end{cases}$$

If the system begins at $x_0 = [h, 0]$ and the horizon is infinite, a positive cost can be avoided if and only if there exists $(\zeta_1, \zeta_2, \ldots)$ such that $h(\zeta_1, \zeta_2, \ldots, \zeta_n) = 1$ for every n, i.e., $J^*([h, 0]) = 0$ if and only if $h \in N(R)$. Therefore, J^* is not Borel-measurable. Over the finite horizon, we have

$$J_{k+1}(x) = T(J_k)(x) = \inf_{u \in C}\{g(x, u) + J_k[f(x, u)]\},$$

and since C is countable and f, g, and J_0 are Borel-measurable, J_k is Borel-measurable for $k = 0, 1, 2, \ldots$. It follows that J_∞ is Borel-measurable.

The equivalent conditions of Proposition 9.16 are not easily verified in practice. We give here some more readily verifiable conditions which imply that $J_\infty = J^*$.

Proposition 9.17 (P)(D)   Assume that there exists a nonnegative integer $\bar{k}$ such that for each $x \in S$, $\lambda \in R$, and $k \geq \bar{k}$, the set

$$U_k(x, \lambda) = \left\{ u \in U(x) \,\middle|\, g(x, u) + \alpha \int J_k(x')t(dx'|x, u) \leq \lambda \right\} \tag{34}$$

is compact in C. Then $\bar{J}_\infty = \bar{J}^*$, $J_\infty = J^*$, and there exists an optimal non-randomized stationary policy for (SM).

Proof   Under (P), we have, for each k, $J_k \leq J_\infty$, so $J_{k+1} = T(J_k) \leq T(J_\infty)$, and letting $k \to \infty$ we obtain

$$J_\infty \leq T(J_\infty). \tag{35}$$

Let $x \in S$ be such that $J_\infty(x) < \infty$. By Lemma 3.1 for $k \geq \bar{k}$ there exists $u_k \in U(x)$ such that

$$J_{k+1}(x) = g(x, u_k) + \alpha \int J_k(x')t(dx'|x, u_k).$$

Since $J_k \leq J_{k+1} \leq \cdots \leq J_\infty$, it follows that for $k \geq \bar{k}$

$$g(x, u_i) + \alpha \int J_k(x')t(dx'|x, u_i) \leq g(x, u_i) + \alpha \int J_i(x')t(dx'|x, u_i)$$

$$= J_{i+1}(x) \leq J_\infty(x) \qquad \forall i \geq k.$$

Therefore, $\{u_i | i \geq k\} \subset U_k[x, J_\infty(x)]$ for every $k \geq \bar{k}$. Since $U_k[x, J_\infty(x)]$ is compact, all limit points of the sequence $\{u_i | i \geq k\}$ belong to $U_k[x, J_\infty(x)]$, and at least one such limit point exists. It follows that if $\bar{u}$ is a limit point of the sequence $\{u_i | i \geq \bar{k}\}$, then

$$\bar{u} \in \bigcap_{k=\bar{k}}^{\infty} U_k[x, J_\infty(x)].$$

Therefore, for all $k \geq \bar{k}$,

$$J_\infty(x) \geq g(x, \bar{u}) + \alpha \int J_k(x') t(dx' | x, \bar{u}) \geq J_{k+1}(x).$$

Letting $k \to \infty$ and using the monotone convergence theorem, we obtain

$$J_\infty(x) = g(x, \bar{u}) + \alpha \int J_\infty(x') t(dx' | x, \bar{u}) \geq T(J_\infty)(x) \tag{36}$$

for all $x \in S$ such that $J_\alpha(x) < \infty$. We also have that (36) holds if $J_\alpha(x) = \infty$, and thus it holds for all $x \in S$. From (35) and (36) we see that $J_\infty = T(J_\infty)$ and conditions (a)–(d) of Proposition 9.16 must hold. In particular, we have from (35) and (36) that for every $x \in S$, there exists $\bar{u} \in U(x)$ such that

$$J^*(x) = g(x, \bar{u}) + \alpha \int J^*(x') t(dx' | x, \bar{u}) = T(J^*)(x).$$

The existence of an optimal nonrandomized stationary policy for (SM) follows from Corollary 9.12.1.

Under (D), conditions (a)–(d) of Proposition 9.16 hold by Proposition 9.14. If we replace g by $g + b$, we obtain a model satisfying (P). This new model also satisfies the hypotheses of the proposition, so there exists an optimal nonrandomized stationary policy for it. This policy is optimal for the original (D) model as well.   Q.E.D.

Corollary 9.17.1 (P)(D)   Assume that the set $U(x)$ is finite for each $x \in S$. Then $\bar{J}_\infty = \bar{J}^*$, $J_\infty = J^*$, and there exists an optimal nonrandomized stationary policy for (SM). In fact, if C is finite and g and Γ are Borel-measurable, then J^* is Borel-measurable and there exists a Borel-measurable optimal nonrandomized stationary policy for (SM).

Corollary 9.17.2 (P)(D)   Suppose conditions (a)–(e) of Definition 8.7 (the lower semicontinuous model) are satisfied. Then $\bar{J}_\infty = \bar{J}^*$, $J_\infty = J^*$, J^* is lower semicontinuous, and there exists a Borel-measurable optimal nonrandomized stationary policy for (SM).

Proof   From the proof of Proposition 8.6, we see that J_k is lower semi-continuous for $k = 1, 2, \ldots$, as are the functions

$$\hat{K}_k(x, u) = \begin{cases} g(x, u) + \alpha \int J_k(x') t(dx' | x, u) & \text{if } (x, u) \in \Gamma, \\ \infty & \text{if } (x, u) \notin \Gamma. \end{cases} \tag{37}$$

For $\lambda \in R$ and k fixed, the lower level set

$$\{(x,u) \in SC | \hat{K}_k(x,u) \le \lambda\} \subset \Gamma$$

is closed, so for each fixed $x \in S$

$$U_k(x,\lambda) = \{u \in C | \hat{K}_k(x,u) \le \lambda\}$$

is compact. Proposition 9.17 can now be invoked, and it remains only to prove that the optimal nonrandomized stationary policy whose existence is guaranteed by that proposition can be chosen to be Borel-measurable. This will follow from Proposition 9.12 and the proof of Proposition 8.6 once we show that $J_\infty = J^*$ is lower semicontinuous. Under (P), $J_k \uparrow J^*$, so

$$\{x \in S | J^*(x) \le \lambda\} = \bigcap_{k=0}^{\infty} \{x \in S | J_k(x) \le \lambda\}$$

is closed, and J^* is lower semicontinuous. Under (D),

$$J_k - b \sum_{j=k}^{\infty} \alpha^k \uparrow J^*,$$

so a similar argument can be used to show that J^* is lower semicontinuous.
                                                                        Q.E.D.

By using the argument used to prove Corollary 8.6.1, we also have the following.

Corollary 9.17.3   The conclusions of Corollary 9.17.2 hold if instead of assuming that C is compact and each Γ^j is closed in Definition 8.7, we assume that each Γ^j is compact.

Proposition 9.17 and its corollaries provide conditions under which the dynamic programming algorithm can be used in the (P) and (D) models to generate J^*. It is also possible to use the dynamic programming algorithm to generate an optimal stationary policy, as is indicated by the next proposition.

Proposition 9.18   (P)(D)   Suppose that either $U(x)$ is finite for each $x \in S$ or else conditions (a)–(e) of Definition 8.7 hold. Then for each $k \ge 0$ there exists a universally measurable $\mu_k : S \to C$ such that $\mu_k(x) \in U(x)$ for every $x \in S$ and

$$T_{\mu_k}(J_k) = T(J_k). \tag{38}$$

If $\{\mu_k\}$ is a sequence of such functions, then for each $x \in S$ the sequence $\{\mu_k(x)\}$ has at least one accumulation point. If $\mu : S \to C$ is universally measurable, $\mu(x)$ is an accumulation point of $\{\mu_k(x)\}$ for each $x \in S$ such that $J^*(x) < \infty$, and $\mu(x) \in U(x)$ for each $x \in S$ such that $J^*(x) = \infty$, then $\pi = (\mu, \mu, \ldots)$ is an optimal stationary policy for (SM).

Proof If $U(x)$ is finite for each $x \in S$, then the sets $U_k(x, \lambda)$ of (34) are compact for all $k \geq 0$, $x \in S$, and $\lambda \in R$. The proof of Corollary 9.17.2 shows that these sets are also compact under conditions (a)–(e) of Definition 8.7. The existence of functions $\mu_k : S \to C$ satisfying (38) such that $\mu_k(x) \in U(x)$ for every $x \in S$ is a consequence of Lemma 3.1 and Proposition 7.50.

Under (P) we see from the proof of Proposition 9.17 that $\{\mu_k(x)\}$ has at least one accumulation point for each $x \in S$ such that $J^*(x) < \infty$ and every accumulation point of $\{\mu_k(x)\}$ is in $U(x)$. If $\mu : S \to C$ is universally measurable and $\mu(x)$ is an accumulation point of $\{\mu_k(x)\}$ for each $x \in S$ such that $J^*(x) < \infty$, then from (35), (36), and Proposition 9.17 we have

$$J^*(x) = g[x, \mu(x)] + \alpha \int J^*(x')t(dx'|x, \mu(x)) \tag{39}$$

for all $x \in S$ such that $J^*(x) < \infty$. If $\mu(x) \in U(x)$ for all $x \in S$ such that $J^*(x) = \infty$, then

$$J^*(x) = T(J^*)(x) \leq g[x, \mu(x)] + \alpha \int J^*(x')t(dx'|x, \mu(x)) \leq \infty = J^*(x) \tag{40}$$

for all $x \in S$ such that $J^*(x) = \infty$. From (39) and (40) we have $J^* = T_\mu(J^*)$, and the policy $\pi = (\mu, \mu, \ldots)$ is optimal by Proposition 9.12.

Under (D) we can replace g by $g + b$ to obtain a model satisfying (P) and the hypotheses of the proposition. The conclusions of the proposition are valid for this new model, so they are valid for the original (D) model as well.   Q.E.D.

A slightly stronger version of Proposition 9.18 can be found in [S12].

Corollary 9.18.1 If conditions (b)–(e) of Definition 8.7 hold and if each Γ^j of condition (b) is compact, then the conclusions of Proposition 9.18 hold.

9.6 Existence of ε-Optimal Policies

We have characterized stationary optimal policies and given conditions under which optimal policies exist. We turn now to the existence of ε-optimal policies. For fixed $x \in S$, by definition there is a policy which is ε-optimal at x. We would like to know how this collection of policies, each of which is ε-optimal at a single point, can be pieced together to form a single policy which is ε-optimal at every point. There is a related question concerning optimal policies. If at each point there is a policy which is optimal at that point, is it possible to find an optimal policy? Answers to these questions are provided by the next two propositions.

Proposition 9.19 (P)(D) For each $\varepsilon > 0$, there exists an ε-optimal non-randomized Markov policy for (SM), and if $\alpha < 1$, it can be taken to be stationary. If for each $x \in S$ there exists a policy for (SM) which is optimal at x, then there exists an optimal nonrandomized stationary policy.

Proof   Choose $\varepsilon > 0$ and $\varepsilon_k > 0$ such that $\sum_{k=0}^{\infty} \alpha^k \varepsilon_k = \varepsilon$. If $\alpha < 1$, let $\varepsilon_k = (1 - \alpha)\varepsilon$ for every k. By Proposition 7.50, there are universally measurable functions $\mu_k : S \to C$, $k = 0, 1, \ldots$, such that $\mu_k(x) \in U(x)$ for every $x \in S$ and

$$T_{\mu_k}(J^*) \le J^* + \varepsilon_k.$$

If $\alpha < 1$, we choose all the μ_k to be identical. Then

$$(T_{\mu_{k-1}} T_{\mu_k})(J^*) \le T_{\mu_{k-1}}(J^*) + \alpha \varepsilon_k \le J^* + \varepsilon_{k-1} + \alpha \varepsilon_k.$$

Continuing this process, we have

$$(T_{\mu_0} T_{\mu_1} \cdots T_{\mu_k})(J^*) \le J^* + \sum_{j=0}^{k} \alpha^j \varepsilon_j \le J^* + \varepsilon,$$

and, letting $k \to \infty$, we obtain

$$\lim_{k \to \infty} (T_{\mu_0} T_{\mu_1} \cdots T_{\mu_k})(J^*) \le J^* + \varepsilon.$$

Under (P) we have

$$J_\pi = \lim_{k \to \infty} (T_{\mu_0} T_{\mu_1} \cdots T_{\mu_k})(J_0) \le \lim_{k \to \infty} (T_{\mu_0} T_{\mu_1} \cdots T_{\mu_k})(J^*), \qquad (41)$$

so $\pi = (\mu_0, \mu_1, \ldots)$ is ε-optimal. Under (D),

$$J_0 \le J^* + [b/(1 - \alpha)],$$
$$(T_{\mu_0} T_{\mu_1} \cdots T_{\mu_k})(J_0) \le (T_{\mu_0} T_{\mu_1} \cdots T_{\mu_k})(J^* + [b/(1 - \alpha)])$$
$$= [\alpha^{k+1} b/(1 - \alpha)] + (T_{\mu_0} T_{\mu_1} \cdots T_{\mu_k})(J^*),$$

so (41) is valid and $\pi = (\mu_0, \mu_1, \ldots)$ is ε-optimal. This proves the first part of the proposition.

Suppose that for each $x \in S$ there is a policy for (SM) which is optimal at x. Fix x and let $\pi = (\mu_0, \mu_1, \ldots)$ be a policy which is optimal at x. By Proposition 9.1, we may assume without loss of generality that π is Markov. By Lemma 8.4(b) and (c), we have

$$J^*(x) = J_\pi(x)$$
$$= \lim_{k \to \infty} (T_{\mu_0} T_{\mu_1} \cdots T_{\mu_k})(J_0)(x)$$
$$= T_{\mu_0} \left[\lim_{k \to \infty} (T_{\mu_1} \cdots T_{\mu_k})(J_0) \right](x)$$
$$\ge T_{\mu_0}(J^*)(x) \ge T(J^*)(x) = J^*(x).$$

Consequently,

$$T_{\mu_0}(J^*)(x) = T(J^*)(x).$$

This implies that the infimum in the expression

$$\inf_{u \in U(x)} \left\{ g(x, u) + \alpha \int J^*(x')t(dx'|x, u) \right\}$$

is achieved. Since x is arbitrary, Corollary 9.12.1 implies the existence of an optimal nonrandomized stationary policy.   Q.E.D.

Proposition 9.20 (N)   For each $\varepsilon > 0$, there exists an ε-optimal nonrandomized semi-Markov policy for (SM). If for each $x \in S$ there exists a policy for (SM) which is optimal at x, then there exists a semi-Markov (randomized) optimal policy.

Proof   Under (N) we have $J_k \downarrow J^*$ (Proposition 9.14), so, given $\varepsilon > 0$, the analytically measurable sets

$$A_k = \{x \in S | J^*(x) > -\infty, J_k(x) \leq J^*(x) + \varepsilon/2\}$$
$$\cup \{x \in S | J^*(x) = -\infty, J_k(x) \leq -(2 + \varepsilon^2)/2\varepsilon\}$$

converge up to S as $k \to \infty$. By Proposition 8.3, for each k there exists a k-stage nonrandomized semi-Markov policy π^k such that for every $x \in S$

$$J_{k, \pi^k}(x) \leq \begin{cases} J_k(x) + (\varepsilon/2) & \text{if } J_k(x) > -\infty, \\ -1/\varepsilon & \text{if } J_k(x) = -\infty. \end{cases}$$

Then for $x \in A_k$ we have either $J^*(x) > -\infty$ and

$$J_{k, \pi^k}(x) \leq J_k(x) + (\varepsilon/2) \leq J^*(x) + \varepsilon,$$

or else $J^*(x) = -\infty$. If $J^*(x) = -\infty$, then either $J_k(x) = -\infty$ and

$$J_{k, \pi^k}(x) \leq -1/\varepsilon,$$

or else $J_k(x) > -\infty$ and

$$J_{k, \pi^k}(x) \leq J_k(x) + (\varepsilon/2) \leq -[(2 + \varepsilon^2)/2\varepsilon] + (\varepsilon/2) = -1/\varepsilon.$$

Choose any $\mu \in U(C|S)$ and define $\hat{\pi}^k = (\mu_0^k, \ldots, \mu_{k-1}^k, \mu, \mu, \ldots)$, where $\pi^k = (\mu_0^k, \ldots, \mu_{k-1}^k)$. For every $x \in A_k$, we have

$$J_{\hat{\pi}^k}(x) \leq J_{k, \pi^k}(x) \leq \begin{cases} J^*(x) + \varepsilon & \text{if } J^*(x) > -\infty, \\ -1/\varepsilon & \text{if } J^*(x) = -\infty, \end{cases}$$

so $\hat{\pi}^k$ is a nonrandomized semi-Markov policy which is ε-optimal for every $x \in A_k$. The policy π defined to be $\hat{\pi}^k$ when the initial state is in A_k, but not in A_j for any $j < k$, is semi-Markov, nonrandomized, and ε-optimal at every $x \in \bigcup_{k=1}^{\infty} A_k = S$.

Suppose now that for each $x \in S$ there exists a policy π^x for (SM) which is optimal at x. Let $\bar{\pi}^x$ be a policy for (DM) which corresponds to π^x, and let $(p_x, q_0^x, q_1^x, \ldots)$ be the sequence generated from p_x by $\bar{\pi}^x$ via (10) and (11). If $G: \Delta \to [-\infty, 0]$ is defined by (15), then we have from Proposition 9.3 that

$$J^*(x) = J_{\pi^x}(x) = \bar{J}_{\bar{\pi}^x}(p_x) = G(p_x, q_0^x, q_1^x, \ldots). \tag{42}$$

We have from Proposition 9.5 and (16) that

$$J^*(x) = \bar{J}^*(p_x) = \inf_{(q_0, q_1, \ldots) \in \Delta_{p_x}} G(p_x, q_0, q_1, \ldots). \tag{43}$$

Therefore the infimum in (43) is attained for every $p_x \in \bar{S}$, where $\bar{S} = \{p_y | y \in S\}$, so by Proposition 7.50, there exists a universally measurable selector $\psi: \bar{S} \to P(SC)P(SC) \cdots$ such that $\psi(p_x) \in \Delta_{p_x}$ and

$$J^*(x) = \bar{J}^*(p_x) = G[p_x, \psi(p_x)] \qquad \forall p_x \in \bar{S}.$$

Let $\delta: S \to \bar{S}$ be the homeomorphism $\delta(x) = p_x$ and let $\varphi(x) = \psi[\delta(x)]$. Then φ is universally measurable, $\varphi(x) \in \Delta_{p_x}$, and

$$J^*(x) = G[p_x, \varphi(x)] \qquad \forall x \in S. \tag{44}$$

Denote

$$\varphi(x) = [q_0(d(x_0, u_0)|x), q_1(d(x_1, u_1)|x), \ldots].$$

For each $k \geq 0$, $q_k(d(x_k, u_k)|x)$ is a universally measurable stochastic kernel on $S_k C_k$ given S, and by Proposition 7.27 and Lemma 7.28(a), (b), $q_k(d(x_k, u_k)|x)$ can be decomposed into its marginal $p_k(dx_k|x)$, which is a universally measurable stochastic kernel on S_k given S, and a universally measurable stochastic kernel $\mu_k(du_k|x, x_k)$ on C_k given SS_k. Since $p_0(dx_0|x) = p_x(dx_0)$, the stochastic kernel $\mu_0(du_0|x, x_0)$ is arbitrary except when $x = x_0$. Set

$$\bar{\mu}_0(du_0|x) = \mu_0(du_0|x, x) \qquad \forall x \in S.$$

The sequence $\pi = (\bar{\mu}_0, \mu_1, \mu_2, \ldots)$ is a randomized semi-Markov policy for (SM). From (7) of Chapter 8, we see that for each $x \in S$

$$q_k(\pi, p_x) = q_k(d(x_k, u_k)|x) \qquad \forall x \in S, \quad k = 0, 1, \ldots.$$

From (5), (15), and (44), we have

$$J_\pi(x) = G[p_x, q_0(\pi, p_x), q_1(\pi, p_x), \ldots] = J^*(x) \qquad \forall x \in S,$$

so π is optimal.   Q.E.D.

Although randomized polcies may be considered inferior and are avoided in practice, under (N) as posed here they cannot be disregarded even in deterministic problems, as the following example demonstrates.

EXAMPLE 3 (St. Petersburg paradox) Let $S = \{0, 1, 2, \ldots\}$, $C = \{0, 1\}$, $U(x) = C$ for every $x \in S$, $\alpha = 1$,

$$f(x, u) = \begin{cases} x + 1 & \text{if } u = 1, \quad x \neq 0, \\ 0 & \text{otherwise,} \end{cases}$$

$$g(x, u) = \begin{cases} -2^x & \text{if } x \neq 0, \quad u = 0, \\ 0 & \text{otherwise.} \end{cases}$$

Beginning in state one, any nonrandomized policy either increases the state by one indefinitely and incurs no nonzero cost or else, after k increases, jumps the system to zero at a cost of -2^{k+1}, where it remains at no further cost. Thus $J^*(1) = -\infty$, but this cost is not achieved by any nonrandomized policy. On the other hand, the randomized stationary policy which jumps the system to zero with probability $\frac{1}{2}$ when the state x is nonzero yields an expected cost of $-\infty$ and is optimal at every $x \in S$.

The one-stage cost g in Example 3 is unbounded, but by a slight modification an example can be constructed in which g is bounded and the only optimal policies are randomized. If one stipulates that J^* must be finite, it may be possible to restrict attention to nonrandomized policies in Proposition 9.20. This is an unsolved problem.

If (SM) is lower semicontinuous, then Proposition 9.19 can be strengthened, as Corollary 9.17.2 shows. Similarly, if (SM) is upper semicontinuous, a stronger version of Proposition 9.20 can be proved.

Proportional 9.21 Assume (SM) satisfies conditions (a)–(d) of Definition 8.8 (the upper semicontinuous model).

(D) For each $\varepsilon > 0$, there exists a Borel-measurable, ε-optimal, nonrandomized, stationary policy.

(N) For each $\varepsilon > 0$, there exists a Borel-measurable, ε-optimal, nonrandomized, semi-Markov policy.

Under both (D) and (N), J^* is upper semicontinuous.

Proof Under (D) and (N) we have $\lim_{k \to \infty} J_k = J^*$ (Proposition 9.14), and each J_k is upper semicontinuous (Proposition 8.7). By an argument similar to that used in the proof of Corollary 9.17.2, J^* is upper semicontinuous.

By using Proposition 7.34 in place of Proposition 7.50, the proof of Proposition 9.19 can be modified to show the existence of a Borel-measurable, ε-optimal, nonrandomized, stationary policy under (D). By using Proposition 8.7 in place of Proposition 8.3, the proof of Proposition 9.20 can be modified to show the existence of a Borel-measurable, ε-optimal, nonrandomized, semi-Markov policy under (N).   Q.E.D.

Chapter 10

The Imperfect State Information Model

In the models of Chapters 8 and 9 the current state of the system is known to the controller at each stage. In many problems of practical interest, however, the controller has instead access only to imperfect measurements of the system state. This chapter is devoted to the study of models relating to such situations. In our analysis we will encounter nonstationary versions of the models of Chapters 8 and 9. We will show in the next section that nonstationary models can be reduced to stationary ones by appropriate reformulation. We will thus be able to obtain nonstationary counterparts to the results of Chapters 8 and 9.

10.1 Reduction of the Nonstationary Model—State Augmentation

The finite horizon stochastic optimal control model of Definition 8.1 and the infinite horizon stochastic optimal control model of Definition 9.1 are said to be *stationary*, i.e., the data defining the model does not vary from stage to stage. In this section we define a nonstationary model and show how it can be reduced to a stationary one by augmenting the state with the time index.

242

We combine the treatments of the finite and infinite horizon models. Thus when $N = \infty$ and notation of the form $S_0, S_1, \ldots, S_{N-1}$ or $k = 0, \ldots, N - 1$ appears, we take this to mean $S_0, S_1, \ldots$ and $k = 0, 1, \ldots$, respectively.

Definition 10.1   A *nonstationary stochastic optimal control model*, denoted by (NSM), consists of the following objects:

N *Horizon.* A positive integer or ∞.

$S_k, k = 0, \ldots, N - 1$ *State spaces.* For each k, S_k is a nonempty Borel space.

$C_k, k = 0, \ldots, N - 1$ *Control spaces.* For each k, C_k is a nonempty Borel space.

$U_k, k = 0, \ldots, N - 1$ *Control constraints.* For each k, U_k is a function from S_k to the set of nonempty subsets of C_k, and the set

$$\Gamma_k = \{(x_k, u_k) | x_k \in S_k, u_k \in U_k(x_k)\} \tag{1}$$

is analytic in $S_k C_k$.

$W_k, k = 0, \ldots, N - 1$ *Disturbance spaces.* For each k, W_k is a nonempty Borel space.

$p_k(dw_k | x_k, u_k), \ k = 0, \ldots, N - 1$ *Disturbance kernels.* For each k, $p_k(dw_k | x_k, u_k)$ is a Borel-measurable stochastic kernel on W_k given $S_k C_k$.

$f_k, k = 0, \ldots, N - 2$ *System functions.* For each k, f_k is a Borel-measurable function from $S_k C_k W_k$ to S_{k+1}.

α *Discount factor.* A positive real number.

$g_k, k = 0, \ldots, N - 1$ *One-stage cost functions.* For each k, g_k is a lower semianalytic function from Γ_k to R^*.

We envision a system which begins at some $x_k \in S_k$ and moves successively through state spaces $S_{k+1}, S_{k+2}, \ldots$ and, if $N < \infty$, finally terminates in S_{N-1}. A policy governing such a system evolution is a sequence $\pi^k = (\mu_k, \mu_{k+1}, \ldots, \mu_{N-1})$, where each μ_j is a universally measurable stochastic kernel on C_j given $S_k C_k \cdots C_{j-1} S_j$ satisfying

$$\mu_j(U_j(x_j) | x_k, u_k, \ldots, u_{j-1}, x_j) = 1$$

for every $(x_k, u_k, \ldots, u_{j-1}, x_j)$. Such a policy is called a *k-originating policy* and the collection of all k-originating policies will be denoted by Π^k. The concepts of *semi-Markov, Markov, nonrandomized* and $\mathscr{F}$-*measurable* policies are analogous to those of Definitions 8.2 and 9.2. The set Π^0 is also written as Π', and the subset of Π' consisting of all Markov policies is denoted by Π.

Define the Borel-measurable *state transition stochastic kernels* by

$$t_k(\underline{S}_{k+1} | x_k, u_k)$$
$$= p_k(\{w_k \in W_k | f_k(x_k, u_k, w_k) \in \underline{S}_{k+1}\} | x_k, u_k) \qquad \forall \underline{S}_{k+1} \in \mathscr{B}_{S_{k+1}}.$$

Given a probability measure $p_k \in P(S_k)$ and a policy $\pi^k = (\mu_k, \ldots, \mu_{N-1}) \in \Pi^k$, define for $j = k, k+1, \ldots, N-1$

$$
\begin{aligned}
q_j(\pi^k, p_k)(\underline{S}_j \underline{C}_j) = & \int_{S_k} \int_{C_k} \cdots \int_{C_{j-1}} \int_{\underline{S}_j} \mu_j(\underline{C}_j | x_k, u_k, \ldots, u_{j-1}, x_j) \\
& \times t_{j-1}(dx_j | x_{j-1}, u_{j-1}) \\
& \times \mu_{j-1}(du_{j-1} | x_k, u_k, \ldots, u_{j-2}, x_{j-1}) \cdots \mu_k(du_k | x_k) p_k(dx_k) \\
& \qquad\qquad\qquad\qquad \forall \underline{S}_j \in \mathcal{B}_{S_j}, \quad \underline{C}_j \in \mathcal{B}_{C_j}. \qquad (2)
\end{aligned}
$$

There is a unique probability measure $q_j(\pi^k, p_k) \in P(S_j C_j)$ satisfying (2).

If the horizon N is finite, we treat (NSM) only under one of the following assumptions:

$$
\int_{S_j C_j} g_j^-(x_j, u_j) \, dq_j(\pi^k, p_{x_k}) < \infty \qquad \forall \pi^k \in \Pi^k, \quad x_k \in S_k, \quad k \le j \le N-1, \quad (\mathrm{F}^+)
$$

$$
k = 0, \ldots, N-1.
$$

$$
\int_{S_j C_j} g_j^+(x_j, u_j) \, dq_j(\pi^k, p_{x_k}) < \infty \qquad \forall \pi^k \in \Pi^k, \quad x_k \in S_k, \quad k \le j \le N-1, \quad (\mathrm{F}^-)
$$

$$
k = 0, \ldots, N-1.
$$

If $N = \infty$, we treat (NSM) only under one of the assumptions:

(P) $0 \le g_k(x_k, u_k)$ for every $(x_k, u_k) \in \Gamma_k$, $k = 0, \ldots, N-1$.

(N) $g_k(x_k, u_k) \le 0$ for every $(x_k, u_k) \in \Gamma_k$, $k = 0, \ldots, N-1$.

(D) $0 < \alpha < 1$, and for some $b \in R$, $-b \le g_k(x_k, u_k) \le b$ for every $(x_k, u_k) \in \Gamma_k$, $k = 0, \ldots, N-1$.

As in Chapters 8 and 9, the symbols (F^+), (F^-), (P), (N), and (D) will be used to indicate when a result is valid under the appropriate assumption.

We define the *k-originating cost corresponding to* π^k at $x_k \in S_k$ to be

$$
J_{\pi^k}(x_k, k) = \sum_{j=k}^{N-1} \alpha^j \int_{S_j C_j} g_j(x_j, u_j) \, dq_j(\pi^k, p_{x_k}),
$$

and the *k-originating optimal cost* at $x_k \in S_k$ to be

$$
J^*(x_k, k) = \inf_{\pi^k \in \Pi^k} J_{\pi^k}(x_k, k). \qquad (3)
$$

A policy $\pi \in \Pi^0$ is ε-*optimal* at $x_0 \in S_0$ if

$$
J_\pi(x_0, 0) \le \begin{cases} J^*(x_0, 0) + \varepsilon & \text{if } J^*(x_0, 0) > -\infty, \\ -1/\varepsilon & \text{if } J^*(x_0, 0) = -\infty. \end{cases}
$$

The policy π is *optimal* at x_0 if $J_\pi(x_0, 0) = J^*(x_0, 0)$. We say $\pi \in \Pi^0$ is ε-*optimal* (*optimal*) if it is ε-optimal (optimal) at every $x_0 \in S_0$. Let $\{\varepsilon_n\}$ be a sequence of positive numbers with $\varepsilon_n \downarrow 0$. A sequence of policies $\{\pi_n\} \subset \Pi^0$ is said to exhibit $\{\varepsilon_n\}$-*dominated convergence to optimality* if

$$\lim_{n \to \infty} J_{\pi_n}(x_0, 0) = J^*(x_0, 0) \qquad \forall x_0 \in S_0,$$

and for $n = 2, 3, \ldots$

$$J_{\pi_n}(x_0, 0) \leq \begin{cases} J^*(x_0, 0) + \varepsilon_n & \text{if} \quad J^*(x_0, 0) > -\infty, \\ J_{\pi_{n-1}}(x_0, 0) + \varepsilon_n & \text{if} \quad J^*(x_0, 0) = -\infty. \end{cases}$$

Definition 10.2   Let a nonstationary stochastic optimal control model as defined by Definition 10.1 be given. The corresponding *stationary stochastic optimal control model*, denoted by (SSM), consists of the following objects. (T is both a terminal state and the only control available at that state. If $N = \infty$, the introduction of T is unnecessary.):

$S = \bigcup_{k=0}^{N-1} \{(x_k, k) | x_k \in S_k\} \cup \{T\}$ *State space.*

$C = \bigcup_{k=0}^{N-1} \{(u_k, k) | u_k \in C_k\} \cup \{T\}$ *Control space.*

U *Control constraint.* A function from S to the set of nonempty subsets of C defined by $U(x_k, k) = \{(u_k, k) | u_k \in U_k(x_k)\}$, $U(T) = \{T\}$.

$W = \bigcup_{k=0}^{N-1} \{(w_k, k) | w_k \in W_k\}$ *Disturbance space.*

$p(dw|x, u)$ *Disturbance kernel.* If $\underline{W}_k \in \mathscr{B}_{W_k}$, we define

$$p[\{(w_k, k) | w_k \in \underline{W}_k\} | (x_k, k), (u_k, k)] = p_k(\underline{W}_k | x_k, u_k). \tag{4}$$

f *System function.* We define for $k = 0, \ldots, N - 2$

$$f[(x_k, k), (u_k, k), (w_k, k)] = [f_k(x_k, u_k, w_k), k + 1], \tag{5}$$

and for the remaining two stages

$$f[(x_{N-1}, N - 1), (u_{N-1}, N - 1), (w_{N-1}, N - 1)] = T, \tag{6}$$

$$f(T, T, w) = T. \tag{7}$$

α *Discount factor.*

g *One-stage cost function.* We define

$$g[(x_k, k), (u_k, k)] = g_k(x_k, u_k), \tag{8}$$

$$g(T, T) = 0. \tag{9}$$

N *Horizon.*

Consider the mapping $\varphi_k : S_k \to S$ given by $\varphi_k(x_k) = (x_k, k)$. We endow S with the topology that makes each φ_k a homeomorphism, and we endow C and W with similar topologies. The spaces S, C, and W are Borel. The set

$$\Gamma = \{(x, u) | x \in S, u \in U(x)\}$$

$$= \bigcup_{k=0}^{N-1} \{[(x_k, k), (u_k, k)] | (x_k, u_k) \in \Gamma_k\} \cup \{(T, T)\}$$

is analytic, and g defined on Γ by (8) and (9) is lower semianalytic. The disturbance kernel $p(dw|x,u)$ is not defined on all of SC by (4), but it is defined on a Borel subset of SC containing $\Gamma - \{(T,T)\}$, which is all that is necessary. Likewise, the system function f is not defined on all of SCW by (5)–(7), but the set of points where it is not defined has probability zero under any policy governing the system evolution. Both $p(dw|x,u)$ and f are Borel-measurable on their domains. Thus (SSM) is a special case of the stochastic optimal control model of Definition 8.1 $(N < \infty)$ or Definition 9.1 $(N = \infty)$.

If $N < \infty$, the (F^+) and (F^-) assumptions on (SSM) are given in Section 8.1. These are equivalent to the respective (F^+) and (F^-) assumptions on (NSM) given earlier in this section. If $N = \infty$, the (P), (N), and (D) assumptions on (SSM) of Definition 9.1 are equivalent to the respective (P), (N), and (D) assumptions on (NSM) given earlier in this section.

The reader can verify that there is a correspondence of policies between (NSM) and (SSM), and the optimal cost at $(x_k, k) \in S$ for (SSM) is $J^*(x_k, k)$ given by (3). Because of these facts, results already proved for (SSM) with either a finite or infinite horizon have immediate counterparts for (NSM). An illustration of this is the nonstationary optimality equation.

Proposition 10.1 (P)(N)(D)   Let $J^*(x_k, k)$ be defined by (3). For fixed k, $J^*(x_k, k)$ is lower semianalytic on S_k, and

$$J^*(x_k, k) = \inf_{u_k \in U_k(x_k)} \left\{ g_k(x_k, u_k) + \alpha \int_{S_{k+1}} J^*(x_{k+1}, k+1) t_k(dx_{k+1}|x_k, u_k) \right\}.$$

We do not list all the results for (NSM) that can be obtained from (SSM). The reader may verify, for example, that the existence results of Propositions 8.3 and 8.4, are valid for (NSM) in exactly the form stated. From Propositions 9.19 and 9.20 we conclude that, under (P) and (D), an ε-optimal nonrandomized Markov policy exists for (NSM), while under (N), an ε-optimal nonrandomized semi-Markov policy exists. In what follows, we make use of these results and reference only their stationary versions.

10.2   Reduction of the Imperfect State Information Model—Sufficient Statistics

Before defining the imperfect state information model, we give without proof some of the standard properties of conditional expectations and probabilities we will be using. For a detailed treatment, see Ash [A1]. Throughout this discussion, $(\Omega, \mathscr{F}, P)$ is a probability space and X is an extended real-valued random variable on Ω for which either $E[X^+]$ or $E[X^-]$ is finite.

If $\mathscr{D} \subset \mathscr{F}$ is a σ-algebra on Ω, then the *expectation of X conditioned on $\mathscr{D}$* is any $\mathscr{D}$-measurable, extended real-valued, random variable $E[X|\mathscr{D}](\cdot)$

on Ω which satisfies

$$\int_D X(\omega)P(d\omega) = \int_D E[X|\mathcal{D}](\omega)P(d\omega) \qquad \forall D \in \mathcal{D}.$$

It can be shown that at least one such random variable exists. Any such random variable will be called a *version* of $E[X|\mathcal{D}]$. If $X(\omega) \geq b$ for some $b \in R$ and every $\omega \in \Omega$, then it can be shown that for any version $E[X|\mathcal{D}](\cdot)$ the random variable $\hat{E}[X|\mathcal{D}](\cdot)$ defined by

$$\hat{E}[X|\mathcal{D}](\omega) = \max\{E[X|\mathcal{D}](\omega), b\},$$

is also a version of $E[X|\mathcal{D}]$. If $\mathscr{E} \subset \mathcal{D}$ is a collection of sets which is closed under finite intersections and generates the σ-algebra $\mathcal{D}$ and if Y is an extended real-valued, $\mathcal{D}$-measurable, random variable satisfying

$$\int_D X(\omega)P(d\omega) = \int_D Y(\omega)P(d\omega) \qquad \forall D \in \mathscr{E}, \tag{10}$$

then Y satisfies (10) for every $D \in \mathcal{D}$, and Y is a version of $E[X|\mathcal{D}]$. If $\mathscr{E} \subset \mathcal{D}$ is a σ-algebra, then

$$E\{E[X|\mathcal{D}]|\mathscr{E}\}(\omega) = E[X|\mathscr{E}](\omega) \tag{11}$$

for P almost every ω.

Suppose now that $(\Omega_1, \mathscr{F}_1)$ and $(\Omega_2, \mathscr{F}_2)$ are measurable spaces and $Y_1 : \Omega \to \Omega_1$ and $Y_2 : \Omega \to \Omega_2$ are measurable. Let $g : \Omega_1 \Omega_2 \to R^*$ be measurable and satisfy either $E[g^+(Y_1, Y_2)] < \infty$ or $E[g^-(Y_1, Y_2)] < \infty$. We define

$$E[X|Y_1](\omega) = E[X|\mathscr{F}(Y_1)](\omega),$$

where

$$\mathscr{F}(Y_1) = \{Y_1^{-1}(F)|F \in \mathscr{F}_1\}.$$

We define for $y_1 \in \Omega_1$

$$E[X|Y_1 = y_1] = E[X|Y_1](\omega(y_1)),$$

where $\omega(y_1)$ is any element of $Y_1^{-1}(\{y_1\})$. Since $E[X|Y_1]$ is $\mathscr{F}(Y_1)$-measurable, it is constant on $Y_1^{-1}(\{y_1\})$, and this definition makes sense. Note that $E[X|Y_1 = y_1]$ is a function of y_1, not of ω. We have for any $y_1 \in Y_1$

$$E[g(Y_1, Y_2)|Y_1 = y_1] = E[g(y_1, Y_2)] \tag{12}$$

for P almost every y_1. We use the phrase "for P almost every y_1" to indicate that, in this case,

$$P(\{\omega \in \Omega|(12) \text{ fails when } y_1 = Y_1(\omega)\}) = 0.$$

For $F \in \mathscr{F}_2$, define

$$P[Y_2 \in F|Y_1](\omega) = E[\chi_F(Y_2)|Y_1](\omega),$$
$$P[Y_2 \in F|Y_1 = y_1] = E[\chi_F(Y_2)|Y_1 = y_1].$$

Suppose $t(dy_2|y_1)$ is a stochastic kernel on $(\Omega_2, \mathscr{F}_2)$ given Ω_1 such that for every $F \in \mathscr{F}_2$

$$P[Y_2 \in F | Y_1 = y_1] = t(F|y_1)$$

for P almost every y_1. Then (12) can be extended to

$$E[g(Y_1, Y_2)|Y_1 = y_1] = E[g(y_1, Y_2)] = \int g(y_1, y_2) t(dy_2|y_1) \qquad (13)$$

for P almost every y_1. We will find (11) and (13) particularly useful in our treatment of the imperfect state information model. They will be used without reference to this discussion.

Definition 10.3 The *imperfect state information stochastic optimal control model* (ISI) is the ten-tuple $(S, C, (U_0, \ldots, U_{N-1}), Z, \alpha, g, t, s_0, s, N)$ described as follows:

S, C, α, g, t *State space, control space, discount factor, one-stage cost function, and state transition kernel* as given in Definition 8.1 and (3) of Chapter 8. We assume that g is defined on all of SC.

Z *Observation space.* A nonempty Borel space.

$U_k, k = 0, \ldots, N-1$ *Control constraints.* Define for $k = 0, \ldots, N-1$,

$$I_k = Z_0 C_0 \cdots C_{k-1} Z_k. \qquad (14)$$

An element of I_k is called a *kth information vector*. For each k, U_k is a mapping from I_k to the set of nonempty subsets of C such that

$$\Gamma_k = \{(i_k, u)|i_k \in I_k, u \in U_k(i_k)\} \qquad (15)$$

is analytic.

s_0 *Initial observation kernel.* A Borel-measurable stochastic kernel on Z given S.

s *Observation kernel.* A Borel-measurable stochastic kernel on Z given CS.

N *Horizon.* A positive integer or ∞.

For the sake of simplicity, we have eliminated the system function, disturbance space, and disturbance kernel from the model definition. In what follows, our notation will generally indicate a finite N. If $N = \infty$, the appropriate interpretation is required.

The system moves stochastically from state x_k to state x_{k+1} via the state transition kernel $t(dx_{k+1}|x_k, u_k)$ and generates cost at each stage of $g(x_k, u_k)$. The observation z_{k+1} is stochastically generated via the observation kernel $s(dz_{k+1}|u_k, x_{k+1})$ and added to the past observations and controls $(z_0, u_0, \ldots, z_k, u_k)$ to form the $(k+1)$st information vector $i_{k+1} = (z_0, u_0, \ldots, z_k, u_k, z_{k+1})$. The first information vector $i_0 = (z_0)$ is generated by the initial observation kernel $s_0(dz_0|x_0)$, and the initial state x_0 has some given initial distribution p. The goal is to choose u_k dependent on the kth information vector i_k so as to minimize

$$E\left\{\sum_{k=0}^{N-1} \alpha^k g(x_k, u_k)\right\}.$$

Definition 10.4   A *policy* for (ISI) is a sequence $\pi = (\mu_0, \ldots, \mu_{N-1})$ such that, for each k, $\mu_k(du_k|p; i_k)$ is a universally measurable stochastic kernel on C given $P(S)I_k$ satisfying

$$\mu_k(U_k(i_k)|p; i_k) = 1 \qquad \forall (p; i_k) \in P(S)I_k.$$

If for each p, k, and i_k, $\mu_k(du_k|p; i_k)$ assigns mass one to some point in C, π is *nonrandomized*.

The concepts of Markov and semi-Markov policies are of no use in (ISI), since the initial distribution, past observations, and past controls are of genuine value in estimating the current state. Thus we expect policies to depend on the initial distribution p and the total information vector. In the remainder of this chapter, Π will denote the set of all policies in (ISI).

Just as we denote the set of all sequences of the form $(z_0, u_0, \ldots, u_{k-1}, z_k) \in ZC \cdots CZ$ by I_k and call these sequences the kth information vectors, we find it notationally convenient to denote the set of all sequences of the form $(x_0, z_0, u_0, \ldots, x_k, z_k, u_k) \in SZC \cdots SZC$ by H_k and call these sequences the kth *history vectors*. Except for u_k, the kth information vector is that portion of the kth history vector known to the controller at the kth stage. Given $p \in P(S)$ and $\pi = (\mu_0, \ldots, \mu_{N-1}) \in \Pi$, by Proposition 7.45 there is a sequence of consistent probability measures $P_k(\pi, p)$ on H_k, $k = 0, \ldots, N-1$, defined on measurable rectangles by

$$P_k(\pi, p)(\underline{S}_0\underline{Z}_0\underline{C}_0 \cdots \underline{S}_k\underline{Z}_k\underline{C}_k)$$

$$= \int_{\underline{S}_0}\int_{\underline{Z}_0}\int_{\underline{C}_0} \cdots \int_{\underline{S}_k}\int_{\underline{Z}_k} \mu_k(\underline{C}_k|p; z_0, u_0, \ldots, u_{k-1}, z_k)s(dz_k|u_{k-1}, x_k)$$

$$\times t(dx_k|u_{k-1}, x_{k-1}) \cdots \mu_0(du_0|p; z_0)s_0(dz_0|x_0)p(dx_0). \qquad (16)$$

Definition 10.5   Given $p \in P(S)$, a policy $\pi = (\mu_0, \ldots, \mu_{N-1}) \in \Pi$, and a positive integer $K \le N$, the K-*stage cost corresponding* to π at p is

$$J_{K,\pi}(p) = \int_{H_{K-1}} \left[\sum_{k=0}^{K-1} \alpha^k g(x_k, u_k)\right] dP_{K-1}(\pi, p). \qquad (17)$$

If $N < \infty$, the *cost corresponding* to π is $J_{N,\pi}$, and we assume either

$$\int_{H_{N-1}} \left[\sum_{k=0}^{N-1} \alpha^k g^-(x_k, u_k)\right] dP_{N-1}(\pi, p) < \infty \qquad \forall \pi \in \Pi, \quad p \in P(S) \quad (\mathrm{F}^+)$$

or

$$\int_{H_{N-1}} \left[\sum_{k=0}^{N-1} \alpha^k g^+(x_k, u_k) \right] dP_{N-1}(\pi, p) < \infty \qquad \forall \pi \in \Pi, \quad p \in P(S). \quad (F^-)$$

If $N = \infty$, the *cost corresponding to* π is $J_\pi = \lim_{K \to \alpha} J_{K, \pi}$, and to ensure that this limit is a well-defined extended real number, we impose one of the following conditions:

(P) $0 \le g(x, u)$ for every $(x, u) \in SC$.

(N) $g(x, u) \le 0$ for every $(x, u) \in SC$.

(D) $0 < \alpha < 1$, and for some $b \in R$, $-b \le g(x, u) \le b$ for every $(x, u) \in SC$.

The *optimal cost* at p is

$$J_N^*(p) = \inf_{\pi \in \Pi} J_{N, \pi}(p).$$

The concepts of *optimality at* p, *optimality*, ε-*optimality at* p, and ε-*optimality* of policies are analogous to those given in Definition 8.3.

If $N < \infty$ and (F^+) or (F^-) holds, then by Lemma 7.11(b)

$$J_{N, \pi}(p) = \sum_{k=0}^{N-1} \alpha^k \int_{H_k} g(x_k, u_k) dP_k(\pi, p) \qquad \forall \pi \in \Pi, \quad p \in P(S). \qquad (18)$$

If $N = \infty$ and (P), (N), or (D) holds, then

$$J_\pi(p) = \sum_{k=0}^{\infty} \alpha^k \int_{H_k} g(x_k, u_k) dP_k(\pi, p) \qquad \forall \pi \in \Pi, \quad p \in P(S). \qquad (19)$$

To aid in the analysis of (ISI), we introduce the idea of a statistic sufficient for control. This statistic is defined in such a way that knowledge of its values is sufficient to control the model.

Definition 10.6   A *statistic* for the model (ISI) is a sequence $(\eta_0, \ldots, \eta_{N-1})$ of Borel-measurable functions $\eta_k : P(S)I_k \to Y_k$, where Y_k is a nonempty Borel space, $k = 0, \ldots, N - 1$. The statistic $(\eta_0, \ldots, \eta_{N-1})$ is *sufficient for control* provided:

(a)   For each k, there exists an analytic set $\hat{\Gamma}_k \subset Y_k C$ such that $\text{proj}_{Y_k}(\hat{\Gamma}_k) = Y_k$ and for every $p \in P(S)$

$$\Gamma_k = \{(i_k, u) | [\eta_k(p; i_k), u] \in \hat{\Gamma}_k\}, \qquad (20)$$

where Γ_k is defined by (15). We define

$$\hat{U}_k(y_k) = (\hat{\Gamma}_k)_{y_k}. \qquad (21)$$

(b)   There exist Borel-measurable stochastic kernels $\hat{t}_k(dy_{k+1} | y_k, u_k)$ on Y_{k+1} given $Y_k C$ such that for every $p \in P(S)$, $\pi \in \Pi$, $\underline{Y}_{k+1} \in \mathscr{B}_{Y_{k+1}}$, $k = 0, \ldots,$

$N - 2$, we have

$$P_{k+1}(\pi, p)[\eta_{k+1}(p; i_{k+1}) \in \underline{Y}_{k+1} | \eta_k(p; i_k) = \bar{y}_k, u_k = \bar{u}_k] = \hat{t}_k(\underline{Y}_{k+1} | \bar{y}_k, \bar{u}_k) \quad (22)$$

for $P_k(\pi, p)$ almost every $(\bar{y}_k, \bar{u}_k)$.[†]

(c)   There exist lower semianalytic functions $\hat{g}_k : \hat{\Gamma}_k \to [-\infty, \infty]$ satisfying for every $p \in P(S)$, $\pi \in \Pi$, $k = 0, \ldots, N - 1$,

$$E[g(x_k, u_k) | \eta_k(p; i_k) = \bar{y}_k, u_k = \bar{u}_k] = \hat{g}_k(\bar{y}_k, \bar{u}_k) \quad (23)$$

for $P_k(\pi, p)$ almost every $(\bar{y}_k, \bar{u}_k)$, where the expectation is with respect to $P_k(\pi, p)$.

Condition (a) of Definition 10.6 guarantees that the control constraint set $U_k(i_k)$ can be recovered from $\eta_k(p; i_k)$. Indeed, from (15), (20), and (21), we have for any $p \in P(S)$, $i_k \in I_k$, $k = 0, \ldots, N - 1$,

$$U_k(i_k) = \hat{U}_k[\eta_k(p; i_k)]. \quad (24)$$

If $U_k(i_k) = C$ for every $i_k \in I_k$, $k = 0, \ldots, N - 1$, then condition (a) is satisfied with $\hat{\Gamma}_k = Y_k C$. This is the case of no control constraint. Condition (b) guarantees that the distribution of y_{k+1} depends only on the values of y_k and u_k. This is necessary in order for the variables y_k to form the states of a stochastic optimal control model of the type considered in Section 10.1. Condition (c) guarantees that the cost corresponding to a policy can be computed from the distributions induced on the (y_k, u_k) pairs.

We temporarily postpone discussion on the existence and the nature of particular statistics sufficient for control, and consider first a perfect state information model corresponding to model (ISI) and a given sufficient statistic.

Definition 10.7   Let the model (ISI) and a statistic sufficient for control $(\eta_0, \ldots, \eta_{N-1})$ be given. The *perfect state information stochastic optimal control model*, denoted by (PSI), consists of the following (we use the notation of Definitions 10.3 and 10.6):

Y_k, $k = 0, \ldots, N - 1$ *State spaces.*
C *Control space.*
$\hat{U}_k$, $k = 0, \ldots, N - 1$ *Control constraints.*
α *Discount factor.*
$\hat{g}_k$, $k = 0, \ldots, N - 1$ *One-stage cost functions.*
$\hat{t}_k$, $k = 0, \ldots, N - 2$ *State transition kernels.*
N *Horizon.*

[†] In this context "for $P_k(\pi, p)$ almost every $(\bar{y}_k, \bar{u}_k)$" means that the set $\{(x_0, z_0, u_0, \ldots, x_k, z_k, u_k) \in H_k |$ (22) holds when $\bar{y}_k = \eta_k(p; i_k), \bar{u}_k = u_k\}$ has $P_k(\pi, p)$-measure one.

Thus defined, (PSI) is a nonstationary stochastic optimal control model in the sense of Definition 10.1.[†] The definitions of policies and cost functions for (PSI) are given in Section 10.1. We will use ($\hat{\ }$) to denote these objects in (PSI). For example, $\hat{\Pi}'$ is the set of all (0-originating) policies and $\hat{\Pi}$ is the set of all Markov (0-originating) policies for (PSI). If $\hat{\pi} = (\hat{\mu}_0, \ldots, \hat{\mu}_{N-1})$ is a policy for (PSI), then by (24) and Proposition 7.44 the sequence

$$(\hat{\mu}_0[du_0|\eta_0(p;i_0)], \ldots, \hat{\mu}_{N-1}[du_{N-1}|\eta_0(p;i_0), u_0, \ldots, u_{N-2}, \eta_{N-1}(p;i_{N-1})]),$$

where

$$i_k = (z_0, u_0, \ldots, u_{k-1}, z_k), \qquad k = 0, \ldots, N-1, \tag{25}$$

is a policy for (ISI). We call this policy $\hat{\pi}$ also, and can regard $\hat{\Pi}'$ as a subset of Π in this sense. If $\hat{\pi}$ is a nonrandomized policy for (PSI), then it is also nonrandomized when considered as a policy for (ISI). We will see in Proposition 10.2 that $\hat{\pi}$ results in the same cost for both (PSI) and (ISI).

Define $\varphi: P(S) \to P(Y_0)$ by

$$\varphi(p)(\underline{Y}_0) = \int_S s_0(\{z_0|\eta_0(p;z_0) \in \underline{Y}_0\}|x_0)p(dx_0) \qquad \forall \underline{Y}_0 \in \mathscr{B}_{Y_0} \tag{26}$$

Thus defined, $\varphi(p)$ is the distribution of the initial state y_0 in (PSI) when the initial state x_0 in (ISI) has distribution p. By Corollary 7.26.1, for every $\underline{Y}_0 \in \mathscr{B}_{Y_0}$ the mapping

$$\psi_{\underline{Y}_0}(x_0, p) = s_0(\{z_0|\eta_0(p;z_0) \in \underline{Y}_0\}|x_0)$$

is Borel-measurable. Define a Borel-measurable stochastic kernel on S given $P(S)$ by $q(dx_0|p) = p(dx_0)$. Then (26) can be written as

$$\varphi(p)(\underline{Y}_0) = \int \psi_{\underline{Y}_0}(x_0, p)q(dx_0|p).$$

It follows from Propositions 7.26 and 7.29 that φ is Borel-measurable. For $p \in P(S)$, define the mapping $V_{p,k}: H_k \to Y_0 C_0 \cdots Y_k C_k$ by

$$V_{p,k}(x_0, z_0, u_0, \ldots, x_k, z_k, u_k) = [\eta_0(p;i_0), u_0, \ldots, \eta_k(p;i_k), u_k], \tag{27}$$

where (25) holds. For $q \in P(Y_0)$ and $\hat{\pi} = (\hat{\mu}_0, \ldots, \hat{\mu}_{N-1}) \in \hat{\Pi}'$, there is a sequence of consistent probability measures $\hat{P}_k(\hat{\pi}, q)$ generated on $Y_0 C_0 \cdots Y_k C_k$, $k = 0, \ldots, N-1$, defined on measurable rectangles by

$$\hat{P}_k(\hat{\pi}, q)(\underline{Y}_0\underline{C}_0 \cdots \underline{Y}_k\underline{C}_k) = \int_{\underline{Y}_0} \int_{\underline{C}_0} \cdots \int_{\underline{Y}_k} \hat{\mu}_k(\underline{C}_k|y_0, u_0, \ldots, u_{k-1}, y_k)$$

$$\times \hat{t}_{k-1}(dy_k|y_{k-1}, u_{k-1}) \cdots \hat{\mu}_0(du_0|y_0)q(dy_0). \tag{28}$$

[†] The disturbance spaces, disturbance kernels, and system functions in (PSI) can be taken to be $W_k = Y_{k+1}$, $p_k(dw_k|y_k, u_k) = \hat{t}_k(dy_{k+1}|y_k, u_k)$, and $f_k(y_k, u_k, w_k) = w_k$, respectively.

For a Markov policy $\hat{\pi} \in \hat{\Pi}$, these objects are related to the probability measures $P_k(\hat{\pi}, p)$ defined by (16) in the following manner.

Lemma 10.1 Suppose $p \in P(S)$ and $\hat{\pi} \in \hat{\Pi}$. Then for $k = 0, \ldots, N-1$ and for every Borel set $B \subset Y_0 C_0 \cdots Y_k C_k$, we have

$$P_k(\hat{\pi}, p)[V_{p,k}^{-1}(B)] = \hat{P}_k[\hat{\pi}, \varphi(p)](B). \tag{29}$$

Proof It suffices to prove that if $\underline{Y}_0 \in \mathscr{B}_{Y_0}$, $\underline{C}_0 \in \mathscr{B}_{C_0}, \ldots, \underline{Y}_k \in \mathscr{B}_{Y_k}$, $\underline{C}_k \in \mathscr{B}_{C_k}$, then

$$P_k(\hat{\pi}, p)(\{\eta_0(p; i_0) \in \underline{Y}_0, u_0 \in \underline{C}_0, \ldots, \eta_k(p; i_k) \in \underline{Y}_k, u_k \in \underline{C}_k\})$$
$$= \hat{P}_k[\hat{\pi}, \varphi(p)](\underline{Y}_0 \underline{C}_0 \cdots \underline{Y}_k \underline{C}_k).^{\dagger} \tag{30}$$

For $k = 0$, (30) follows from (16), (26), and (28). If (29) holds for some $k < N$, then using (16), (22), (28), and (29), we obtain

$$P_{k+1}(\hat{\pi}, p)(\{\eta_0(p; i_0) \in \underline{Y}_0, u_0 \in \underline{C}_0, \ldots, \eta_{k+1}(p; i_{k+1}) \in \underline{Y}_{k+1}, u_{k+1} \in \underline{C}_{k+1}\})$$

$$= \int_{\{\eta_0(p; i_0) \in \underline{Y}_0, u_0 \in \underline{C}_0, \ldots, \eta_k(p; i_k) \in \underline{Y}_k, u_k \in \underline{C}_k\}} \int_{\underline{Y}_{k+1}} \hat{\mu}_{k+1}(\underline{C}_{k+1}|y_{k+1})$$
$$\times \hat{t}_k(dy_{k+1}|\eta_k(p; i_k), u_k) \, dP_k(\hat{\pi}, p)$$

$$= \int_{\underline{Y}_0 \underline{C}_0 \cdots \underline{Y}_k \underline{C}_k} \int_{\underline{Y}_{k+1}} \hat{\mu}_{k+1}(\underline{C}_{k+1}|y_{k+1}) \hat{t}_k(dy_{k+1}|y_k, u_k) \, d\hat{P}_k[\hat{\pi}, \varphi(p)]$$

$$= \hat{P}_{k+1}[\hat{\pi}, \varphi(p)](\underline{Y}_0 \underline{C}_0 \cdots \underline{Y}_{k+1} \underline{C}_{k+1}). \quad \text{Q.E.D.}$$

As noted earlier, (PSI) is a model of the type considered in Section 10.1. The (F^+) and (F^-) conditions of Section 10.1, when specialized to the (PSI) model, will be denoted by $(\hat{F}^+)$ and $(\hat{F}^-)$, respectively. These conditions are not to be confused with the (F^+) and (F^-) conditions for the ISI model given in this section. In a particular problem it is often possible to see the relationship between these finiteness conditions on the two models. In the general case, the relationship is unclear. We point out, however, that if g is bounded below or above, then (F^+) or (F^-) is satisfied for (ISI), respectively, and given any statistic sufficient for control, the corresponding $\hat{g}_k$ can be chosen so as to be bounded below or above, respectively. If a particular result holds when we assume (F^+) on the (ISI) model and $(\hat{F}^+)$ on the (PSI) model, the notation $(F^+, \hat{F}^+)$ will appear. The notation $(F^-, \hat{F}^-)$ has a similar meaning.

---

† In this context, we define

$$\{\eta_0(p; i_0) \in \underline{Y}_0, u_0 \in \underline{C}_0, \ldots, \eta_k(p; i_k) \in \underline{Y}_k, u_k \in \underline{C}_k\}$$
$$= \{(x_0, z_0, u_0, \ldots, x_k, z_k, u_k) | \eta_0(p; i_0) \in \underline{Y}_0, u_0 \in \underline{C}_0, \ldots, \eta_k(p; i_k) \in \underline{Y}_k, u_k \in \underline{C}_k\}.$$

where $i_j = (z_0, u_0, \ldots, u_{j-1}, z_j)$. We will often use this notation to indicate a set which depends on functions of some or all of the components of a Cartesian product.

If $N = \infty$, we consider conditions (P), (N), and (D) for (ISI) and the corresponding conditions ($\hat{P}$), ($\hat{N}$), and ($\hat{D}$) for (PSI). In this case, however, if (P) holds for (ISI) and lower semianalytic functions $\hat{g}_k : \hat{\Gamma}_k \to [-\infty, \infty]$ satisfying (23) exist, there is no loss of generality in assuming that $\hat{g}_k \geq 0$ for every k, i.e., ($\hat{P}$) holds for (PSI). Likewise, if (N) or (D) holds for (ISI), we may assume without loss of generality that ($\hat{N}$) or ($\hat{D}$), respectively, holds for (PSI). As in the finite horizon case, we adopt the notation (P, $\hat{P}$), (N, $\hat{N}$), and (D, $\hat{D}$) to indicate which assumptions are sufficient for a result to hold.

From Section 10.1, we have that when ($\hat{F}^+$), ($\hat{F}^-$), ($\hat{P}$), ($\hat{N}$), or ($\hat{D}$) holds, then the (0-originating) *cost corresponding to a policy* $\hat{\pi}$ for (PSI) at $y \in Y_0$ is

$$\hat{J}_{N,\hat{\pi}}(y) = \sum_{k=0}^{N-1} \alpha^k \int_{Y_0 C_0 \cdots Y_k C_k} \hat{g}_k(y_k, u_k) \, d\hat{P}_k(\hat{\pi}, p_y), \tag{31}$$

where N may be infinite. The (0-originating) *optimal cost* for (PSI) at $y \in Y_0$ is

$$\hat{J}_N^*(y) = \inf_{\hat{\pi} \in \hat{\Pi}'} \hat{J}_{N,\hat{\pi}}(y). \tag{32}$$

The remainder of this section is devoted to establishing relations between costs, optimal costs, and optimal and nearly optimal policies for the (ISI) and (PSI) models.

Proposition 10.2 $(F^+, \hat{F}^+)(F^-, \hat{F}^-)(P, \hat{P})(N, \hat{N})(D, \hat{D})$ For every $p \in P(S)$ and $\hat{\pi} \in \hat{\Pi}$, we have

$$J_{N,\hat{\pi}}(p) = \int_{Y_0} \hat{J}_{N,\hat{\pi}}(y_0) \varphi(p)(dy_0). \tag{33}$$

Proof   From (31), (28), (23), (18), (19), and Lemma 10.1, we have

$$\int_{Y_0} \hat{J}_{N,\hat{\pi}}(y) \varphi(p)(dy) = \sum_{k=0}^{N-1} \alpha^k \int_{Y_0} \int_{Y_0 C_0 \cdots Y_k C_k} \hat{g}_k(y_k, u_k) \, d\hat{P}_k(\hat{\pi}, p_y) \varphi(p)(dy)$$

$$= \sum_{k=0}^{N-1} \alpha^k \int_{Y_0 C_0 \cdots Y_k C_k} \hat{g}_k(y_k, u_k) \, d\hat{P}_k[\hat{\pi}, \varphi(p)]$$

$$= \sum_{k=0}^{N-1} \alpha^k \int_{H_k} g(x_k, u_k) \, dP_k(\hat{\pi}, p) = J_{N,\hat{\pi}}(p),$$

where the (F^+) or (F^-) assumption is used to interchange integration and summation when $N < \infty$, and the monotone or bounded convergence theorem is used when $N = \infty$.   Q.E.D.

Corollary 10.2.1 $(F^+, \hat{F}^+)(F^-, \hat{F}^-)(P, \hat{P})(N, \hat{N})(D, \hat{D})$ For every $p \in P(S)$, we have

$$J_N^*(p) \leq \int_{Y_0} \hat{J}_N^*(y_0) \varphi(p)(dy_0). \tag{34}$$

Proof   The function $\hat{J}_N^*(y_0)$ is lower semianalytic, so the integral in (34) is defined. From Proposition 10.2, we have

$$J_N^*(p) = \inf_{\pi \in \Pi} J_{N,\pi}(p) \leq \inf_{\hat{\pi} \in \hat{\Pi}} \int_{Y_0} \hat{J}_{N,\hat{\pi}}(y_0)\varphi(p)(dy_0),$$

so it suffices to show that

$$\inf_{\hat{\pi} \in \hat{\Pi}} \int_{Y_0} \hat{J}_{N,\hat{\pi}}(y_0)\varphi(p)(dy_0) = \int_{Y_0} \hat{J}_N^*(y_0)\varphi(p)(dy_0). \tag{35}$$

This follows from Lemma 8.6 and Corollary 9.5.2.   Q.E.D.

We wish now to establish a relationship similar to (33) between the optimal cost functions for (ISI) and (PSI). In light of Corollary 10.2.1, it suffices to show that given any policy for (ISI), a policy for (PSI) can be found which does at least as well. This is formalized in the next lemma, and the analog of (33) is given as part of Proposition 10.3.

Lemma 10.2 $(F^+, \hat{F}^+)(F^-, \hat{F}^-)(P, \hat{P})(N, \hat{N})(D, \hat{D})$ Given $p \in P(S)$ and $\pi \in \Pi$, there exists $\hat{\pi} \in \hat{\Pi}$ such that

$$J_{N,\pi}(p) = \int_{Y_0} \hat{J}_{N,\hat{\pi}}(y_0)\varphi(p)(dy_0). \tag{36}$$

Proof   Let $p \in P(S)$ and $\pi = (\mu_0, \dots, \mu_{N-1}) \in \Pi$ be given. For $k = 0, \dots, N-1$, let $Q_k(\pi, p)$ be the probability measure on $Y_k C_k$ defined on measurable rectangles to be

$$Q_k(\pi, p)(\underline{Y_k}\underline{C_k}) = P_k(\pi, p)(\{\eta_k(p; i_k) \in \underline{Y_k}, u_k \in \underline{C_k}\}). \tag{37}$$

There exists a Borel-measurable stochastic kernel $\hat{\mu}_k(du_k|y_k)$ on C_k given Y_k such that for every Borel set $B \subset Y_k C_k$ we have

$$Q_k(\pi, p)(B) = \int_{Y_k C_k} \hat{\mu}_k(B_{y_k}|y_k)\, dQ_k(\pi, p). \tag{38}$$

In particular,

$$\begin{aligned}
1 &= P_k(\pi, p)(\{(i_k, u_k) \in \Gamma_k\}) \\
&= P_k(\pi, p)(\{[\eta_k(p; i_k), u_k] \in \hat{\Gamma}_k\}) \\
&= Q_k(\pi, p)(\hat{\Gamma}_k) = \int_{Y_k C_k} \hat{\mu}_k(\hat{U}_k(y_k)|y_k)\, dQ_k(\pi, p),
\end{aligned}$$

so, altering $\hat{\mu}_k(du_k|y_k)$ on a set of measure zero if necessary, we may assume that (38) holds and $\hat{\mu}_k(\hat{U}_k(y_k)|y_k) = 1$ for every $y_k \in Y_k$. Let $\hat{\pi} = (\hat{\mu}_0, \dots, \hat{\mu}_{N-1})$. Then $\hat{\pi}$ is a Markov policy for (PSI).

We show by induction that for $\underline{Y_k} \in \mathscr{B}_{Y_k}$, $\underline{C_k} \in \mathscr{B}_C$, $k = 0, \dots, N-1$,

$$Q_k(\pi, p)(\underline{Y_k}\underline{C_k}) = \hat{P}_k[\hat{\pi}, \varphi(p)](\{y_k \in \underline{Y_k}, u_k \in \underline{C_k}\}). \tag{39}$$

We see from (26) and (37) that the marginal of $Q_0(\pi, p)$ on Y_0 is $\varphi(p)$. Equation (39) for $k = 0$ follows from (28) and (38). Assume that (39) holds for k. From (38), (37), (22), and the induction hypothesis, we have

$$Q_{k+1}(\pi, p)(\underline{Y}_{k+1}\underline{C}_{k+1})$$

$$= \int_{\underline{Y}_{k+1}\underline{C}_{k+1}} \hat{\mu}_{k+1}(\underline{C}_{k+1}|y_{k+1}) dQ_{k+1}(\pi, p)$$

$$= \int_{\{\eta_{k+1}(p; i_{k+1}) \in \underline{Y}_{k+1}\}} \hat{\mu}_{k+1}(\underline{C}_{k+1}|\eta_{k+1}(p; i_{k+1})) dP_{k+1}(\pi, p)$$

$$= \int_{H_k} \int_{\underline{Y}_{k+1}} \hat{\mu}_{k+1}(\underline{C}_{k+1}|y_{k+1}) \hat{t}_k(dy_{k+1}|\eta_k(p; i_k), u_k) dP_k(\pi, p)$$

$$= \int_{Y_k C_k} \int_{\underline{Y}_{k+1}} \hat{\mu}_{k+1}(\underline{C}_{k+1}|y_{k+1}) \hat{t}_k(dy_{k+1}|y_k, u_k) dQ_k(\pi, p)$$

$$= \int_{Y_0 C_0 \cdots Y_k C_k} \int_{\underline{Y}_{k+1}} \hat{\mu}_{k+1}(\underline{C}_{k+1}|y_{k+1}) \hat{t}_k(dy_{k+1}|y_k, u_k) d\hat{P}_k[\hat{\pi}, \varphi(p)]$$

$$= P_{k+1}[\hat{\pi}, \varphi(p)](\{y_{k+1} \in \underline{Y}_{k+1}, u_k \in \underline{C}_{k+1}\}).$$

Taken together, (37) and (39) imply that for $\underline{Y}_k \in \mathscr{B}_{Y_k}, \underline{C}_k \in \mathscr{B}_C, k = 0, \ldots, N - 1$, we have

$$P_k(\pi, p)(\{\eta_k(p; i_k) \in \underline{Y}_k, u_k \in \underline{C}_k\}) = \hat{P}_k[\hat{\pi}, \varphi(p)](\{y_k \in \underline{Y}_k, u_k \in \underline{C}_k\}). \quad (40)$$

If (40) is used in place of Lemma 10.1, the proof of Proposition 10.2 can now be used to prove (36).   Q.E.D.

Definition 10.8   Given $q \in P(Y_0)$ and $\varepsilon > 0$, a policy $\hat{\pi} \in \hat{\Pi}'$ is said to be *weakly q-ε-optimal* if

$$\int_{Y_0} \hat{J}_{N, \hat{\pi}}(y_0) q(dy_0) \leq \begin{cases} \int\int_{Y_0} \hat{J}_N^*(y_0) q(dy_0) + \varepsilon & \text{if } \int_{Y_0} \hat{J}_N^*(y_0) q(dy_0) > -\infty, \\ -1/\varepsilon & \text{if } \int_{Y_0} \hat{J}_N^*(y_0) q(dy_0) = -\infty. \end{cases}$$

The policy $\hat{\pi}$ is said to be *q-optimal* if $q(\{y_0 \in Y_0 | \hat{J}_{N, \hat{\pi}}(y_0) = \hat{J}_N^*(y_0)\}) = 1$.

Equation (35) shows that given any $p \in P(S)$ and $\varepsilon > 0$, a weakly $\varphi(p)$-ε-optimal Markov policy exists. The next proposition shows that such a policy is ε-optimal at p when considered as a policy in (ISI).

Proposition 10.3 $(F^+, \hat{F}^+)(F^-, \hat{F}^-)(P, \hat{P})(N, \hat{N})(D, \hat{D})$ We have

$$J_N^*(p) = \int_{Y_0} \hat{J}_N^*(y_0) \varphi(p)(dy_0) \qquad \forall p \in P(S). \quad (41)$$

Furthermore, if $\hat{\pi}$ is optimal, $\varphi(p)$-optimal, or weakly $\varphi(p)$-ε-optimal for (PSI), then $\hat{\pi}$ is optimal, optimal at p, or ε-optimal at p, respectively, for (ISI). If $\hat{\pi}$ is ε-optimal for (PSI) and $(F^+, \hat{F}^+)$, $(P, \hat{P})$, or $(D, \hat{D})$ holds, then $\hat{\pi}$ is also ε-optimal for (ISI).

Proof   Equation (41) follows from Corollary 10.2.1 and Lemma 10.2. Let $\hat{\pi}$ be ε-optimal for (PSI). It is clear that under $(P, \hat{P})$ and $(D, \hat{D})$, we have

$$\hat{J}_N^*(y_0) > -\infty \qquad \forall y_0 \in Y_0, \tag{42}$$

so

$$\hat{J}_{N, \hat{\pi}}(y_0) \le \hat{J}_N^*(y_0) + \varepsilon \qquad \forall y_0 \in Y_0. \tag{43}$$

Under $(F^+, \hat{F}^+)$, (42) follows from Lemma 8.3 and Proposition 8.2, so again (43) holds. We have from (41) and Proposition 10.2 that

$$
\begin{aligned}
J_{N, \hat{\pi}}(p) &= \int_{Y_0} \hat{J}_{N, \hat{\pi}}(y_0)\varphi(p)(dy_0) \\
&\le \int_{Y_0} \hat{J}_N^*(y_0)\varphi(p)(dy_0) + \varepsilon \\
&= J_N^*(p) + \varepsilon \qquad \forall p \in P(S),
\end{aligned}
$$

so $\hat{\pi}$ is ε-optimal for (ISI). The remainder of the proposition follows from (41) and Proposition 10.2.   Q.E.D.

We shall show shortly that a statistic sufficient for control always exists, and indeed, in many cases it can be chosen so that (PSI) is stationary. The existence of such a statistic for (ISI) and the consequent existence of the corresponding model (PSI) enable us to utilize the results of Chapters 8 and 9. For example, we have the following proposition.

Proposition 10.4 $(F^+, \hat{F}^+)(F^-, \hat{F}^-)(P, \hat{P})(N, \hat{N})(D, \hat{D})$ If $(\eta_0, \ldots, \eta_{N-1})$ is a statistic sufficient for control for (ISI), then for every $\varepsilon > 0$, there exists an ε-optimal nonrandomized policy for (ISI) which depends on $i_k = (z_0, u_0, \ldots, u_{k-1}, z_k)$ only through $\eta_k(p; i_k)$, i.e., has the form

$$\pi = (\mu_0[p; \eta_0(p; i_0)], \ldots, \mu_{N-1}[p; \eta_{N-1}(p; i_{N-1})]). \tag{44}$$

Under $(F^+, \hat{F}^+)$, $(P, \hat{P})$, or $(D, \hat{D})$, we may choose this ε-optimal policy to have the simpler form

$$\hat{\pi} = (\hat{\mu}_0[\eta_0(p; i_0)], \ldots, \hat{\mu}_{N-1}[\eta_{N-1}(p; i_{N-1})]). \tag{45}$$

Proof   Under $(F^+, \hat{F}^+)$, $(P, \hat{P})$, or $(D, \hat{D})$, there exists an ε-optimal, non-randomized, Markov policy $\hat{\pi} = (\hat{\mu}_0, \ldots, \hat{\mu}_{N-1})$ for (PSI) (Propositions 8.3 and 9.19). This policy $\hat{\pi}$ is ε-optimal for (ISI) by Proposition 10.3, and the second part of the proposition is proved.

Assume $(F^-, \hat{F}^-)$ holds and let $\{\varepsilon_n\}$ be a sequence of positive numbers with $\sum_{n=1}^{\infty} \varepsilon_n < \infty$ and $\varepsilon_n \downarrow 0$. Let $\hat{\pi}_n = (\hat{\mu}_0^n, \ldots, \hat{\mu}_{N-1}^n)$ be a sequence of non-randomized Markov policies for (PSI) exhibiting $\{\varepsilon_n\}$-dominated convergence to optimality (Proposition 8.4). By Proposition 10.2 and the $(F^-, \hat{F}^-)$

assumption, we have

$$\int_{Y_0} \hat{J}_{N,\hat{\pi}_n}(y_0)\varphi(p)(dy_0) = J_{N,\hat{\pi}_n}(p) < \infty \qquad \forall p \in P(S).$$

Since

$$\hat{J}_{N,\hat{\pi}_n}(y_0) + \sum_{k=n+1}^{\infty} \varepsilon_k \downarrow \hat{J}_N^*(y_0) \qquad \forall y_0 \in Y_0,$$

we have

$$\lim_{n\to\infty} \int_{Y_0} \hat{J}_{N,\hat{\pi}_n}(y_0)\varphi(p)(dy_0) = \int_{Y_0} \hat{J}^*(y_0)\varphi(p)(dy_0) \qquad \forall p \in P(S).$$

Let $\varepsilon > 0$ be given and let $n(p)$ be the smallest positive integer n for which

$$\int_{Y_0} \hat{J}_{N,\hat{\pi}_n}(y_0)\varphi(p)(dy_0) \leq \begin{cases} \int_{Y_0} \hat{J}^*(y_0)\varphi(p)(dy_0) + \varepsilon \\ \qquad \text{if } \int_{Y_0} \hat{J}^*(y_0)\varphi(p)(dy_0) > -\infty, \\ -1/\varepsilon \\ \qquad \text{if } \int_{Y_0} \hat{J}^*(y_0)\varphi(p)(dy_0) = -\infty. \end{cases}$$

Define $\mu_k(p; y_k) = \hat{\mu}_k^{n(p)}(y_k)$, $k = 0, \ldots, N-1$. Then by Propositions 10.2 and 10.3, π given by (44) is an ε-optimal nonrandomized policy for (ISI).

Assume $(N, \hat{N})$ holds. Consider the nonstationary stochastic optimal control model (NPSI) for which the initial state space is $P(Y_0)$, the initial control space is a singleton set $\{u_0\}$, the initial cost function is $g_0(q, u_0) = 0$ for every $q \in P(Y_0)$, and the initial transition kernel is given by $t(dy_0|q, u_0) = q(dy_0)$ for every $q \in P(Y_0)$. For $k \geq 0$, the $(k+1)$st state and control spaces, control constraint, cost function, and transition kernel are Y_k, C, $\hat{U}_k$, $\hat{g}_k$, and $\hat{t}_k(dy_{k+1}|y_k, u_k)$ of (PSI), respectively. The discount factor is α and the horizon is infinite. By definition, the optimal cost for (NPSI) at $q \in P(Y_0)$ is

$$\inf_{\hat{\pi}\in\hat{\Pi}} \int_{Y_0} \hat{J}_{\hat{\pi}}(y_0) q(dy_0),$$

which, by Corollaries 9.1.1 and 9.5.2, is the same as

$$\int_{Y_0} \hat{J}^*(y_0) q(dy_0).$$

Now (NPSI) has a nonpositive one-stage cost function, so, by Proposition 9.20, for each $\varepsilon > 0$ there exists an ε-optimal, nonrandomized, semi-Markov policy

$$\bar{\pi} = (\bar{\mu}(q), \bar{\mu}_0(q; y_0), \bar{\mu}_1(q; y_1), \ldots).$$

For fixed $q \in P(Y_0)$, let $\hat{\pi}(q)$ be the policy for (PSI) given by

$$\hat{\pi}(q) = (\bar{\mu}_0(q; y_0), \bar{\mu}_1(q; y_1), \ldots).$$

Then

$$
\int_{Y_0} \hat{J}_{\hat{\pi}(q)}(y_0)q(dy_0) \leq
\begin{cases}
\int_{Y_0} \hat{J}^*(y_0)q(dy_0) + \varepsilon & \text{if } \int_{Y_0} \hat{J}^*(y_0)q(dy_0) > -\infty, \\[2mm]
-1/\varepsilon & \text{if } \int_{Y_0} \hat{J}^*(y_0)q(dy_0) = -\infty,
\end{cases}
$$

i.e., $\hat{\pi}(q)$ is weakly q-ε-optimal for (PSI). By Proposition 10.3, the policy π defined by (44), where $\mu_k(p; y_k) = \bar{\mu}(\varphi(p); y_k)$, is ε-optimal for (ISI).   Q.E.D.

The other specific results which can be derived for (ISI) from Chapters 8 and 9 are obvious and shall not be exhaustively listed. We content ourselves with describing the dynamic programming algorithm over a finite horizon.

By Proposition 8.2, the dynamic programming algorithm has the following form under $(\mathrm{F}^+, \hat{\mathrm{F}}^+)$ or $(\mathrm{F}^-, \hat{\mathrm{F}}^-)$, where we assume for notational simplicity that (PSI) is stationary:

$$
\hat{J}_0^*(y) = 0 \qquad \forall y \in Y, \tag{46}
$$

$$
\hat{J}_{k+1}^*(y) = \inf_{u \in \hat{U}(y)} \{\hat{g}(y, u) + \alpha \int \hat{J}_k^*(y')\hat{t}(dy'|y, u)\}, \qquad k = 0, \ldots, N-1. \tag{47}
$$

If the infimum in (47) is achieved for every y and $k = 0, \ldots, N-1$, then there exist universally measurable functions $\hat{\mu}_k: Y \to C$ such that for every y and $k = 0, \ldots, N-1$, $\hat{\mu}_k(y) \in \hat{U}(y)$ and $\hat{\mu}_k(y)$ achieves the infimum in (47). Then $\hat{\pi} = (\hat{\mu}_0, \ldots, \hat{\mu}_{N-1})$ is optimal in (PSI) (Proposition 8.5), so $\hat{\pi}$ is optimal in (ISI) as well (Proposition 10.3).

If $(\mathrm{F}^+, \hat{\mathrm{F}}^+)$ holds and the infimum in (47) is not achieved for every y and $k = 0, \ldots, N-1$, the dynamic programming algorithm (46) and (47) can still be used in the manner of Proposition 8.3 to construct an ε-optimal, nonrandomized, Markov policy $\hat{\pi}$ for (PSI). We see from Proposition 10.3 that $\hat{\pi}$ is an ε-optimal policy for (ISI) as well.

In many cases, $\eta_{k+1}(p; i_{k+1})$ is a function of $\eta_k(p; i_k)$, u_k, and z_{k+1}. The computational procedure in such a case is to first construct $(\hat{\mu}_0, \ldots, \hat{\mu}_{N-1})$ via (46) and (47), then compute $y_0 = \eta_0(p; i_0)$ from the initial distribution and the initial observation, and apply control $u_0 = \hat{\mu}_0(y_0)$. Given y_k, u_k, and z_{k+1}, compute y_{k+1} and apply control $u_{k+1} = \hat{\mu}_{k+1}(y_{k+1})$, $k = 0, \ldots, N-2$. In this way the information contained in $(p; i_k)$ has been condensed into y_k. This condensation of information is the historical motivation for statistics sufficient for control.

10.3   Existence of Statistics Sufficient for Control

Turning to the question of the existence of a statistic sufficient for control, it is not surprising to discover that the sequence of identity mappings on $P(S)I_k$, $k = 0, \ldots, N-1$, is such an object (Proposition 10.6). Although this represents no condensation of information, it is sufficient to justify our analysis thus far. We will show that if the constraint sets Γ_k are equal to $I_k C$, $k = 0, \ldots, N - 1$, then the functions mapping $P(S)I_k$ into the distribution of x_k conditioned on $(p; i_k)$, $k = 0, \ldots, N - 1$, constitute a statistic sufficient for control (Proposition 10.5). This statistic has the property that its value at the $(k + 1)$st stage is a function of its value at the kth stage, u_k and z_{k+1} [see (52)], so it represents a genuine condensation of information. It also results in a stationary perfect state information model and, if the conditional distributions can be characterized by a finite set of parameters, it may result in significant computational simplification. This latter condition is the case, for example, if it is possible to show beforehand that all these distributions are Gaussian.

10.3.1   Filtering and the Conditional Distributions of the States

We discuss filtering with the aid of the following basic lemma.

Lemma 10.3   Consider the (ISI) model. There exist Borel-measurable stochastic kernels $r_0(dx_0|p; z_0)$ on S given $P(S)Z$ and $r(dx|p; u, z)$ on S given $P(S)CZ$ which satisfy

$$\int_{\underline{S}_0} s_0(\underline{Z}_0|x_0)p(dx_0) = \int_S \int_{\underline{Z}_0} r_0(\underline{S}_0|p; z_0)s_0(dz_0|x_0)p(dx_0)$$

$$\forall \underline{S}_0 \in \mathscr{B}_S, \quad \underline{Z}_0 \in \mathscr{B}_Z, \quad p \in P(S), \qquad (48)$$

$$\int_{\underline{S}} s(\underline{Z}|u, x)p(dx) = \int_S \int_{\underline{Z}} r(\underline{S}|p; u, z)s(dz|u, x)p(dx)$$

$$\forall \underline{S} \in \mathscr{B}_S, \quad \underline{Z} \in \mathscr{B}_Z, \quad p \in P(S), \quad u \in C. \qquad (49)$$

Proof   For fixed $(p; u) \in P(S)C$, define a probability measure q on SZ by specifying its values on measurable rectangles to be (Proposition 7.28)

$$q(\underline{S}\underline{Z}|p; u) = \int_{\underline{S}} s(\underline{Z}|u, x)p(dx).$$

By Propositions 7.26 and 7.29, $q(d(x, z)|p; u)$ is a Borel-measurable stochastic kernel on SZ given $P(S)C$. By Corollary 7.27.1, this stochastic kernel can be decomposed into its marginal on Z given $P(S)C$ and a Borel-measurable stochastic kernel $r(dx|p; u, z)$ on S given $P(S)CZ$ such that (49) holds. The existence of $r_0(dx_0|p; z_0)$ is proved in a similar manner.   Q.E.D.

It is customary to call p, the given distribution of x_0, the *a priori distribution* of the initial state. After z_0 is observed, the distribution is "up-dated", i.e., the distribution of x_0 conditioned on z_0 is computed. The up-dated distribution is called the *a posteriori distribution* and, as we will show in Lemma 10.4, is just $r_0(dx_0|p; z_0)$. At the kth stage, $k \geq 1$, we will have some a priori distribution p'_k of x_k based on $i_{k-1} = (z_0, u_0, \ldots, u_{k-2}, z_{k-1})$. Control

u_{k-1} is applied, some z_k is observed, and an a posteriori distribution of x_k conditioned on (i_{k-1}, u_{k-1}, z_k) is computed. We will show that this distribution is just $r(dx|p'_k; u_{k-1}, z_k)$. The process of passing from an a priori to an a posteriori distribution in this manner is called *filtering*, and it is formalized next.

Consider the function $\bar{f} : P(S)C \to P(S)$ defined by

$$\bar{f}(p, u)(\underline{S}) = \int t(\underline{S}|x, u)p(dx) \qquad \forall \underline{S} \in \mathcal{B}_S. \tag{50}$$

Equation (50) is called the *one-stage prediction equation*. If x_k has an a posteriori distribution p_k and the control u_k is chosen, then the a priori distribution of x_{k+1} is $\bar{f}(p_k, u_k)$. The mapping $\bar{f}$ is Borel-measurable (Propositions 7.26 and 7.29).

Given a sequence $i_k \in I_k$ such that $i_{k+1} = (i_k, u_k, z_{k+1})$, $k = 0, \ldots, N-2$, and given $p \in P(S)$, define recursively

$$p_0(p; i_0) = r_0(dx_0|p; z_0), \tag{51}$$

$$p_{k+1}(p; i_{k+1}) = r(dx|\bar{f}[p_k(p; i_k), u_k]; u_k, z_{k+1}), \qquad k = 0, \ldots, N-2. \tag{52}$$

Note that for each k, $p_k : P(S)I_k \to P(S)$ is Borel-measurable.

Equations (48)–(52) are called the *filtering equations* corresponding to the (ISI) model. For a given initial distribution and policy, they generate the conditional distribution of the state given the current information, as the following lemma shows,

Lemma 10.4 Let the model (ISI) be given. For any $p \in P(S)$, $\pi = (\mu_0, \ldots, \mu_{N-1}) \in \Pi$ and $\underline{S}_k \in \mathcal{B}_S$, we have

$$P_k(\pi, p)[x_k \in \underline{S}_k | i_k] = p_k(p; i_k)(\underline{S}_k) \tag{53}$$

for $P_k(\pi, p)$ almost every i_k, $k = 0, \ldots, N-1$.

Proof[†] We proceed by induction. For any $\underline{S}_0 \in \mathcal{B}_S$ and $\underline{Z}_0 \in \mathcal{B}_Z$, we have from (51), (16), and (48), that

$$\int_{\{z_0 \in \underline{Z}_0\}} p_0(p; z_0)(\underline{S}_0) \, dP_0(\pi, p) = \int_{\{z_0 \in \underline{Z}_0\}} r_0(\underline{S}_0|p; z_0) \, dP_0(\pi, p)$$

$$= \int_S \int_{\underline{Z}_0} r_0(\underline{S}_0|p; z_0) s_0(dz_0|x_0) p(dx_0)$$

$$= \int_{\underline{S}_0} s_0(\underline{Z}_0|x_0) p(dx_0)$$

$$= P_0(\pi, p)(\{x_0 \in \underline{S}_0, z_0 \in \underline{Z}_0\}). \tag{54}$$

Equation (53) for $k = 0$ follows from (54) and the definition of conditional probability.

[†] In this and subsequent proofs, the reader may find the discussion concerning conditional expectations and probabilities at the beginning of Section 10.2 helpful.

Assume now that (53) holds for k. For any $\underline{I}_k \in \mathscr{B}_{I_k}$, $\underline{C}_k \in \mathscr{B}_C$, $\underline{Z}_{k+1} \in \mathscr{B}_Z$ and $\underline{S}_{k+1} \in \mathscr{B}_S$, we have from (16), the induction hypothesis, Fubini's theorem, (50), (52), and (49) that

$$\int_{\{i_k \in \underline{I}_k, u_k \in \underline{C}_k, z_{k+1} \in \underline{Z}_{k+1}\}} p_{k+1}(p; i_k, u_k, z_{k+1})(\underline{S}_{k+1}) \, dP_{k+1}(\pi, p)$$

$$= \int_{\{i_k \in \underline{I}_k\}} \int_{\underline{C}_k} \int_{S_{k+1}} \int_{\underline{Z}_{k+1}} p_{k+1}(p; i_k, z_{k+1})(\underline{S}_{k+1}) s(dz_{k+1}|u_k, x_{k+1})$$
$$\times \, t(dx_{k+1}|x_k, u_k)\mu_k(du_k|p; i_k) \, dP_k(\pi, p)$$

$$= \int_{\{i_k \in \underline{I}_k\}} \int_{S_k} \int_{\underline{C}_k} \int_{S_{k+1}} \int_{\underline{Z}_{k+1}} p_{k+1}(p; i_k, u_k, z_{k+1})(\underline{S}_{k+1}) s(dz_{k+1}|u_k, x_{k+1})$$
$$\times \, t(dx_{k+1}|x_k, u_k)\mu_k(du_k|p; i_k)[p_k(p; i_k)(dx_k)] \, dP_k(\pi, p)$$

$$= \int_{\{i_k \in \underline{I}_k\}} \int_{\underline{C}_k} \int_{S_k} \int_{S_{k+1}} \int_{\underline{Z}_{k+1}} p_{k+1}(p; i_k, u_k, z_{k+1})(\underline{S}_{k+1}) s(dz_{k+1}|u_k, x_{k+1})$$
$$\times \, t(dx_{k+1}|x_k, u_k)[p_k(p; i_k)(dx_k)]\mu_k(du_k|p; i_k) \, dP_k(\pi, p)$$

$$= \int_{\{i_k \in \underline{I}_k\}} \int_{\underline{C}_k} \int_{S_{k+1}} \int_{\underline{Z}_{k+1}} r(\underline{S}_{k+1}|\tilde{f}[p_k(p; i_k), u_k]; u_k, z_{k+1})$$
$$\times \, s(dz_{k+1}|u_k, x_{k+1})\tilde{f}[p_k(p; i_k), u_k](dx_{k+1})\mu_k(du_k|p; i_k) \, dP_k(\pi, p)$$

$$= \int_{\{i_k \in \underline{I}_k\}} \int_{\underline{C}_k} \int_{S_{k+1}} s(\underline{Z}_{k+1}|u_k, x_{k+1})\tilde{f}[p_k(p; i_k), u_k](dx_{k+1})$$
$$\times \, \mu_k(du_k|p; i_k) \, dP_k(\pi, p)$$

$$= \int_{\{i_k \in \underline{I}_k\}} \int_{\underline{C}_k} \int_{S_k} \int_{S_{k+1}} s(\underline{Z}_{k+1}|u_k, x_{k+1}) t(dx_{k+1}|x_k, u_k)[p_k(p; i_k)(dx_k)]$$
$$\times \, \mu_k(du_k|p; i_k) \, dP_k(\pi, p)$$

$$= \int_{\{i_k \in \underline{I}_k\}} \int_{S_k} \int_{\underline{C}_k} \int_{S_{k+1}} s(\underline{Z}_{k+1}|u_k, x_{k+1}) t(dx_{k+1}|x_k, u_k)\mu_k(du_k|p; i_k)$$
$$\times \, [p_k(p; i_k)(dx_k)] \, dP_k(\pi, p)$$

$$= \int_{\{i_k \in \underline{I}_k\}} \int_{\underline{C}_k} \int_{S_{k+1}} s(\underline{Z}_{k+1}|u_k, x_{k+1}) t(dx_{k+1}|x_k, u_k)\mu_k(du_k|p; i_k) \, dP_k(\pi, p)$$

$$= P_{k+1}(\pi, p)(\{i_k \in \underline{I}_k, u_k \in \underline{C}_k, x_{k+1} \in \underline{S}_{k+1}, z_{k+1} \in \underline{Z}_{k+1}\}). \qquad (55)$$

It follows from (55) and the definition of conditional probability that

$$P_{k+1}(\pi, p)[x_{k+1} \in \underline{S}_{k+1}|i_{k+1}] = p_{k+1}(p; i_{k+1})(\underline{S}_{k+1})$$

for $P_{k+1}(\pi, p)$ almost every i_k, and the induction step is completed.   Q.E.D.

Proposition 10.5 Consider the (ISI) model and assume that $U_k(x) = C$ for every $x \in S$ and $k = 0, \ldots, N - 1$. Then the sequence $[p_0(p; i_0), \ldots, p_{N-1}(p; i_{N-1})]$ defined by (51) and (52) is a statistic sufficient for control, and the resulting perfect state information model is stationary.

Proof Let Y_k in Definition 10.6 be $P(S)$, $k = 0, \ldots, N - 1$. We have already seen that the mappings $p_k : P(S)I_k \to P(S)$ are Borel-measurable, so $(p_0, \ldots, p_{N-1})$ is a statistic. Condition (a) of Definition 10.6 is satisfied with $\hat{\Gamma}_k = P(S)C$, $k = 0, \ldots, N - 1$.

For $y \in P(S)$, $u \in C$ and $\underline{Y} \in \mathscr{B}_{P(S)}$, define

$$\underline{Z}(y, u, \underline{Y}) = \{z \in Z | r[dx | \bar{f}(y, u); u, z] \in \underline{Y}\},$$

$$\hat{t}(\underline{Y} | y, u) = \int_S \int_S s[\underline{Z}(y, u, \underline{Y}) | u, x'] t(dx' | x, u) y(dx). \tag{56}$$

Note that $\underline{Z}(y, u, \underline{Y})$ is the (y, u)-section of the inverse image of $\underline{Y}$ under a Borel-measurable function. The stochastic kernel

$$\lambda(\underline{Z} | x, u) = \int_S s(\underline{Z} | u, x') t(dx' | x, u)$$

is Borel-measurable by Propositions 7.26 and 7.29, so the stochastic kernel

$$\Lambda(\underline{Z} | y, u) = \int_S \int_S s(\underline{Z} | u, x') t(dx' | x, u) y(dx) = \int_S \lambda(\underline{Z} | x, u) y(dx)$$

is Borel-measurable by the same propositions. It follows from Proposition 7.26 and Corollary 7.26.1 that $\hat{t}(dy' | y, u)$ is a Borel-measurable stochastic kernel on $P(S)$ given $P(S)C$. For $\pi \in \Pi$, $p \in P(S)$, $\underline{Y} \in \mathscr{B}_{P(S)}$ and $k = 0, \ldots, N - 2$, we have from Lemma 10.4

$$P_{k+1}(\pi, p)[p_{k+1}(p; i_{k+1}) \in \underline{Y} | p_k(p; i_k) = \bar{y}_k, u_k = \bar{u}_k]$$
$$= P_{k+1}(\pi, p)[z_{k+1} \in \underline{Z}(\bar{y}_k, \bar{u}_k, \underline{Y}) | p_k(p; i_k) = \bar{y}_k, u_k = \bar{u}_k]$$
$$= E\{P_{k+1}(\pi, p)[z_{k+1} \in \underline{Z}(\bar{y}_k, \bar{u}_k, \underline{Y}) | i_k, u_k] | p_k(p; i_k) = \bar{y}_k, u_k = \bar{u}_k\}$$
$$= E\left\{ \int_S \int_S s[\underline{Z}(\bar{y}_k, \bar{u}_k, \underline{Y}) | u_k, x_{k+1}] t(dx_{k+1} | x_k, u_k) \right.$$
$$\left. \times [p_k(p; i_k)(dx_k)] | p_k(p; i_k) = \bar{y}_k, u_k = \bar{u}_k \right\}$$
$$= \hat{t}(\underline{Y} | \bar{y}_k, \bar{u}_k)$$

for $P_k(\pi, p)$ almost every $(\bar{y}_k, \bar{u}_k)$, where the expectations are with respect to $P_{k+1}(\pi, p)$. Thus (22) is satisfied.

For $\pi \in \Pi$, $p \in P(S)$, and $k = 0, \ldots, N - 1$, we have from Lemma 10.4

$$E[g(x_k, u_k) | p_k(p; i_k) = \bar{y}_k, u_k = \bar{u}_k]$$
$$= E\{E[g(x_k, u_k) | i_k, u_k] | p_k(p; i_k) = \bar{y}_k, u_k = \bar{u}_k\}$$
$$= E\left\{ \int_S g(x_k, u_k) p_k(p; i_k)(dx_k) | p_k(p; i_k) = \bar{y}_k, u_k = \bar{u}_k \right\}$$
$$= \int_S g(x_k, \bar{u}_k) \bar{y}_k(dx_k) \tag{57}$$

for $P_k(\pi, p)$ almost every $(\bar{y}_k, \bar{u}_k)$, where the expectations are with respect to $P_k(\pi, p)$. The function $\hat{g}: P(S)C \to R^*$ defined by

$$\hat{g}(\bar{y}, \bar{u}) = \int_S g(x, \bar{u})\bar{y}(dx) \tag{58}$$

is lower semianalytic (Proposition 7.48), and, by (57), $\hat{g}$ satisfies (23).   Q.E.D.

If the horizon is finite, then the transition kernel $\hat{t}$ and the one-stage cost function $\hat{g}$ defined by (56) and (58) can be substituted in the dynamic programming algorithm (46)–(47) to compute the optimal cost function $\hat{J}_N^*$ for (PSI). The optimal cost function J_N^* for (ISI) can then be determined from (41). If the horizon is infinite, in the limit the dynamic programming algorithm (46)–(47) yields $\hat{J}^*$ under $(\hat{N})$ and $(\hat{D})$ and under $(\hat{P})$ in some cases (Propositions 9.14 and 9.17). The determination of J^* from $\hat{J}^*$ is again accomplished by using (41).

10.3.2   The Identity Mappings

Proposition 10.6   Let the model (ISI) be given. The sequence of identity mappings on $P(S)I_k$, $k = 0, \ldots, N - 1$, is a statistic sufficient for control.

Proof   Let Y_k in Definition 10.6 be $P(S)I_k$, $k = 0, \ldots, N - 1$, and let η_k be the identity mapping on $P(S)I_k$. Then $(\eta_0, \ldots, \eta_{N-1})$ is a statistic. Condition (a) of Definition 10.6 is satisfied with $\hat{\Gamma}_k = P(S)\Gamma_k$, $k = 0, \ldots, N - 1$.

If $\underline{Y}_{k+1} \in \mathscr{B}_{P(S)I_{k+1}}$, $\bar{y}_k \in P(S)I_k$, and $\bar{u}_k \in C_k$, we adopt the notation

$$(\underline{Y}_{k+1})_{(\bar{y}_k, \bar{u}_k)} = \{z_{k+1} \in Z \mid (\bar{p}; \bar{z}_0, \bar{u}_0, \ldots, \bar{u}_{k-1}, \bar{z}_k, \bar{u}_k, z_{k+1}) \in \underline{Y}_{k+1}\},$$

where $\bar{y}_k = (\bar{p}; \bar{z}_0, \bar{u}_0, \ldots, \bar{u}_{k-1}, \bar{z}_k)$. Using this notation, we define for $k = 0, \ldots, N - 2$ the stochastic kernel $\hat{t}_k(dy_{k+1} \mid \bar{y}_k, \bar{u}_k)$ on $P(S)I_{k+1}$ given $P(S)I_kC$ by

$$\hat{t}_k(\underline{Y}_{k+1} \mid \bar{y}_k, \bar{u}_k) = \int_{S_{k+1}} s[(\underline{Y}_{k+1})_{(\bar{y}_k, \bar{u}_k)} \mid \bar{u}_k, x_{k+1}] t(dx_{k+1} \mid x_k, \bar{u}_k) p_k(\bar{y}_k)(dx_k)$$

$$\forall \underline{Y}_{k+1} \in \mathscr{B}_{P(S)I_{k+1}}, \tag{59}$$

where $p_k(\bar{y}_k)$ is given by (51) and (52). By an argument similar to that used in Proposition 10.5, it can be shown that $\hat{t}_k$ is Borel-measurable. For $p \in P(S)$, $\pi \in \Pi$, $\underline{Y}_{k+1} \in \mathscr{B}_{P(S)I_{k+1}}$, and $k = 0, \ldots, N - 2$, we have from Lemma 10.4

$$P_{k+1}(\pi, p)[\eta_{k+1}(p; i_{k+1}) \in \underline{Y}_{k+1} \mid \eta_k(p; i_k) = \bar{y}_k, u_k = \bar{u}_k]$$
$$= P_{k+1}(\pi, p)[(\bar{y}_k, \bar{u}_k, z_{k+1}) \in \underline{Y}_{k+1}]$$
$$= \int_{S_{k+1}} s[(\underline{Y}_{k+1})_{(\bar{y}_k, \bar{u}_k)} \mid \bar{u}_k, x_{k+1}] t(dx_{k+1} \mid x_k, \bar{u}_k) p_k(\bar{y}_k)(dx_k)$$
$$= \hat{t}_k(\underline{Y}_{k+1} \mid \bar{y}_k, \bar{u}_k),$$

for $P_k(\pi, p)$ almost every $(\bar{y}_k, \bar{u}_k)$, so (22) is satisfied.

For $k = 0, \ldots, N - 1$, define $\hat{g}_k : P(S)I_k C \to R^*$ by

$$\hat{g}_k(\bar{y}_k, \bar{u}_k) = \int_{S_k} g(x_k, \bar{u}_k) p_k(\bar{y}_k)(dx_k). \tag{60}$$

By Proposition 7.48, $\hat{g}_k$ is lower semianalytic for each k. For $p \in P(S)$, $\pi \in \Pi$, and $k = 0, \ldots, N - 1$, we have from Lemma 10.4

$$E[g(x_k, u_k)|\eta(p; i_k) = \bar{y}_k, u_k = \bar{u}_k] = \int_{S_k} g(x_k, \bar{u}_k) p_k(\bar{y}_k)(dx_k)$$

$$= \hat{g}_k(\bar{y}_k, \bar{u}_k)$$

for $P_k(\pi, p)$ almost every $(\bar{y}_k, \bar{u}_k)$, where the expectation is with respect to $P_k(\pi, p)$, so (23) is satisfied.   Q.E.D.

The transition kernels $\hat{t}_k$ and one-stage cost functions $\hat{g}_k$ defined by (59) and (60) can be used in the nonstationary version of the dynamic programming algorithm (46)–(47). See the discussion following Proposition 10.5.

Chapter 11

Miscellaneous

11.1 Limit-Measurable Policies

In this section we strengthen the results of Section 7.7 concerning universally measurable functions. In particular, we show that these results are still valid if limit-measurable functions (Definitions B.2 and B.3) are used in place of universally measurable functions. This allows us to replace all the results on the existence of universally measurable policies in Chapters 8 and 9 by stronger results on the existence of limit-measurable policies.

We now rework the main results of Section 7.7 with the aid of the concepts and results of Appendix B.

Proposition 11.1 Let X, Y, and Z be Borel spaces, $D \in \mathscr{L}_X$, and $E \in \mathscr{L}_Y$. Suppose $f : D \to Y$ and $g : E \to Z$ are limit-measurable and $f(D) \subset E$. Then the composition $g \circ f$ is limit-measurable.

Proof This follows from Corollary B.11.1.   Q.E.D.

Corollary 11.1.1 Let X and Y be Borel spaces, let $f : X \to Y$ be a function, and let $q(dy|x)$ be a stochastic kernel on Y given X such that, for each x, $q(dy|x)$ assigns probability one to the point $f(x) \in Y$. Then $q(dy|x)$ is limit-measurable if and only if f is limit-measurable.

Proof See the proof of Corollary 7.44.3.   Q.E.D.

Proposition 11.2   Let X and Y be Borel spaces and let $q(dy|x)$ be a stochastic kernel on Y given X. The following statements are equivalent:

(a)   The stochastic kernel $q(dy|x)$ is limit-measurable.

(b)   For every $B \in \mathscr{B}_Y$, the mapping $\lambda_B : X \to R$ defined by

$$\lambda_B(x) = q(B|x) \qquad (1)$$

is limit-measurable.

(c)   For every $Q \in \mathscr{L}_Y$, the mapping λ_Q of (1) is limit-measurable.

Proof   We prove (a) $\Rightarrow$ (c) $\Rightarrow$ (b) $\Rightarrow$ (a). Suppose (a) holds and $Q \in \mathscr{L}_Y$. Now $\lambda_Q = \theta_Q \circ \gamma$, where $\gamma : X \to P(Y)$ is given by

$$\gamma(x) = q(dy|x) \qquad (2)$$

and $\theta_Q : P(Y) \to R$ is given by

$$\theta_Q(q) = q(Q). \qquad (3)$$

We have assumed that γ is limit-measurable, and θ_Q is limit-measurable by Proposition B.12. Therefore (c) holds.

It is clear that (c) $\Rightarrow$ (b). Suppose now that (b) holds. Then

$$\sigma\left[\bigcup_{B \in \mathscr{B}_Y} \lambda_B^{-1}(\mathscr{B}_R)\right] \subset \mathscr{L}_X.$$

Letting γ and θ_B be defined by (2) and (3), we have from Proposition 7.25

$$\gamma^{-1}[\mathscr{B}_{P(Y)}] = \gamma^{-1}\left[\sigma\left(\bigcup_{B \in \mathscr{B}_Y} \theta_B^{-1}(\mathscr{B}_R)\right)\right]$$

$$= \sigma\left[\bigcup_{B \in \mathscr{B}_Y} \gamma^{-1}(\theta_B^{-1}(\mathscr{B}_R))\right] = \sigma\left[\bigcup_{B \in \mathscr{B}_Y} \lambda_B^{-1}(\mathscr{B}_R)\right] \subset \mathscr{L}_X,$$

so $q(dy|x)$ is limit-measurable.   Q.E.D.

Proposition 11.3   Let X and Y be Borel spaces and let $f : XY \to R^*$ be limit-measurable. Let $q(dy|x)$ be a limit-measurable stochastic kernel on Y given X. Then the mapping $\lambda : X \to R^*$ defined by

$$\lambda(x) = \int f(x, y) q(dy|x)$$

is limit-measurable.

Proof   The mapping $\delta(x) = p_x$ is continuous (Corollary 7.21.1), as is the mapping $\sigma : P(X)P(Y) \to P(XY)$ defined by $\sigma(p, q) = pq$, where pq is the product measure (Lemma 7.12). Suppose $Q \in \mathcal{L}_{XY}$ and $f = \chi_Q$. For every $x \in X$,

$$\lambda(x) = [p_x q(dy|x)](Q) = \theta_Q(\sigma[\delta(x), \gamma(x)]), \tag{4}$$

where γ and θ_Q are given by (2) and (3). Since all the functions on the right-hand side of (4) are limit-measurable, λ is limit-measurable. It follows that λ is limit-measurable when f is a limit-measurable simple function. The extension to the general limit-measurable, extended real-valued function f is straightforward.   Q.E.D.

Corollary 11.3.1   Let X be a Borel space and let $f : X \to R^*$ be limit-measurable. Then the function $\theta_f : P(X) \to R^*$ defined by

$$\theta_f(p) = \int f \, dp$$

is limit-measurable.

We have the following sharpened version of the selection theorem for lower semianalytic functions.

Proposition 11.4   Let X and Y be Borel spaces, $D \subset XY$ an analytic set, and $f : D \to R^*$ a lower semianalytic function. Define $f^* : \text{proj}_X(D) \to R^*$ by

$$f^*(x) = \inf_{y \in D_x} f(x, y).$$

The set

$$I = \{x \in \text{proj}_X(D) | \text{ for some } y_x \in D_x, f(x, y_x) = f^*(x)\}$$

is limit-measurable, and for every $\varepsilon > 0$ there exists a limit-measurable function $\varphi : \text{proj}_X(D) \to Y$ such that $\text{Gr}(\varphi) \subset D$ and for all $x \in \text{proj}_X(D)$

$$f[x, \varphi(x)] = f^*(x) \qquad \text{if} \quad x \in I,$$

$$f[x, \varphi(x)] \leq \begin{cases} f^*(x) + \varepsilon & \text{if} \quad x \notin I, \quad f^*(x) > -\infty, \\ -1/\varepsilon & \text{if} \quad x \notin I, \quad f^*(x) = -\infty. \end{cases}$$

Proof   The proof is the same as in Proposition 7.50(b), except that at the points where Corollary 7.44.2 is invoked to say that the composition of analytically measurable functions is universally measurable, we use Proposition 11.1 to say that the composition is limit-measurable.   Q.E.D.

By the remark following Corollary B.11.1, we see that *I and the selector obtained in Proposition 11.4 are in fact $\mathcal{L}_X^2$-measurable.* This remark further suggests that the constructions in Chapters 8 and 9 of optimal and ε-optimal policies can be done more carefully by keeping track of the minimal $\mathscr{L}_S^z$ with respect to which policies and costs are measurable. We do this to some extent in the next section, but do not pursue this matter to any great length.

Propositions 11.1–11.4 are sufficient to allow us to replace every reference to a "(universally measurable) policy" in Chapters 8 and 9 by the words "limit-measurable policy." It does not matter which class of policies is considered when defining J_N^* and J^*; the proof of Proposition 8.1 together with Proposition 11.5 given below can be used to show that these functions are determined by the analytically measurable Markov policies alone. Corollary 11.1.1 tells us that the nonrandomized limit-measurable policies are just the set of sequences of limit-measurable functions from state to control which satisfy the control constraint (cf. Definition 8.2). This fact and Proposition 11.2 are needed for the proof of the limit-measurable counterpart of Lemma 8.2. From Proposition 11.3 we can deduce that the cost corresponding to a limit-measurable policy is limit-measurable (cf. Definitions 8.3 and 9.3). This fact was used, for example, in proving that under (F^-) a nonrandomized, semi-Markov, ε-optimal policy exists (Proposition 8.3). Proposition 11.4 allows limit-measurable ε-optimal and optimal selection. The ε-optimal selection property for universally measurable functions is used in practically every proof in Chapters 8 and 9. The exact selection property is used in showing the existence under certain conditions of optimal policies (Propositions 8.5, 9.19, and 9.20).

11.2 Analytically Measurable Policies

Some of the existence results of Chapters 8 and 9 can be sharpened to state the existence of ε-optimal analytically measurable policies. This is due to Proposition 7.50(a) and the following propositions. Proposition 11.5 is the analog of Corollary 7.44.3 for universally measurable policies and of Corollary 11.1.1 for limit-measurable policies.

Proposition 11.5 Let X and Y be Borel spaces, let $f: X \to Y$ be a function, and let $q(dy|x)$ be a stochastic kernel on Y given X such that, for each x, $q(dy|x)$ assigns probability one to the point $f(x) \in Y$. Then $q(dy|x)$ is analytically measurable if and only if f is analytically measurable.

Proof We sharpen the proof of Corollary 7.44.3. Let $\gamma(x) = q(dy|x)$ and $\delta(y) = p_y$, so that $\gamma = \delta \circ f$ and $f = \delta^{-1} \circ \gamma$. Now δ is a homeomorphism from Y to $\bar{Y} = \{p_y | y \in Y\}$, so δ and $\delta^{-1}: \bar{Y} \to Y$ are both Borel-measurable. If f is analytically measurable and $C \in \mathscr{B}_{P(Y)}$, then

$$\gamma^{-1}(C) = f^{-1}[\delta^{-1}(C)] \in \mathscr{A}_X$$

because $\delta^{-1}(C) \in \mathscr{B}_Y$. If γ is analytically measurable and $B \in \mathscr{B}_Y$, then

$$f^{-1}(B) = \gamma^{-1}[\delta(B)] \in \mathscr{A}_X$$

because $\delta(B) \in \mathscr{B}_{P(Y)}$.   Q.E.D.

Proposition 11.6 Let X and Y be Borel spaces and let $q(dy|x)$ be a stochastic kernel on Y given X. The following statements are equivalent:

(a)   The stochastic kernel $q(dy|x)$ is analytically measurable.
(b)   For every $B \in \mathscr{B}_Y$, the mapping $\lambda_B : X \to R$ defined by

$$\lambda_B(x) = q(B|x) \tag{5}$$

is analytically measurable.

Proof   Assume (a) holds and define $\gamma(x) = q(dy|x)$. Then for $B \in \mathscr{B}_Y$, $C \in \mathscr{B}_R$, and $\theta_B : P(Y) \to R$ defined by (3), we have

$$\lambda_B^{-1}(C) = \gamma^{-1}[\theta_B^{-1}(C)] \in \mathscr{A}_X$$

because $\theta_B^{-1}(C) \in \mathscr{B}_{P(Y)}$ (Proposition 7.25). Therefore (b) holds.

If (b) holds, we can show that (a) holds by the same argument used in the proof of Proposition 11.2.   Q.E.D.

We know from Corollary B.11.1 that the composition of analytically measurable functions need not be analytically measurable, so the cost corresponding to an analytically measurable policy for a stochastic optimal control model may not be analytically measurable. To see this, just write out explicitly the cost corresponding to a two-stage, nonrandomized, Markov, analytically measurable policy (cf. Definition 8.3).

A review of Chapters 8 and 9 shows the following. Proposition 8.3 is still valid if the word "policy" is replaced by "analytically measurable policy," except that under (F⁻) an analytically measurable, nonrandomized, semi-Markov, ε-optimal policy is not guranteed to exist. However, an analytically measurable nonrandomized ε-optimal policy can be shown to exist if $g \leq 0$ [B12]. The proof of the existence of a sequence of nonrandomized Markov policies exhibiting $\{\varepsilon_n\}$-dominated convergence to optimality (Proposition 8.4) breaks down at the point where we assume that a sequence of one-stage policies $\{\mu_0^n\}$ exists for which

$$T_{\mu_0^n}(J_0) \leq T_{\mu_0^{n-1}}(J_0).$$

This occurs because $T_{\mu_0^{n-1}}(J_0)$ may not be analytically measurable. In the first sentence of Proposition 9.19, the word "policy" can be replaced by "analytically measurable policy." The ε-optimal part of Proposition 9.20 depends on the (F⁻) part of Proposition 8.3, so it cannot be strengthened in this way. Under assumption (N), an analytically measurable, nonrandomized, ε-optimal policy can be shown to exist [B12], but it is unknown whether this policy can be taken to be semi-Markov. The results of Chapters 8 and 9 relating to existence of universally measurable optimal policies depend on the exact selection property of Proposition 7.50(b). Since this property is not available for analytically measurable functions, we cannot use the same arguments to infer existence of optimal analytically measurable policies.

11.3 Models with Multiplicative Cost

In this section we revisit the stochastic optimal control model with a multiplicative cost functional first encountered in Section 2.3.4. We pose the finite horizon model in Borel spaces and state the results which are obtainable by casting this Borel space model in the generalized framework of Chapter 6. This does not permit a thorough treatment of the type already given to the model with additive cost in Chapters 8 and 9, but it does yield some useful results and illustrates how the generalized abstract model of Chapter 6 can be applied. The reader can, of course, use the mathematical theory of Chapter 7 to analyze the model with multiplicative cost directly under conditions more general than those given here.

We set up the *Borel model with multiplicative cost*. Let the state space S, the control space C, and the disturbance space W be Borel spaces. Let the control constraint U mapping S into the set of nonempty subsets of C be such that

$$\Gamma = \{(x, u) | x \in S, u \in U(x)\}$$

is analytic. Let the disturbance kernel $p(dw|x, u)$ and the system function $f : SCW \to S$ be Borel-measurable. Let the one-stage cost function g be Borel-measurable, and assume that there exists a $b \in R$ such that $0 \le g(x, u, w) \le b$ for all $x \in S$, $u \in U(x)$, $w \in W$. Let the horizon N be a positive integer.

In the framework of Section 6.1, we define $\tilde{F}$ to be the set of extended real-valued, universally measurable functions on S and F^* to be the set of functions in $\tilde{F}$ which are lower semianalytic. We let $\tilde{M}$ be the set of universally measurable functions from S to C with graph in Γ. Define $H : SC\tilde{F} \to [0, \infty]$ by

$$H(x, u, J) = \int_W g(x, u, w) J[f(x, u, w)] p(dw|x, u),$$

where we define $0 \cdot \infty = \infty \cdot 0 = 0 \cdot (-\infty) = (-\infty) \cdot 0 = 0$. We take $J_0 : S \to R^*$ to be identically one. Then Assumptions A.1–A.4, $\tilde{F}$.2, and the Exact Selection Assumption of Section 6.1 hold. (Assumption A.2 follows from Lemma 7.30(4) and Propositions 7.47 and 7.48. Assumption A.4 follows from Proposition

7.50.) From Propositions 6.1(a), 6.2(a), and 6.3 we have the following results, where the notation of Section 6.1 is used.

Proposition 11.7 In the finite horizon Borel model with multiplicative cost, we have

$$J_N^* = T^N(J_0),$$

and for every $\varepsilon > 0$ there exists an N-stage ε-optimal (Markov) policy. A policy $\pi^* = (\mu_0^*, \ldots, \mu_{N-1}^*)$ is uniformly N-stage optimal if and only if $(T_{\mu_k^*} T^{N-k-1})(J_0) = T^{N-k}(J_0)$, $k = 0, \ldots, N - 1$, and such a policy exists if and only if the infimum in the relation

$$T^{k+1}(J_0)(x) = \inf_{u \in U(x)} H[x, u, T^k(J_0)]$$

is attained for each $x \in S$ and $k = 0, \ldots, N - 1$. A sufficient condition for this infimum to be attained is for the set

$$U_k(x, \lambda) = \{u \in U(x) | H[x, u, T^k(J_0)] \leq \lambda\}$$

to be compact for each $x \in S$, $\lambda \in R$, and $k = 0, \ldots, N - 1$.

Appendix A

The Outer Integral

Throughout this appendix, $(X, \mathcal{B}, p)$ is a probability space. Unless otherwise specified, f, g, and h are functions from X to $[-\infty, \infty]$.

Definition A.1 If $f \geq 0$, the *outer integral* of f with respect to p is defined by

$$\int^* f \, dp = \inf \left\{ \int g \, dp \,\middle|\, f \leq g, g \text{ is } \mathcal{B}\text{-measurable} \right\}. \tag{1}$$

If f is arbitrary, define

$$\int^* f \, dp = \int^* f^+ \, dp - \int^* f^- \, dp, \tag{2}$$

where

$$f^+(x) = \max\{0, f(x)\}, \qquad f^-(x) = \max\{0, -f(x)\},$$

and we set $\infty - \infty = \infty$.

Lemma A.1 If $f \geq 0$, then there exists a $\mathcal{B}$-measurable g with $g \geq f$, such that

$$\int^* f \, dp = \int g \, dp. \tag{3}$$

273

Proof   Choose $g_n \geq f$, g_n $\mathscr{B}$-measurable, so that

$$\int g_n \, dp \downarrow \int^* f \, dp.$$

We assume without loss of generality that $g_1 \geq g_2 \geq \cdots$. Let $g = \lim_{n \to \infty} g_n$. Then $g \geq f$, g is $\mathscr{B}$-measurable, and (3) holds.   Q.E.D.

Lemma A.2   If $f \geq 0$ and $h \geq 0$, then

$$\int^* (f + h) \, dp \leq \int^* f \, dp + \int^* h \, dp. \qquad (4)$$

If either f or h is $\mathscr{B}$-measurable, then equality holds in (4).

Proof   Suppose $g_1 \geq f$, $g_2 \geq f$, g_1 and g_2 are $\mathscr{B}$-measurable, and $\int^* f \, dp = \int g_1 \, dp$, $\int^* h \, dp = \int g_2 \, dp$. Then $g_1 + g_2 \geq f + h$ and (4) follows from (1).

Suppose h is $\mathscr{B}$-measurable and $\int h \, dp < \infty$. [If $\int h \, dp = \infty$, equality is easily seen to hold in (4).] Suppose $f + h \leq g$, where g is $\mathscr{B}$-measurable and

$$\int^* (f + h) \, dp = \int g \, dp.$$

Then $f \leq g - h$ and $g - h$ is $\mathscr{B}$-measurable, so

$$\int^* f \, dp \leq \int g \, dp - \int h \, dp,$$

which implies

$$\int^* f \, dp + \int h \, dp \leq \int^* (f + h) \, dp.$$

Therefore equality holds in (4).   Q.E.D.

We provide an example to show that strict inequality can occur in (4), even if $f + h$ is $\mathscr{B}$-measurable. For this and subsequent examples we will need the following observation: For any $E \subset X$,

$$\int^* \chi_E \, dp = p^*(E), \qquad (5)$$

where $p^*(E)$ is p-outer measure defined by

$$p^*(E) = \inf\{p(B) | E \subset B, B \in \mathscr{B}\}$$

and χ_E is the indicator function of E defined by

$$\chi_E(x) = \begin{cases} 1 & \text{if } x \in E, \\ 0 & \text{if } x \notin E. \end{cases}$$

To verify (5), note that if $\chi_E \leq g$ and g is $\mathscr{B}$-measurable, then $\{x | g(x) \geq 1\}$ is a $\mathscr{B}$-measurable set containing E and consequently

$$\int g \, dp \geq p^*(E).$$

Definition A.1 implies

$$\int^* \chi_E \, dp \geq p^*(E). \tag{6}$$

On the other hand, if $\{B_n\}$ is a sequence of $\mathscr{B}$-measurable sets with $E \subset B_n$ and $p(B_n) \downarrow p^*(E)$, then $p(\bigcap_{n=1}^{\infty} B_n) = p^*(E)$. By construction, $\chi_{\bigcap_{n=1}^{\infty} B_n} \geq \chi_E$. But $\chi_{\bigcap_{n=1}^{\infty} B_n}$ is $\mathscr{B}$-measurable, and

$$\int \chi_{\bigcap_{n=1}^{\infty} B_n} dp = p^*(E).$$

The reverse of inequality (6) follows. Note that the preceding argument shows that for any set E, there exists a set $B \in \mathscr{B}$ such that $E \subset B$ and $p(B) = p^*(E)$.

EXAMPLE 1   Let $X = [0, 1]$, let $\mathscr{B}$ be the Borel σ-algebra, and let p be Lebesgue measure restricted to $\mathscr{B}$. Let $E \subset X$ be a set for which $p^*(E) = p^*(X - E) = 1$ (see [H1, Section 16, Theorem E]). Then

$$\int (\chi_E + \chi_{X-E}) \, dp = \int 1 \, dp = 1,$$

$$\int^* \chi_E \, dp + \int^* \chi_{X-E} \, dp = 2,$$

and strict inequality holds in (4).

Lemma A.2 cannot be extended to (possibly negative) bounded functions, even if h is $\mathscr{B}$-measurable, as the following example demonstrates.

EXAMPLE 2   Let $(X, \mathscr{B}, p)$ and E be as before. Let $f = \chi_E - \chi_{X-E}$, $h = 1$. Then

$$\int^* (f + h) \, dp = \int^* 2\chi_E \, dp = 2,$$

$$\int^* f \, dp + \int h \, dp = \int^* \chi_E \, dp - \int^* \chi_{X-E} \, dp + 1 = 1.$$

Lemma A.3

(a) If $f \leq g$, then $\int^* f \, dp \leq \int^* g \, dp$.

(b) If $\varepsilon > 0$ and $f \leq g \leq f + \varepsilon$, then

$$\int^* f \, dp \leq \int^* g \, dp \leq \int^* f \, dp + 2\varepsilon. \tag{7}$$

(c)   If $\int^* f^+ \, dp < \infty$ or $\int^* f^- \, dp < \infty$, then

$$\int^* (-f) \, dp = -\int^* f \, dp. \tag{8}$$

(d)   If $A, B \in \mathcal{B}$ are disjoint, then for any f

$$\int^* \chi_{A \cup B} f \, dp = \int^* \chi_A f \, dp + \int^* \chi_B f \, dp. \tag{9}$$

(e)   If $E \subset X$ satisfies $p^*(E) = 0$, then for any f

$$\int^* f \, dp = \int^* \chi_{X-E} f \, dp.$$

(f)   If $p^*(\{x | f(x) = \infty\}) > 0$, then for every g, $\int^* (g + f) \, dp = \infty$.

(g)   If $p^*(\{x | f(x) = -\infty\}) > 0$, then for every g either $\int^* (g + f) \, dp = \infty$ or $\int^* (g + f) \, dp = -\infty$.

Proof   (a)   If $f \leq g$, then $f^+ \leq g^+$ and $f^- \geq g^-$. By (1),

$$\int^* f^+ \, dp \leq \int^* g^+ \, dp, \qquad \int^* f^- \, dp \geq \int^* g^- \, dp.$$

The result follows from (2).

(b)   In light of (a), it remains only to show that

$$\int^* (f + \varepsilon) \, dp \leq \int^* f \, dp + 2\varepsilon. \tag{10}$$

For $g_1 \geq f^+$, g_1 $\mathcal{B}$-measurable, and

$$\int^* f^+ \, dp = \int g_1 \, dp,$$

we have

$$(f + \varepsilon)^+ \leq g_1 + \varepsilon$$

so

$$\int^* (f + \varepsilon)^+ \, dp \leq \int g_1 \, dp + \varepsilon = \int^* f^+ \, dp + \varepsilon. \tag{11}$$

For $g_2 \geq (f + \varepsilon)^-$, g_2 $\mathcal{B}$-measurable, and

$$\int^* (f + \varepsilon)^- \, dp = \int g_2 \, dp,$$

we have

$$g_2 + \varepsilon \geq (f + \varepsilon)^- + \varepsilon = \max\{f^- - \varepsilon, 0\} + \varepsilon \geq f^-,$$

so

$$\varepsilon + \int^* (f + \varepsilon)^- \, dp = \varepsilon + \int g_2 \, dp = \int (g_2 + \varepsilon) \, dp \geq \int^* f^- \, dp. \tag{12}$$

Combine (11) and (12) to conclude (10).

(c)   We have

$$\int^{*}(-f)\,dp = \int^{*}(-f)^{+}\,dp - \int^{*}(-f)^{-}\,dp$$

$$= \int^{*} f^{-}\,dp - \int^{*} f^{+}\,dp = -\left[\int^{*} f^{+}\,dp - \int^{*} f^{-}\,dp\right]$$

$$= -\int^{*} f\,dp,$$

where the assumption that $\int^{*} f^{+}\,dp < \infty$ or $\int^{*} f^{-}\,dp < \infty$ is necessary for the next to last equality.

(d)   Suppose $f \geq 0$. Let g be a $\mathscr{B}$-measurable function with $g \geq \chi_{A \cup B} f$ and

$$\int^{*} \chi_{A \cup B} f\,dp = \int g\,dp.$$

Then $\chi_A g \geq \chi_A f$, $\chi_B g \geq \chi_B f$, so

$$\int^{*} \chi_{A \cup B} f\,dp = \int \chi_A g\,dp + \int \chi_B g\,dp$$

$$\geq \int^{*} \chi_A f\,dp + \int^{*} \chi_B f\,dp. \tag{13}$$

Now suppose $g_1 \geq \chi_A f$, $g_2 \geq \chi_B f$ are $\mathscr{B}$-measurable and

$$\int g_1\,dp = \int^{*} \chi_A f\,dp, \qquad \int g_2\,dp = \int^{*} \chi_B f\,dp.$$

Then $g_1 + g_2 \geq \chi_{A \cup B} f$, so

$$\int^{*} \chi_A f\,dp + \int^{*} \chi_B f\,dp = \int (g_1 + g_2)\,dp$$

$$\geq \int^{*} \chi_{A \cup B} f\,dp. \tag{14}$$

Combine (13) and (14) to conclude (9) for $f \geq 0$. The extension to arbitrary f is straightforward.

(e)   Suppose $f \geq 0$. Choose $B \in \mathscr{B}$ with $p(B) = p^{*}(E) = 0$, $B \supset E$. By (d),

$$\int^{*} f\,dp = \int^{*} \chi_{X-B} f\,dp \leq \int^{*} \chi_{X-E} f\,dp \leq \int^{*} f\,dp.$$

Hence $\int^{*} f\,dp = \int^{*} \chi_{X-E} f\,dp$. The extension to arbitrary f is straightforward.

(f)   We have $(g + f)^{+}(x) = \infty$ if $f(x) = \infty$, so that

$$p^{*}(\{x | (g + f)^{+}(x) = \infty\}) > 0.$$

Hence $\int^{*}(g + f)^{+}\,dp = \infty$, and it follows that $\int^{*}(g + f)\,dp = \infty$.

(g)   Consider the sets $E = \{x | f(x) = -\infty\}$ and $E_g = \{x | f(x) = -\infty, g(x) < \infty\}$. If $p^{*}(E_g) = 0$, then

$$p^{*}(E - E_g) = p^{*}(E - E_g) + p^{*}(E_g) \geq p^{*}(E) > 0.$$

Since we have $f(x) + g(x) = \infty$ for $x \in E - E_g$, it follows from (f) that $\int^*(g + f)\,dp = \infty$. If $p^*(E_g) > 0$, then $p^*(\{x|(g + f)^-(x) = \infty\}) \geq p^*(E_g) > 0$ and hence, by (f), $\int^*(g + f)^-\,dp = \infty$. Hence, if $\int^*(g + f)^+\,dp = \infty$, then $\int^*(g + f)\,dp = \infty$, while if $\int^*(g + f)^+\,dp < \infty$, then $\int^*(g + f)\,dp = -\infty$.

Q.E.D.

The bound given in (7) is the sharpest possible. To see this, let f be as defined in Example 2, $g = f + 1$, and $\varepsilon = 1$. Despite these pathologies of outer integration, there is a monotone convergence theorem, which we now prove.

Proposition A.1 If $\{f_n\}$ is a sequence of nonnegative functions and $f_n \uparrow f$, then

$$\int^* f_n\,dp \uparrow \int^* f\,dp. \tag{15}$$

If $\{f_n\}$ is a sequence of nonpositive functions and $f_n \downarrow f$, then

$$\int^* f_n\,dp \downarrow \int^* f\,dp.$$

Proof We prove the first statement of the theorem. The second follows from the first and Lemma A.3(c). Assume $f_n \geq 0$ and $f_n \uparrow f$. Let $\{g_n\}$ be a sequence of $\mathscr{B}$-measurable functions such that $g_n \geq f_n$ and

$$\int^* f_n\,dp = \int g_n\,dp. \tag{16}$$

If, for some n, $\int g_n\,dp = \int^* f_n\,dp = \infty$, then (15) is assured. If not, then for every n,

$$\int g_n\,dp < \infty. \tag{17}$$

Suppose (17) holds for every n and for some n,

$$p(\{x|g_n(x) > g_{n+1}(x)\}) > 0.$$

Then since $g_{n+1} \geq f_{n+1} \geq f_n$, we have that $\bar{g}$ defined by

$$\bar{g}(x) = \begin{cases} g_n(x) & \text{if } g_n(x) \leq g_{n+1}(x), \\ g_{n+1}(x) & \text{if } g_n(x) > g_{n+1}(x), \end{cases}$$

satisfies $g_n \geq \bar{g} \geq f_n$ everywhere and $\bar{g} < g_n$ on a set of positive measure. This contradicts (16). We may therefore assume without loss of generality that (17) holds and $g_1 \leq g_2 \cdots$. Let $g = \lim_{n \to \infty} g_n$. Then $g \geq f$ and

$$\lim_{n \to \infty} \int^* f_n\,dp = \lim_{n \to \infty} \int g_n\,dp = \int g\,dp \geq \int^* f\,dp.$$

But $f_n \leq f$ for every n, so the reverse inequality holds as well.   Q.E.D.

One might hope that if $\{f_n\}$ is a sequence of functions which are bounded below and $f_n \uparrow f$, then (15) remains valid. This is not the case, as the following example shows.

EXAMPLE 3   Let $X = [0, 1)$, $\mathscr{B}$ be the Borel σ-algebra, and p be Lebesgue measure restricted to $\mathscr{B}$. Define an equivalence relation $\sim$ on X by

$$x \sim y \Leftrightarrow x - y \text{ is rational.}$$

Let F_0 be constructed by choosing one representative from each equivalence class. Let $Q = \{q_0, q_1, \ldots\}$ be an enumeration of the rationals in $[0, 1)$ with $q_0 = 0$ and define

$$F_k = \{x + q_k [\text{mod } 1] \,|\, x \in F_0\} = F_0 + q_k [\text{mod } 1] \qquad k = 0, 1, \ldots .$$

Then $F_0, F_1, \ldots$ is a sequence of disjoint sets with

$$\bigcup_{k=0}^{\infty} F_k = [0, 1). \tag{18}$$

If for some $n < \infty$, we have $p^*(\bigcup_{k=n}^{\infty} F_k) < 1$, then $E = \bigcup_{k=0}^{n-1} F_k$ contains a $\mathscr{B}$-measurable set with measure $\delta > 0$. For $k = 1, \ldots, n - 1$, let $q_k = r_k/s_k$, where r_k and s_k are integers and r_k/s_k is reduced to lowest terms. Let $\{p_1, p_2, \ldots\}$ be a sequence of prime numbers such that

$$\max_{1 \leq k \leq n-1} s_k < p_1 < p_2 < \cdots$$

Then the sets $E, E + p_1^{-1} [\text{mod } 1], E + p_2^{-1} [\text{mod } 1], \ldots$ are disjoint, and by the translation invariance of p, each contains a $\mathscr{B}$-measurable set with measure $\delta > 0$. It follows that $[0, 1)$ must contain a $\mathscr{B}$-measurable set of infinite measure. This contradiction implies

$$p^*\left(\bigcup_{k=n}^{\infty} F_k\right) = 1 \tag{19}$$

for every n. Define

$$f_n = -\chi_{\bigcup_{k=n}^{\infty} F_k}, \qquad n = 0, 1, \ldots .$$

Then $f_n \uparrow 0$, but (5) and (19) imply that for every n

$$\int^* f_n \, dp = -1.$$

By a change of sign in Example 3, we see that the second part of Theorem A.1 cannot be extended to functions which are bounded above unless additional conditions are imposed. We impose such conditions in order to prove a corollary.

Corollary A.1.1 Let $\{\varepsilon_n\}$ be a sequence of positive numbers with $\sum_{n=1}^{\infty} \varepsilon_n < \infty$. Let $\{f_n\}$ be a sequence with

$$\lim_{n \to \infty} f_n = f, \tag{20}$$

$$f \leq f_n, \qquad n = 1, 2, \ldots, \tag{21}$$

$$f_n(x) \leq f(x) + \varepsilon_n \qquad \text{if} \quad f(x) > -\infty, \tag{22}$$

$$f_n(x) \leq f_{n-1}(x) + \varepsilon_n \qquad \text{if} \quad f(x) = -\infty, \quad n = 2, 3, \ldots, \tag{23}$$

$$\int^* f_1 \, dp < \infty. \tag{24}$$

Then

$$\lim_{n \to \infty} \int^* f_n \, dp = \int^* f \, dp. \tag{25}$$

Proof From (20) we have $\lim_{n \to \infty} f_n^+ = f^+$ and $\lim_{n \to \infty} f_n^- = f^-$. Now $\inf_{k \geq n} f_k^- \leq f_n^- \leq f^-$ and $\inf_{k \geq n} f_k^- \uparrow f^-$ as $n \to \infty$. By Proposition A.1,

$$\int^* f^- \, dp = \lim_{n \to \infty} \int^* \inf_{k \geq n} f_k^- \, dp \leq \lim_{n \to \infty} \int^* f_n^- \, dp \leq \int^* f^- \, dp,$$

so

$$\lim_{n \to \infty} \int^* f_n^- \, dp = \int^* f^- \, dp. \tag{26}$$

Let $A = \{x \mid f(x) = -\infty\}$. If $p^*(A) = 0$, then (21), (22), (24), and Lemmas A.3(b) and (e) imply

$$\int^* f^+ \, dp \leq \int^* f_n^+ \, dp \leq 2\varepsilon_n + \int^* f^+ \, dp < \infty,$$

so

$$\lim_{n \to \infty} \int^* f_n^+ \, dp = \int^* f^+ \, dp < \infty. \tag{27}$$

Combine (26) and (27) to conclude (25). If $p^*(A) > 0$, then $\int^* f^- \, dp = -\infty$ and (26) will imply (25) provided that

$$\int^* f^+ \, dp < \infty \tag{28}$$

and

$$\limsup_{n \to \infty} \int^* f_n^+ \, dp < \infty. \tag{29}$$

Conditions (21) and (24) imply (28). Conditions (21)–(23) imply for every $x \in X$

$$f_n(x) \leq f_{n-1}(x) + \varepsilon_n, \qquad n = 2, 3, \ldots,$$

so

$$\int^* f_n^+ \, dp \le 2\varepsilon_n + \int^* f_{n-1}^+ \, dp$$

and

$$\int^* f_n^+ \, dp \le 2 \sum_{k=2}^n \varepsilon_k + \int^* f_1^+ \, dp.$$

The finiteness of $\sum_{k=2}^\infty \varepsilon_k$ and (24) imply (29).   Q.E.D.

Appendix B

Additional Measurability Properties of Borel Spaces

This appendix supplements Section 7.6. The notation and terminology used here is the same as in that section and, in most cases, is defined in Section 7.1.

B.1 Proof of Proposition 7.35(e)

Our first task is to give a proof of Proposition 7.35(e). To do this, we introduce the space $N^* = \{1, 2, \ldots\} \cup \{\infty\}$ with the topology induced by the metric

$$d(x, y) = \left| \frac{1}{x} - \frac{1}{y} \right|,$$

where we define $1/\infty = 0$. Let $\mathscr{N}^* = N^* N^* \cdots$ with the product topology. The space $\mathscr{N}$ of sequences of positive integers is a topological subspace of $\mathscr{N}^*$. The space $\mathscr{N}^*$ is compact by Tychonoff's theorem, while $\mathscr{N}$ is not. If $(X, \mathscr{P})$ and $(Y, \mathscr{Q})$ are paved spaces, we denote by $\mathscr{P}\mathscr{Q}$ the paving of XY:

$$\mathscr{P}\mathscr{Q} = \{PQ \,|\, P \in \mathscr{P}, Q \in \mathscr{Q}\}. \tag{1}$$

Proposition B.1 Let $(X, \mathscr{P})$ be a paved space and $\mathscr{K}$ the collection of compact subsets of $\mathscr{N}^*$. Then the projection on X of a set in $\mathscr{S}(\mathscr{P}\mathscr{K})$ is in $\mathscr{S}(\mathscr{P})$. Conversely, every set in $\mathscr{S}(\mathscr{P})$ is the projection on X of some set in $[(\mathscr{P}\mathscr{K})_\sigma]_\delta$.

282

Proof Let S be a Suslin scheme for $\mathcal{P}\mathcal{K}$. Then for every $s \in \Sigma$, $S(s)$ has the form $S(s) = S_1(s)S_2(s)$, where $S_1(s) \in \mathcal{P}$ and $S_2(s) \in \mathcal{K}$. Now

$$
\begin{aligned}
N(S) &= \bigcup_{z \in \mathcal{N}} \bigcap_{s < z} S(s) \\
&= \bigcup_{z \in \mathcal{N}} \bigcap_{s < z} [S_1(s)S_2(s)] \\
&= \bigcup_{z \in \mathcal{N}} \left\{ \left[\bigcap_{s < z} S_1(s) \right] \left[\bigcap_{s < z} S_2(s) \right] \right\},
\end{aligned}
$$

so

$$
\mathrm{proj}_X[N(S)] = \bigcup_{z \in A} \bigcap_{s < z} S_1(s),
$$

where

$$
A = \left\{ z \in \mathcal{N} \,\Big|\, \bigcap_{s < z} S_2(s) \neq \varnothing \right\}.
$$

Since each $S_2(s)$ is compact, we have

$$
A = \left\{ (\zeta_1, \zeta_2, \ldots) \in \mathcal{N} \,\Big|\, \bigcap_{k=1}^{n} S_2(\zeta_1, \zeta_2, \ldots, \zeta_k) \neq \varnothing \; \forall n \right\}.
$$

Define a Suslin scheme R for $\mathcal{P}$ by

$$
R(\zeta_1, \ldots, \zeta_n) = \begin{cases} S_1(\zeta_1, \ldots, \zeta_n) & \text{if } \bigcap_{k=1}^{n} S_2(\zeta_1, \ldots, \zeta_k) \neq \varnothing, \\ \varnothing & \text{otherwise.} \end{cases}
$$

Then

$$
\begin{aligned}
\mathrm{proj}_X[N(S)] &= \bigcup_{z \in A} \bigcap_{s < z} S_1(s) \\
&= \bigcup_{z \in \mathcal{N}} \bigcap_{s < z} R(s) = N(R),
\end{aligned}
$$

so $\mathrm{proj}_X[N(S)] \in \mathcal{S}(\mathcal{P})$.

For the second part of the proposition, suppose S is a Suslin scheme for $\mathcal{P}$. Define a Suslin scheme R for $\mathcal{K}$ by

$$
R(\sigma_1, \ldots, \sigma_n) = \{ (\zeta_1, \zeta_2, \ldots) \in \mathcal{N}^* \mid \zeta_1 = \sigma_1, \ldots, \zeta_n = \sigma_n \}.
$$

For fixed $z_0 \in \mathcal{N}$, we have $\bigcap_{s < z_0} R(s) = \{z_0\}$, so

$$
\begin{aligned}
\bigcap_{s < z_0} [S(s)R(s)] &= \left[\bigcap_{s < z_0} S(s) \right] \left[\bigcap_{s < z_0} R(s) \right] \\
&= \left\{ (x, z_0) \,\Big|\, x \in \bigcap_{s < z_0} S(s) \right\}. \qquad (2)
\end{aligned}
$$

Therefore,

$$N(S) = \bigcup_{z \in \mathcal{N}} \bigcap_{s < z} S(s)$$

$$= \bigcup_{z \in \mathcal{N}} \text{proj}_X \left\{ \bigcap_{s < z} [S(s)R(s)] \right\}$$

$$= \text{proj}_X \left\{ \bigcup_{z \in \mathcal{N}} \bigcap_{s < z} [S(s)R(s)] \right\},$$

and it remains only to show that

$$\bigcup_{z \in \mathcal{N}} \bigcap_{s < z} [S(s)R(s)] \in [(\mathscr{P}\mathscr{K})_\sigma]_\delta. \qquad (3)$$

If we can show that

$$\bigcup_{z \in \mathcal{N}} \bigcap_{s < z} [S(s)R(s)] = \bigcap_{k=1}^{\infty} \bigcup_{s \in \Sigma_k} [S(s)R(s)], \qquad (4)$$

where Σ_k is the set of elements in Σ having k components, then (3) will follow. Let $x \in X$ and $z_0 = (\zeta_1^0, \zeta_2^0, \dots) \in \mathcal{N}^*$ be given. Suppose

$$(x, z_0) \in \bigcap_{z \in \mathcal{N}} \bigcap_{s < z} [S(s)R(s)].$$

We see from (2) that $z_0 \in \mathcal{N}$ and $(x, z_0) \in \bigcap_{s < z_0}[S(s)R(s)]$, so for every $k \geq 1$, $(x, z_0) \in S(\zeta_1^0, \dots, \zeta_k^0)R(\zeta_1^0, \dots, \zeta_k^0)$. This implies $(x, z_0) \in \bigcap_{k=1}^{\infty} \bigcup_{s \in \Sigma_k}[S(s)R(s)]$, and

$$\bigcup_{z \in \mathcal{N}} \bigcap_{s < z} [S(s)R(s)] \subset \bigcap_{k=1}^{\infty} \bigcup_{s \in \Sigma_k} [S(s)R(s)]. \qquad (5)$$

On the other hand, if $(x, z_0) \in \bigcap_{k=1}^{\infty} \bigcup_{s \in \Sigma_k}[S(s)R(s)]$, then for each $k \geq 1$, $(x, z_0) \in \bigcup_{s \in \Sigma_k}[S(s)R(s)]$. This can happen only if $z_0 \in \mathcal{N}$ and $(x, z_0) \in S(\zeta_1^0, \dots, \zeta_k^0)R(\zeta_1^0, \dots, \zeta_k^0)$. Therefore,

$$(x, z_0) \in \bigcap_{k=1}^{\infty} [S(\zeta_1^0, \dots, \zeta_k^0)R(\zeta_1^0, \dots, \zeta_k^0)]$$

$$= \bigcap_{s < z_0} [S(s)R(s)]$$

$$\subset \bigcup_{z \in \mathcal{N}} \bigcap_{s < z} [S(s)R(s)],$$

which proves the reverse of set containment (5). Equality (4) follows.

Q.E.D.

If $(X, \mathscr{P})$ is a paved space, Y is another space, and $Q \subset Y$, we define a paving of XY by

$$\mathscr{P}Q = \{PQ | P \in \mathscr{P}\}.$$

Lemma B.1 Let $(X, \mathscr{P})$ and $(Y, \mathscr{Q})$ be paved spaces. Then:

(a) $\mathscr{S}(\mathscr{P})Q = \mathscr{S}(\mathscr{P}Q)$ for every $Q \subset Y$;

(b) $\mathscr{S}(\mathscr{P})\mathscr{Q} \subset \mathscr{S}(\mathscr{P}\mathscr{Q})$.

Proof Part (a) is trivial and part (b) follows from (a).   Q.E.D.

We are now in a position to prove part (e) of Proposition 7.35.

Proposition B.2 Let $(X, \mathscr{P})$ be a paved space. Then $\mathscr{S}(\mathscr{P}) = \mathscr{S}[\mathscr{S}(\mathscr{P})]$.

Proof In light of Proposition 7.35(d), we need only prove

$$\mathscr{S}(\mathscr{P}) \supset \mathscr{S}[\mathscr{S}(\mathscr{P})]. \tag{6}$$

Let $\mathscr{N}^*$ and $\mathscr{K}$ be as in Proposition B.1. If $A \in \mathscr{S}[\mathscr{S}(\mathscr{P})]$, then by the second part of Proposition B.1, $A = \text{proj}_X(B)$ for some set $B \in ([\mathscr{S}(\mathscr{P})\mathscr{K}]_\sigma)_\delta$. By Lemma B.1(b) and Proposition 7.35(b) and (c), we have

$$B \in ([\mathscr{S}(\mathscr{P})\mathscr{K}]_\sigma)_\delta \subset ([\mathscr{S}(\mathscr{P}\mathscr{K})]_\sigma)_\delta = \mathscr{S}(\mathscr{P}\mathscr{K}).$$

The first part of Proposition B.1 implies that $A = \text{proj}_X(B) \in \mathscr{S}(\mathscr{P})$ and (6) follows.   Q.E.D.

B.2 Proof of Proposition 7.16

In Proposition 7.16 we stated that Borel spaces X and Y are Borel-isomorphic if and only if they have the same cardinality. A related result is that every uncountable Borel space is Borel-isomorphic to every other uncountable Borel space. We used the latter fact in Proposition 7.27 to assume without loss of generality that the Borel spaces under consideration were actually copies of $(0, 1]$, we used it in Proposition 7.39 to transfer a statement about $\mathscr{N}$ to a statement about any uncountable Borel space, and we will use it again in Proposition B.7 to allow our treatment of the limit σ-algebra to center on the space $\mathscr{N}$. The proofs of Proposition 7.16 and Corollary 7.16.1 depend on the following lemma, which is an immediate consequence of Propositions 7.36 and 7.37. The reader may wish to verify that these propositions depend only on Propositions 7.35, B.1, and B.2, so no circularity is present in the arguments.

Lemma B.2 Let X be a nonempty Borel space. There is a continuous function f from $\mathscr{N}$ onto X.

Define $\mathcal{M}$ to be the set of infinite sequences of zeroes and ones. We can regard $\mathcal{M}$ as the countable product of copies of $\{0,1\}$ and endow it with the product topology, where $\{0,1\}$ has the discrete topology. By Tychonoff's theorem, $\mathcal{M}$ is compact with this topology. It is also metrizable as a complete separable space.

Our proof of Proposition 7.16 consists of three parts. We show first that every uncountable Borel space contains a Borel subset homeomorphic to $\mathcal{M}$, we show second that every uncountable Borel space is isomorphic to a Borel subset of $\mathcal{M}$, and we show finally that these first two facts imply that every uncountable Borel space is isomorphic to $\mathcal{M}$.

Lemma B.3 Let X be an uncountable Borel space. There exists a compact set $K \subset X$ such that $\mathcal{M}$ and K are homeomorphic.

Proof Let $f: \mathcal{N} \to X$ be the continuous, onto function of Lemma B.2. For each $x \in X$, choose an element $z_x \in \mathcal{N}$ such that $x = f(z_x)$. Let $S = \{z_x | x \in X\}$, so that f is a one-to-one function from S onto X. For $z \in S$, if possible choose an open neighborhood $T(z)$ of z such that $S \cap T(z)$ is countable. Let R be the set of all $z \in S$ for which such a $T(z)$ can be found. Since separable metrizable spaces have the Lindelöf property, there exists a countable subset R' of R such that $\bigcup_{z \in R} T(z) = \bigcup_{z \in R'} T(z)$, so

$$R \subset S \cap \left[\bigcup_{z \in R} T(z) \right] = \bigcup_{z \in R'} [S \cap T(z)],$$

and R is countable. Since S is uncountable, $S - R$ must be infinite. Furthermore, if $z \in S - R$, then every open neighborhood of z contains infinitely many points of $S - R$.

Let d be a metric on $\mathcal{N}$ consistent with its topology for which $(\mathcal{N}, d)$ is complete. For $\bar{z} \in \mathcal{N}$, the closed sphere of radius r centered at $\bar{z}$ is the set $\{z \in \mathcal{N} | d(z, \bar{z}) \leq r\}$. The interior of this sphere, denoted $\text{Int}\{z \in \mathcal{N} | d(z, \bar{z}) \leq r\}$, is the set $\{z \in \mathcal{N} | d(z, \bar{z}) < r\}$. Let $z(0)$ and $z(1)$ be distinct points in $S - R$. Then $f[z(0)] \neq f[z(1)]$, so there exist disjoint open neighborhoods U and V of $f[z(0)]$ and $f[z(1)]$ respectively. Let $S(0)$ and $S(1)$ be disjoint closed spheres of radius no greater than one centered at $z(0)$ and $z(1)$ and contained in $f^{-1}(U)$ and $f^{-1}(V)$ respectively. We have that $f[S(0)]$ and $f[S(1)]$ are disjoint. Note also that for every $z \in (S - R) \cap \text{Int } S(0)$, every open neighborhood of z contains infinitely many points of $(S - R) \cap \text{Int } S(0)$, and the same is true of $S(1)$. By the same procedure we can choose distinct points $z(0,0)$ and $z(0,1)$ in $(S - R) \cap \text{Int } S(0)$ and distinct points $z(1,0)$ and $z(1,1)$ in $(S - R) \cap \text{Int } S(1)$, and we can also choose disjoint closed spheres $S(0,0)$, $S(0,1)$, $S(1,0)$ and $S(1,1)$ of radius no greater than $\frac{1}{2}$ centered at $z(0,0)$, $z(0,1)$, $z(1,0)$ and $z(1,1)$, respectively, so that $f[S(0,0)]$, $f[S(0,1)]$, $f[S(1,0)]$ and $f[S(1,1)]$ are all disjoint. We can choose these spheres so that $S(0,0)$ and

$S(0, 1)$ are contained in $S(0)$, while $S(1, 0)$ and $S(1, 1)$ are contained in $S(1)$. At the kth step of this process, we choose a collection of disjoint closed spheres $S(\mu_1, \ldots, \mu_k)$ of radius no greater than $1/k$ centered at distinct points $z(\mu_1, \ldots, \mu_k)$ in $S - R$, where each μ_j is either zero or one. Furthermore, we can choose the spheres so that for each $(\mu_1, \ldots, \mu_{k-1})$

(i) $f[S(\mu_1, \ldots, \mu_{k-1}, 0)] \cap f[S(\mu_1, \ldots, \mu_{k-1}, 1)] = \varnothing$,

(ii) $S(\mu_1, \ldots, \mu_{k-1}, \mu_k) \subset S(\mu_1, \ldots, \mu_{k-1})$, $\quad \mu_k = 0, 1$.

For fixed $m = (\mu_1, \mu_2, \ldots) \in \mathcal{M}$, the sets $\{S(\mu_1, \ldots, \mu_k)\}$ form a decreasing sequence of closed sets with radius converging to zero, so $\{z(\mu_1, \ldots, \mu_k)\}$ is Cauchy and thus has a limit $\varphi(m) \in \bigcap_{k=1}^{\infty} S(\mu_1, \ldots, \mu_k)$.

We show that $\varphi : \mathcal{M} \to \mathcal{N}$ is a homeomorphism. If $(\mu_1, \mu_2, \ldots)$ and $(v_1, v_2, \ldots)$ are distinct elements of $\mathcal{M}$, then for some integer k, we have $\mu_k \neq v_k$. Since $\varphi(\mu_1, \mu_2, \ldots) \in S(\mu_1, \ldots, \mu_k)$, $\varphi(v_1, v_2, \ldots) \in S(v_1, \ldots, v_k)$, and $S(\mu_1, \ldots, \mu_k)$ is disjoint from $S(v_1, \ldots, v_k)$, we see that $\varphi(\mu_1, \mu_2, \ldots) \neq \varphi(v_1, v_2, \ldots)$, so φ is one-to-one. To show φ is continuous, let $\{m_n\}$ be a sequence converging to $m \in \mathcal{M}$. Choose $\varepsilon > 0$ and let k be a positive integer such that $2/k < \varepsilon$. There exists an $\bar{n}$ such that whenever $n \geq \bar{n}$, the elements m_n and $m = (\mu_1, \mu_2, \ldots)$ agree in the first k components, so both $\varphi(m_n)$ and $\varphi(m)$ are in $S(\mu_1, \ldots, \mu_k)$. This implies $d(\varphi(m_n), \varphi(m)) \leq 2/k < \varepsilon$, so φ is continuous. To show that φ^{-1} is continuous, it suffices to show that $\varphi(F)$ is closed in $\varphi(\mathcal{M})$ whenever F is closed in $\mathcal{M}$. This follows from the fact that $\mathcal{M}$ is compact and φ is continuous. Define $\mathcal{N}_1 \subset \mathcal{N}$ to be the compact homeomorphic image of $\mathcal{M}$ under φ.

We now show that $f : \mathcal{N}_1 \to X$ is a homeomorphism. To see that f is one-to-one, choose distinct points z and $\hat{z}$ in $\mathcal{N}_1$. Then there exist distinct points $m = (\mu_1, \mu_2, \ldots)$ and $\hat{m} = (\hat{\mu}_1, \hat{\mu}_2, \ldots)$ in $\mathcal{M}$ such that $z = \varphi(m)$ and $\hat{z} = \varphi(\hat{m})$. For some k, we have $\mu_k \neq \hat{\mu}_k$, so by (i), $f[S(\mu_0, \ldots, \mu_k)] \cap f[S(\hat{\mu}_0, \ldots, \hat{\mu}_k)] = \varnothing$. Since $z \in S(\mu_0, \ldots, \mu_k)$ and $\hat{z} \in S(\hat{\mu}_0, \ldots, \hat{\mu}_k)$, we see that $f(z) \neq f(\hat{z})$, so f is one-to-one. Just as in the case of φ, the continuity of f^{-1} follows from the fact that f is continuous and has a compact domain.

The set $K = f(\mathcal{N}_1)$ is a compact subset of X homeomorphic to $\mathcal{M}$. Q.E.D.

Lemma B.4 Let X be an uncountable Borel space. There exists a Borel subset L of $\mathcal{M}$ such that X and L are Borel-isomorphic.

Proof By definition, X is homeomorphic to a Borel subset B of a complete separable metric space Y. By Urysohn's and Alexandroff's theorems (Propositions 7.2 and 7.3), Y is homeomorphic to a G_δ-subset of the Hilbert cube $\mathcal{H}$, so B and hence X are homeomorphic to a Borel subset of $\mathcal{H}$. It suffices then to show that $\mathcal{H}$ is Borel-isomorphic to a Borel subset of $\mathcal{M}$.

The idea of the proof is this. Each element in $\mathcal{H}$ is a sequence of real numbers in $[0, 1]$. Each of these numbers has a binary expansion, and by mixing all these expansions, we obtain an element in $\mathcal{M}$. Let us first define $\psi:[0,1] \to \mathcal{M}$ which maps a real number into a sequence of zeroes and ones which is its binary expansion. It is easier to define ψ^{-1}, which we define on $\mathcal{M}_1 \cup \{(0,0,0,\ldots)\}$, where

$$\mathcal{M}_1 = \{(\mu_1, \mu_2, \ldots) \in \mathcal{M} \,|\, \mu_k = 1 \text{ for infinitely many } k\}.$$

It is given by

$$\psi^{-1}(\mu_1, \mu_2, \ldots) = \sum_{k=1}^{\infty} \mu_k/2^k,$$

and it is easily verified that ψ^{-1} is one-to-one, continuous, and maps onto $[0,1]$. Since $\mathcal{M} - \mathcal{M}_1$ is countable, the domain of ψ^{-1} is a Borel subset of $\mathcal{M}$, and Proposition 7.15 tells us that ψ is a Borel isomorphism. Since we have not proved Proposition 7.15, we show directly that ψ is Borel-measurable. Consider the collection of sets

$$R(k) = \{(\mu_1, \mu_2, \ldots) \in \mathcal{M} \,|\, \mu_k = 0\}, \qquad k = 1, 2, \ldots,$$
$$\tilde{R}(k) = \{(\mu_1, \mu_2, \ldots) \in \mathcal{M} \,|\, \mu_k = 1\}, \qquad k = 1, 2, \ldots.$$

These sets form a subbase for the topology of $\mathcal{M}$, so by the remark following Definition 7.6, we need only prove that $\psi^{-1}[R(k)]$ and $\psi^{-1}[\tilde{R}(k)]$ are Borel-measurable to conclude that ψ is. Since one of these sets is the complement of the other, we may restrict attention to $\psi^{-1}[R(k)]$. Remembering that the domain of ψ^{-1} is $\mathcal{M}_1 \cup \{0,0,0,\ldots)\}$, we have

$$\psi^{-1}[R(k)] = \left\{ \sum_{j=1}^{\infty} \frac{\mu_j}{2^j} \,\middle|\, (\mu_1, \mu_2, \ldots) \in \mathcal{M}_1, \quad \mu_k = 0 \right\} \cup \{0\},$$

and

$$\left\{ \sum_{j=1}^{\infty} \frac{\mu_j}{2^j} \,\middle|\, (\mu_1, \mu_2, \ldots) \in \mathcal{M}_1, \quad \mu_k = 0 \right\} = \bigcup_{(\mu_1, \ldots, \mu_{k-1})} \left\{ x + \sum_{j=1}^{k-1} \frac{\mu_j}{2^j} \,\middle|\, 0 < x \le \frac{1}{2^k} \right\},$$

which is a finite union of Borel sets.

The proof that $\mathcal{M}\mathcal{M}\cdots$ and $\mathcal{M}$ are homeomorphic is essentially the same one given in Lemma 7.25, and we do not repeat it here. Let θ mapping $\mathcal{M}\mathcal{M}\cdots$ onto $\mathcal{M}$ be a homeomorphism and define $\varphi: \mathcal{H} \to \mathcal{M}$ by

$$\varphi(x_1, x_2, \ldots) = \theta[\psi(x_1), \psi(x_2), \ldots].$$

Then φ is the required Borel-isomorphism.   Q.E.D.

Lemma B.5 If K_1 and L are Borel subsets of $\mathcal{M}$, $K_1 \subset L$, and K_1 is Borel-isomorphic to $\mathcal{M}$, then L is Borel-isomorphic to $\mathcal{M}$.

Proof For Borel subsets A and B of $\mathcal{M}$, we write $A \approx B$ to indicate that A and B are Borel-isomorphic. Note that $A \approx B$ and $B \approx C$ implies

$A \approx C$. Also, if $A_1, A_2, \ldots$ is a sequence of disjoint Borel sets, if $B_1, B_2, \ldots$ is another such sequence, and if $A_i \approx B_i$ for every i, then $\bigcup_{i=1}^{\infty} A_i \approx \bigcup_{i=1}^{\infty} B_i$. We note finally that if $A = A_1 \cup A_2$ and $A \approx B$, then $B = B_1 \cup B_2$, where $A_1 \approx B_1$ and $A_2 \approx B_2$. If A_1 and A_2 are disjoint, then B_1 and B_2 can be taken to be disjoint.

Under the hypotheses of the lemma, let $D_1 = \mathcal{M} - K_1$. Since $\mathcal{M}_1 \approx K_1$ and $\mathcal{M} = K_1 \cup D_1$, there exist disjoint Borel sets K_2 and D_2 such that $K_1 = K_2 \cup D_2$, $K_1 \approx K_2$ and $D_1 \approx D_2$. Since $K_1 \approx K_2$ and $K_1 = K_2 \cup D_2$, there exist disjoint Borel sets K_3 and D_3 such that $K_2 = K_3 \cup D_3$, $K_2 \approx K_3$, and $D_2 \approx D_3$. Continuing in this manner, at the nth step we construct disjoint Borel sets K_n and D_n such that $K_{n-1} = K_n \cup D_n$, $K_{n-1} \approx K_n$, and $D_{n-1} \approx D_n$. Let $K_{\infty} = \bigcap_{n=1}^{\infty} K_n$. Then $\mathcal{M} = K_{\infty} \cup [\bigcup_{n=1}^{\infty} D_n]$, and all the sets on the right side of this equation are disjoint.

Let $A_1 = \mathcal{M} - L$ and $B_1 = L - K_1$. Then A_1 and B_1 are disjoint and $D_1 = A_1 \cup B_1$. For each n, $D_1 \approx D_n$, so $D_n = A_n \cup B_n$, where A_n and B_n are disjoint Borel sets and $A_1 \approx A_n$, $B_1 \approx B_n$. In particular, $A_n \approx A_{n+1}$ for $n = 1, 2, \ldots$, and we have

$$\mathcal{M} = K_{\infty} \cup \left[\bigcup_{n=1}^{\infty} D_n \right] = K_{\infty} \cup \left[\bigcup_{n=1}^{\infty} A_n \right] \cup \left[\bigcup_{n=1}^{\infty} B_n \right]$$

$$\approx K_{\infty} \cup \left[\bigcup_{n=2}^{\infty} A_n \right] \cup \left[\bigcup_{n=1}^{\infty} B_n \right]$$

$$= \left\{ K_{\infty} \cup \left[\bigcup_{n=1}^{\infty} D_n \right] \right\} - A_1 = \mathcal{M} - A_1 = L. \qquad \text{Q.E.D.}$$

We can now prove Proposition 7.16, and the proof clearly shows that Corollary 7.16.1 is also true.

Proposition B.3 Let X and Y be Borel spaces. Then X and Y are isomorphic if and only if they have the same cardinality.

Proof If X and Y are isomorphic, then clearly they must have the same cardinality. If X and Y both have the same finite or countably infinite cardinality, then their Borel σ-algebras are their power sets and any one-to-one onto mapping from one to the other is a Borel-isomorphism.

If X is uncountable, then by Lemma B.4 there exists a Borel isomorphism $\varphi : X \to \mathcal{M}$ such that $L = \varphi(X)$ is a Borel subset of M. By Lemma B.3, X contains a compact set K which is homeomorphic to $\mathcal{M}$, so $\varphi(K)$ is Borel-isomorphic to $\mathcal{M}$ and $\varphi(K) \subset L$. Set $K_1 = \varphi(K)$ and use Lemma B.5 to conclude that L and $\mathcal{M}$ are isomorphic. It follows that X and $\mathcal{M}$ are isomorphic. If Y is uncountable, the same argument shows that Y and $\mathcal{M}$ are isomorphic, so X and Y are isomorphic. Q.E.D.

B.3   An Analytic Set Which Is Not Borel-Measurable

Suslin schemes can be used to generate a strictly increasing sequence of σ-algebras on any given uncountable Borel space X. The first σ-algebra in this sequence is the Borel σ-algebra $\mathscr{B}_X$ and the second is the analytic σ-algebra $\mathscr{A}_X$, and, as a result of the following discussion, we will see that $\mathscr{A}_X$ is strictly larger than $\mathscr{B}_X$. The proof of this depends on a contradiction involving universal functions, which we now introduce.

Let $\mathscr{M}_1$ be the set of sequences of zeroes and ones for which one occurs infinitely many times. If the nonzero components of $m \in \mathscr{M}_1$ are in positions $m_1, m_2, \ldots$, then we can think of m as a mapping from $\mathscr{N}$ to $\mathscr{N}$ defined by

$$m(\zeta_1, \zeta_2, \ldots) = (\zeta_{m_1}, \zeta_{m_2}, \ldots).$$

Definition B.1 Let $\mathscr{P}$ be a paving of $\mathscr{N}$. A *universal function* L for $\mathscr{P}$ is a mapping from $\mathscr{N}$ onto $\mathscr{P}$. If $\mathscr{Q}$ is another paving of $\mathscr{N}$ and

$$\{z \in \mathscr{N} \,|\, z \in L[m(z)]\} \in \mathscr{Q} \qquad \forall m \in \mathscr{M}_1, \tag{7}$$

we say L is *consistent with* $\mathscr{Q}$.

Proposition B.4 Let $\mathscr{G}$ be the collection of open subsets of $\mathscr{N}$. There exists a universal function for $\mathscr{G}$ consistent with $\mathscr{G}$.

Proof The space $\mathscr{N}$ is separable, so its topology has a countable base $\{G(1), G(2), \ldots\}$, where the empty set is included among these basic open sets. Define $L: \mathscr{N} \to \mathscr{G}$ by

$$L(\zeta_1, \zeta_2, \ldots) = \bigcup_{n=1}^{\infty} G(\zeta_n).$$

It is clear that L is a universal function for $\mathscr{G}$. Now choose $m \in \mathscr{M}_1$ and suppose the nonzero components of m are in positions $m_1, m_2, \ldots$. Choose $z_0 = (\zeta_1^0, \zeta_2^0, \ldots)$ in the set

$$\{z \in \mathscr{N} \,|\, z \in L[m(z)]\} = \left\{(\zeta_1, \zeta_2, \ldots) \in \mathscr{N} \,|\, (\zeta_1, \zeta_2, \ldots) \in \bigcup_{k=1}^{\infty} G(\zeta_{m_k})\right\}.$$

Then for some $\bar{k}$, we have $z_0 \in G(\zeta_{m_{\bar{k}}}^0)$. Let

$$U_{\bar{k}}(z_0) = \{(\zeta_1, \zeta_2, \ldots) \in \mathscr{N} \,|\, \zeta_{m_{\bar{k}}} = \zeta_{m_{\bar{k}}}^0\}.$$

Then $G(\zeta_{m_{\bar{k}}}^0) \subset L[m(z)]$ for every $z \in U_{\bar{k}}(z_0)$, so $z \in L[m(z)]$ for every $z \in U_{\bar{k}}(z_0) \cap G(\zeta_{m_{\bar{k}}}^0)$. Therefore $U_{\bar{k}}(z_0) \cap G(\zeta_{m_{\bar{k}}}^0)$ is an open neighborhood of z_0 contained in $\{z \in \mathscr{N} \,|\, z \in L[m(z)]\}$, so this set is open.   Q.E.D.

Given a paved space and a universal function for the paving which satisfies a condition like (7), it is possible to construct similar universal functions for larger pavings. We show first how this is done when the given paving is extended by the use of Suslin schemes.

Proposition B.5   Let $\mathscr{P}$ be a paving for $\mathscr{N}$ and suppose that there exists a universal function for $\mathscr{P}$ consistent with $\mathscr{S}(\mathscr{P})$. Then there exists a universal function for $\mathscr{S}(\mathscr{P})$ consistent with $\mathscr{S}(\mathscr{P})$.

Proof   Fix a partition $\{P_s | s \in \Sigma\}$ of the positive integers into countably many countable sets, and define for each $s \in \Sigma$ a corresponding $m_s = (\mu_1(s), \mu_2(s), \ldots) \in \mathscr{M}_1$ by

$$\mu_k(s) = \begin{cases} 1 & \text{if } k \in P_s, \\ 0 & \text{if } k \notin P_s, \end{cases} \tag{8}$$

Let L be a universal function for $\mathscr{P}$ consistent with $\mathscr{S}(\mathscr{P})$. Define $K : \mathscr{N} \to \mathscr{S}(\mathscr{P})$ by

$$K(z_0) = \bigcup_{z \in \mathscr{N}} \bigcap_{s < z} L[m_s(z_0)]. \tag{9}$$

To show that K is onto, we must show that given any Suslin scheme S for $\mathscr{P}$, there exists $z_0 \in \mathscr{N}$ such that

$$S(s) = L[m_s(z_0)] \qquad \forall s \in \Sigma. \tag{10}$$

If $S : \Sigma \to \mathscr{P}$ is given and $s \in \Sigma$, then $S(s) \in \mathscr{P}$. Since L is a universal function for $\mathscr{P}$, there exists $z_s \in \mathscr{N}$ for which $S(s) = L(z_s)$. If z_0 is chosen so that $m_s(z_0) = z_s$ for every $s \in \Sigma$, then (10) is satisfied, and such a choice of z_0 is possible because $m_s(z_0)$ depends only on the components of z_0 with indices in P_s. Therefore K is a universal function for $\mathscr{S}(\mathscr{P})$.

If $m, n \in \mathscr{M}_1$, then there is an element in $\mathscr{M}_1$, which we denote by mn, such that $(mn)(z) = m[n(z)]$ for every $z \in \mathscr{N}$. In fact, if the nonzero elements of m are $(m_1, m_2, \ldots)$ and the nonzero elements of n are $(n_1, n_2, \ldots)$, then the nonzero elements of mn are $(n_{m_1}, n_{m_2}, \ldots)$. Now suppose $m \in \mathscr{M}_1$. We have

$$\{z_0 \in \mathscr{N} | z_0 \in K[m(z_0)]\} = \left\{ z_0 \in \mathscr{N} \Big| z_0 \in \bigcup_{z \in \mathscr{N}} \bigcap_{s < z} L[(m_s m)(z_0)] \right\}$$

$$= \bigcup_{z \in \mathscr{N}} \bigcap_{s < z} \{z_0 \in \mathscr{N} | z_0 \in L[(m_s m)(z_0)]\},$$

which, since L is consistent with $\mathscr{S}(\mathscr{P})$, is the nucleus of a Suslin scheme for $\mathscr{S}(\mathscr{P})$. It follows from Proposition B.2 that K is consistent with $\mathscr{S}(\mathscr{P})$.

Q.E.D.

Corollary B.5.1   There is a universal function for $\mathscr{S}(\mathscr{F}_{\mathscr{N}})$ consistent with $\mathscr{S}(\mathscr{F}_{\ast})$.

Proof Let $\mathscr{G}$ be the collection of open subsets of $\mathscr{N}$. By Propositions B.4 and B.5, there is a universal function for $\mathscr{S}(\mathscr{G})$ consistent with $\mathscr{S}(\mathscr{G})$, and it remains only to show that $\mathscr{S}(\mathscr{G}) = \mathscr{S}(\mathscr{F}_{\mathscr{N}})$. Since $\mathscr{G} \subset \mathscr{B}_{\mathscr{N}}$, it follows from Proposition 7.36 that $\mathscr{S}(\mathscr{G}) \subset \mathscr{S}(\mathscr{F}_{\mathscr{N}})$. Since every closed subset of $\mathscr{N}$ is a G_δ-set and, by Proposition 7.35, $\mathscr{G}_\delta \subset \mathscr{S}(\mathscr{G})_\delta = \mathscr{S}(\mathscr{G})$, we see that $\mathscr{F}_{\mathscr{N}} \subset \mathscr{S}(\mathscr{G})$. Proposition B.2 implies that $\mathscr{S}(\mathscr{F}_{\mathscr{N}}) \subset \mathscr{S}[\mathscr{S}(\mathscr{G})] = \mathscr{S}(\mathscr{G})$.   Q.E.D.

Corollary B.5.2 Let L be a universal function for $\mathscr{S}(\mathscr{F}_{\mathscr{N}})$ consistent with $\mathscr{S}(\mathscr{F}_{\mathscr{N}})$. The set

$$A_0 = \{z \in \mathscr{N} \,|\, z \in L(z)\} \tag{11}$$

is analytic but not Borel-measurable, and $\mathscr{N} - A_0$ is not analytic.

Proof The set A_0 is analytic because L is consistent with $\mathscr{S}(\mathscr{F}_{\mathscr{N}})$. We have

$$\mathscr{N} - A_0 = \{z \in \mathscr{N} \,|\, z \notin L(z)\}, \tag{12}$$

and if this set is analytic, then there exists $z_0 \in \mathscr{N}$ such that

$$\mathscr{N} - A_0 = L(z_0).$$

If $z_0 \in A_0$, then $z_0 \notin L(z_0)$, and (11) is contradicted. If $z_0 \in \mathscr{N} - A_0$, then $z_0 \in L(z_0)$ and (12) is contradicted. Therefore $\mathscr{N} - A_0$ is not analytic, thus not Borel-measurable, so A_0 is also not Borel-measurable.   Q.E.D.

Proposition B.6 Let X be an uncountable Borel space. There exists an analytic subset A of X such that A is not Borel-measurable and $X - A$ is not analytic.

Proof Let $\varphi : \mathscr{N} \to X$ be a Borel isomorphism from $\mathscr{N}$ onto X (Corollary 7.16.1), and let $A_0 \subset \mathscr{N}$ be as in Corollary B.5.2. Then $A = \varphi(A_0)$ is analytic, but since $\mathscr{N} - A_0 = \varphi^{-1}(X - A)$ is not analytic, neither is $X - A$. It follows that A is not Borel-measurable.   Q.E.D.

B.4 The Limit σ-algebra

We construct a collection of σ-algebras indexed by the countable ordinals, and at the end of this process we arrive at the limit σ-algebra, denoted by $\mathscr{L}_X$. The proofs of many of the properties of $\mathscr{L}_X$, and indeed the definition of $\mathscr{L}_X$, proceed by transfinite induction. We also make frequent use of the fact that if $\{\alpha_n\}$ is a sequence of countable ordinals, then there exists a countable ordinal $\bar{\alpha}$ such that $\alpha_n < \bar{\alpha}$ for every n. In keeping with standard convention, we denote by Ω the first uncountable ordinal.

Definition B.2 Let X be a Borel space and $\mathcal{G}_X$ the collection of open subsets of X. For each countable ordinal α, we define

$$\mathcal{L}_X^0 = \sigma(\mathcal{G}_X), \tag{13}$$

$$\mathcal{L}_X^\alpha = \sigma\left[\mathcal{S}\left(\bigcup_{\beta < \alpha} \mathcal{L}_X^\beta\right)\right]. \tag{14}$$

The *limit σ-algebra* is

$$\mathcal{L}_X = \bigcup_{\alpha < \Omega} \mathcal{L}_X^\alpha. \tag{15}$$

We prove later (Proposition B.10) that $\mathcal{L}_X$ is in fact a σ-algebra. Note that $\mathcal{L}_X^0 = \mathcal{B}_X$ and $\mathcal{L}_X^1 = \mathcal{A}_X$. When X is countable, $\mathcal{B}_X = \mathcal{L}_X^\alpha$ for every $\alpha < \Omega$. If X is uncountable, there is no loss of generality in assuming $X = \mathcal{N}$ when dealing with the σ-algebras $\mathcal{L}_X^\alpha$ and $\mathcal{L}_X$. This is the subject of the next proposition.

Proposition B.7 Let X be an uncountable Borel space and let $\varphi : \mathcal{N} \to X$ be a Borel isomorphism from $\mathcal{N}$ onto X. (Such an isomorphism exists by Corollary 7.16.1.) Then for every $\alpha < \Omega$,

$$\varphi(\mathcal{L}_{\mathcal{N}}^\alpha) = \mathcal{L}_X^\alpha, \qquad \mathcal{L}_{\mathcal{N}}^\alpha = \varphi^{-1}(\mathcal{L}_X^\alpha), \tag{16}$$

and

$$\varphi(\mathcal{L}_{\mathcal{N}}) = \mathcal{L}_X, \qquad \mathcal{L}_{\mathcal{N}} = \varphi^{-1}(\mathcal{L}_X). \tag{17}$$

Proof We prove (16) by transfinite induction. For $\alpha = 0$, (16) clearly holds. If (16) holds for all $\beta < \alpha$, where $\alpha < \Omega$, then we have

$$\varphi\left(\bigcup_{\beta < \alpha} \mathcal{L}_{\mathcal{N}}^\beta\right) = \bigcup_{\beta < \alpha} \mathcal{L}_X^\beta, \qquad \bigcup_{\beta < \alpha} \mathcal{L}_{\mathcal{N}}^\beta = \varphi^{-1}\left(\bigcup_{\beta < \alpha} \mathcal{L}_X^\beta\right).$$

Let S be a Suslin scheme for $\bigcup_{\beta < \alpha} \mathcal{L}_{\mathcal{N}}^\beta$. Then

$$\varphi[N(S)] = N(\varphi \circ S),$$

where

$$(\varphi \circ S)(s) = \varphi[S(s)] \qquad \forall s \in \Sigma.$$

Since $\varphi \circ S$ is a Suslin scheme for $\bigcup_{\beta < \alpha} \mathcal{L}_X^\beta$, we see that

$$\varphi\left[\mathcal{S}\left(\bigcup_{\beta < \alpha} \mathcal{L}_{\mathcal{N}}^\beta\right)\right] \subset \mathcal{S}\left(\bigcup_{\beta < \alpha} \mathcal{L}_X^\beta\right). \tag{18}$$

On the other hand, if R is a Suslin scheme for $\bigcup_{\beta < \alpha} \mathcal{L}_X^\beta$, then

$$N(R) = \varphi[N(\varphi^{-1} \circ R)],$$

where

$$(\varphi^{-1} \circ R)(s) = \varphi^{-1}[R(s)] \qquad \forall s \in \Sigma.$$

This shows that $N(R) \in \varphi[\mathscr{S}(\bigcup_{\beta < \alpha} \mathscr{L}^{\beta}_{\mathscr{N}})]$, which proves the reverse of set containment (18). Therefore,

$$\varphi\left[\mathscr{S}\left(\bigcup_{\beta < \alpha} \mathscr{L}^{\beta}_{\mathscr{N}}\right)\right] = \mathscr{S}\left(\bigcup_{\beta < \alpha} \mathscr{L}^{\beta}_{X}\right). \tag{19}$$

Since φ is one-to-one, we also have

$$\mathscr{S}\left(\bigcup_{\beta < \alpha} \mathscr{L}^{\beta}_{\mathscr{N}}\right) = \varphi^{-1}\left[\mathscr{S}\left(\bigcup_{\beta < \alpha} \mathscr{L}^{\beta}_{X}\right)\right]. \tag{20}$$

Now by (19), $\varphi(\mathscr{L}^{\alpha}_{\mathscr{N}})$ is a σ-algebra containing $\mathscr{S}(\bigcup_{\beta < \alpha} \mathscr{L}^{\beta}_{X})$, so

$$\varphi(\mathscr{L}^{\alpha}_{\mathscr{N}}) \supset \mathscr{L}^{\alpha}_{X}. \tag{21}$$

By (20), $\varphi^{-1}(\mathscr{L}^{\alpha}_{X})$ is a σ-algebra containing $\mathscr{S}(\bigcup_{\beta < \alpha} \mathscr{L}^{\beta}_{\mathscr{N}})$, so

$$\mathscr{L}^{\alpha}_{\mathscr{N}} \subset \varphi^{-1}(\mathscr{L}^{\alpha}_{X}). \tag{22}$$

Since φ is one-to-one, (21) implies

$$\mathscr{L}^{\alpha}_{\mathscr{N}} \supset \varphi^{-1}(\mathscr{L}^{\alpha}_{X}) \tag{23}$$

and (22) implies

$$\varphi(\mathscr{L}^{\alpha}_{\mathscr{N}}) \subset \mathscr{L}^{\alpha}_{X}. \tag{24}$$

Relations (21)–(24) imply (16). Relation (17) follows from (15) and (16).

Q.E.D.

We have already seen that in an uncountable Borel space X, $\mathscr{L}^{0}_{X}$ is properly contained in $\mathscr{L}^{1}_{X}$ (Proposition B.6). We would like to show more generally that if $\beta < \alpha < \Omega$, then $\mathscr{L}^{\beta}_{X}$ is properly contained in $\mathscr{L}^{\alpha}_{X}$. Our method for doing this is to generalize Corollary B.5.1 and then generalize Corollary B.5.2. The following lemmas are a step in this direction. If $\mathscr{P}$ is a paving for a space X, we denote by $\bar{\mathscr{P}}$ the paving

$$\bar{\mathscr{P}} = \mathscr{P} \cup \{X - P | P \in \mathscr{P}\}. \tag{25}$$

Lemma B.6 Let $\mathscr{P}$ be a paving for $\mathscr{N}$ which contains the open subsets of $\mathscr{N}$, and suppose there exists a universal function for $\mathscr{P}$ consistent with $\mathscr{P}$. Then there exists a universal function for $\bar{\mathscr{P}}$ consistent with $\sigma(\mathscr{P})$.

Proof Let L be a universal function for $\mathscr{P}$ consistent with $\mathscr{P}$. Define $K : \mathscr{N} \to \bar{\mathscr{P}}$ by

$$K(\zeta_1, \zeta_2, \ldots) = \begin{cases} L(\zeta_2, \zeta_3, \zeta_4, \ldots) & \text{if } \zeta_1 \text{ is odd,} \\ \mathscr{N} - L(\zeta_2, \zeta_3, \zeta_4, \ldots) & \text{if } \zeta_1 \text{ is even.} \end{cases}$$

It is clear that K is a universal function for $\bar{\mathscr{P}}$. As in the proof of Proposition B.4, choose $m \in \mathscr{M}_1$ and suppose that the nonzero components of m are in positions $m_1, m_2, \ldots$. Then

$$\{z \in \mathcal{N} \mid z \in K[m(z)]\}$$
$$= \{(\zeta_1, \zeta_2, \ldots) \mid \zeta_{m_1} \text{ is odd and } (\zeta_1, \zeta_2, \ldots) \in L(\zeta_{m_2}, \zeta_{m_3}, \ldots)\}$$
$$\cup \{(\zeta_1, \zeta_2, \ldots) \mid \zeta_{m_1} \text{ is even and } (\zeta_1, \zeta_2, \ldots) \notin L(\zeta_{m_2}, \zeta_{m_3}, \ldots)\}$$
$$= \left(\left[\bigcup_{k=1}^{\infty} \{(\zeta_1, \zeta_2, \ldots) \mid \zeta_{m_1} = 2k - 1\} \right] \right.$$
$$\cap \{(\zeta_1, \zeta_2, \ldots) \mid (\zeta_1, \zeta_2, \ldots) \in L(\zeta_{m_2}, \zeta_{m_3}, \ldots)\} \bigg)$$
$$\cup \left(\left[\bigcup_{k=1}^{\infty} \{(\zeta_1, \zeta_2, \ldots) \mid \zeta_{m_1} = 2k\} \right] \right.$$
$$\cap \{(\zeta_1, \zeta_2, \ldots) \mid (\zeta_1, \zeta_2, \ldots) \notin L(\zeta_{m_2}, \zeta_{m_3}, \ldots)\} \bigg). \tag{26}$$

Since L is consistent with $\mathscr{P}$ and $\mathscr{P}$ contains every open set, we have that every set in (26) is in $\sigma(\mathscr{P})$. It follows that K is consistent with $\sigma(\mathscr{P})$.   Q.E.D.

Lemma B.7   Let α be a countable ordinal. For each $\beta < \alpha$, let $\mathscr{P}_\beta$ be a paving for $\mathcal{N}$ which contains the collection $\mathscr{G}$ of open sets, and assume that there exists a universal function L_β for $\mathscr{P}_\beta$ consistent with $\mathscr{P}_\beta$. Then there exists a universal function for $\bigcup_{\beta < \alpha} \mathscr{P}_\beta$ consistent with $\mathscr{S}(\bigcup_{\beta < \alpha} \mathscr{P}_\beta)$.

Proof   The set of ordinals $\{\beta \mid \beta < \alpha\}$ is countable whenever $\alpha < \Omega$, so there exists a partition $\{P(\beta) \mid \beta < \alpha\}$ of the positive integers such that $P(\beta)$ is nonempty for each $\beta < \alpha$. Define a universal function for $\bigcup_{\beta < \alpha} P(\beta)$ by

$$L(\zeta_1, \zeta_2, \ldots) = L_\beta(\zeta_2, \zeta_3, \ldots) \qquad \text{if} \quad \zeta_1 \in P(\beta).$$

Let $m \in \mathscr{M}_1$ have nonzero components $m_1, m_2, \ldots$. Then

$$\{z \in \mathcal{N} \mid z \in L[m(z)]\}$$

$$= \bigcup_{\beta < \alpha} \{(\zeta_1, \zeta_2, \ldots) \mid \zeta_{m_1} \in P(\beta) \text{ and } (\zeta_1, \zeta_2, \ldots) \in L_\beta(\zeta_{m_2}, \zeta_{m_3}, \ldots)\}$$

$$= \bigcup_{\beta < \alpha} [\{(\zeta_1, \zeta_2, \ldots) \mid \zeta_{m_1} \in P(\beta)\} \cap \{(\zeta_1, \zeta_2, \ldots) \mid (\zeta_1, \zeta_2, \ldots) \in L_\beta(\zeta_{m_2}, \zeta_{m_3}, \ldots)\}],$$

and this set is in $\mathscr{S}(\bigcup_{\beta < \alpha} \mathscr{P}_\beta)$ by Proposition 7.35(b), (c), and the fact that each L_β is consistent with $\mathscr{P}_\beta$.   Q.E.D.

Proposition B.8   For each $\alpha < \Omega$, there is a universal function for $\mathscr{S}(\mathscr{L}_{\mathcal{N}}^{\alpha})$ consistent with $\mathscr{S}(\mathscr{L}_{\mathcal{N}}^{\alpha})$.

Proof   For simplicity of notation, we suppress the subscript $\mathcal{N}$. The proof is by transfinite induction. When $\alpha = 0$, the result follows from Corollary B.5.1. Assume now that the result holds for every $\beta < \alpha$, where $\alpha < \Omega$. We prove it for α.

By Lemma B.7 and the induction assumption, there is a universal function for $\bigcup_{\beta < \alpha} \mathcal{S}(\mathcal{L}^\beta)$ consistent with $\mathcal{S}[\bigcup_{\beta < \alpha} \mathcal{S}(\mathcal{L}^\beta)]$. Now

$$\bigcup_{\beta < \alpha} \mathcal{L}^\beta \subset \bigcup_{\beta < \alpha} \mathcal{S}(\mathcal{L}^\beta) \subset \mathcal{S}\left(\bigcup_{\beta < \alpha} \mathcal{L}^\beta\right), \tag{27}$$

and applying $\mathcal{S}$ to both sides of (27) and using Proposition B.2, we obtain

$$\mathcal{S}\left(\bigcup_{\beta < \alpha} \mathcal{L}^\beta\right) = \mathcal{S}\left[\bigcup_{\beta < \alpha} \mathcal{S}(\mathcal{L}^\beta)\right]. \tag{28}$$

From Proposition B.5 and (28) we have the existence of a universal function for $\mathcal{S}(\bigcup_{\beta < \alpha} \mathcal{L}^\beta)$ consistent with $\mathcal{S}(\bigcup_{\beta < \alpha} \mathcal{L}^\beta)$, and Lemma B.6 implies existence of a universal function for $\overline{\mathcal{S}(\bigcup_{\beta < \alpha} \mathcal{L}^\beta)}$ consistent with $\mathcal{L}^\alpha$. From Corollary 7.35.1 we have

$$\mathcal{L}^\alpha = \sigma\left[\mathcal{S}\left(\bigcup_{\beta < \alpha} \mathcal{L}^\beta\right)\right] \subset \mathcal{S}\left[\overline{\mathcal{S}\left(\bigcup_{\beta < \alpha} \mathcal{L}^\beta\right)}\right], \tag{29}$$

so we have a universal function for $\overline{\mathcal{S}(\bigcup_{\beta < \alpha} \mathcal{L}^\beta)}$ consistent with

$$\mathcal{S}\left[\overline{\mathcal{S}\left(\bigcup_{\beta < \alpha} \mathcal{L}^\beta\right)}\right].$$

But from (29),

$$\mathcal{L}^\alpha \subset \mathcal{S}\left[\overline{\mathcal{S}\left(\bigcup_{\beta < \alpha} \mathcal{L}^\beta\right)}\right] \subset \mathcal{S}(\mathcal{L}^\alpha),$$

and applying $\mathcal{S}$ to both sides, we see that

$$\mathcal{S}(\mathcal{L}^\alpha) = \mathcal{S}\left[\overline{\mathcal{S}\left(\bigcup_{\beta < \alpha} \mathcal{L}^\beta\right)}\right]. \tag{30}$$

From Proposition B.5 and (30) we have the existence of a universal function for $\mathcal{S}(\mathcal{L}^\alpha)$ consistent with $\mathcal{S}(\mathcal{L}^\alpha)$.   Q.E.D.

Proposition B.9   Let X be an uncountable Borel space. If $\beta < \alpha < \Omega$, then $\mathcal{L}_X^\beta$ is properly contained in $\mathcal{L}_X^\alpha$.

Proof   We assume without loss of generality that $X = \mathcal{N}$ (Proposition B.7) and suppress the subscript $\mathcal{N}$. It is clear that for $\beta < \alpha$ we have $\mathcal{L}^\beta \subset$

$\mathscr{L}^{\alpha}$. Let L be a universal function for $\mathscr{S}(\mathscr{L}^{\beta})$ consistent with $\mathscr{S}(\mathscr{L}^{\beta})$ and define

$$A = \{z \in \mathcal{N} \mid z \in L(z)\}.$$

Then $A \in \mathscr{S}(\mathscr{L}^{\beta})$. If $\mathcal{N} - A \in \mathscr{S}(\mathscr{L}^{\beta})$, then for some $z_0 \in \mathcal{N}$ we have

$$\mathcal{N} - A = L(z_0).$$

If $z_0 \in A$, then $z_0 \notin L(z_0)$ and a contradiction is reached. If $z_0 \in \mathcal{N} - A$, then $z_0 \in L(z_0)$ and again a contradiction is reached. It follows that $\mathcal{N} - A \notin \mathscr{S}(\mathscr{L}^{\beta})$. But $\mathcal{N} - A \in \mathscr{L}^{\alpha}$, so $\mathscr{L}^{\beta}$ is properly contained in $\mathscr{L}^{\alpha}$.   Q.E.D.

Proposition B.10   Let X be a Borel space. The limit σ-algebra $\mathscr{L}_X$ is contained in $\mathscr{U}_X$ and

$$\mathscr{L}_X = \mathscr{S}(\mathscr{L}_X). \tag{31}$$

Indeed, $\mathscr{L}_X$ is the smallest σ-algebra containing the open subsets of X which satisfies (31).

Proof   The result is trivial if X is countable, so assume that X is uncountable. It is clear that $\varnothing \in \mathscr{L}_X$ and $\mathscr{L}_X$ is closed under complementation, so we need only verify that $\mathscr{L}_X$ is closed under countable unions in order to show that it is a σ-algebra. If $Q_1, Q_2, \ldots$ is a sequence of sets in $\mathscr{L}_X$, then for some $\alpha < \Omega$, we have $Q_k \in \mathscr{L}_X^{\alpha}$ for every k. Then $\bigcup_{k=1}^{\infty} Q_k \in \mathscr{L}_X^{\alpha} \subset \mathscr{L}_X$.

We prove by transfinite induction that $\mathscr{L}_X^{\alpha} \subset \mathscr{U}_X$ for every $\alpha < \Omega$. This is clearly the case if $\alpha = 0$. If $\mathscr{L}_X^{\beta} \subset \mathscr{U}_X$ for every $\beta < \alpha$, where $\alpha < \Omega$, then by Lusin's theorem (Proposition 7.42), $\mathscr{S}(\bigcup_{\beta < \alpha} \mathscr{L}_X^{\beta}) \subset \mathscr{U}_X$. It follows that $\mathscr{L}_X^{\alpha} \subset \mathscr{U}_X$. Therefore $\mathscr{L}_X \subset \mathscr{U}_X$.

We now prove (31). As a result of Proposition 7.35(d), it suffices to prove that $\mathscr{L}_X \supset \mathscr{S}(\mathscr{L}_X)$. Let S be a Suslin scheme for $\mathscr{L}_X$. Since Σ is countable, there exists $\alpha < \Omega$ such that $S(s) \in \mathscr{L}_X^{\alpha}$ for every $s \in \Sigma$. Then $N(S) \in \mathscr{L}_X^{\alpha+1} \subset \mathscr{L}_X$, and (31) is proved.

Suppose $\mathscr{P}$ is a σ-algebra containing the open subsets of X which satisfies $\mathscr{P} = \mathscr{S}(\mathscr{P})$. Clearly, $\mathscr{B}_X = \mathscr{L}_X^0 \subset \mathscr{P}$. If $\mathscr{L}_X^{\beta} \subset \mathscr{P}$ for every $\beta < \alpha$, where $\alpha < \Omega$, then (14) implies that $\mathscr{L}_X^{\alpha} \subset \mathscr{P}$. Therefore $\mathscr{P}$ contains $\mathscr{L}_X$, which must be the smallest σ-algebra containing the open subsets of X and satisfying (31).   Q.E.D.

A major shortcoming of the analytic σ-algebra is that the composition of analytically measurable functions is not necessarily analytically measurable (cf. remarks following Proposition 7.50). However, the composition of limit-measurable functions is limit-measurable. We first give a formal definition of these terms and then prove the preceding statements.

Definition B.3 Let X and Y be Borel spaces, $D \subset X$, and $\mathscr{P}$ a σ-algebra on X. A function $f : D \to Y$ is said to be $\mathscr{P}$-*measurable* if $f^{-1}(B) \in \mathscr{P}$ for every $B \in \mathscr{B}_Y$. If $\mathscr{P} = \mathscr{L}_X$, we say that f is *limit-measurable*. The σ-algebra $\mathscr{P}$ is said to be *closed under composition of functions* if, whenever $f : X \to X$ is $\mathscr{P}$-measurable and $P \in \mathscr{P}$, then $f^{-1}(P) \in \mathscr{P}$.

In Definition B.3 there is no mention of a $\mathscr{P}$-measurable function g mapping X into a Borel space Y with which to compose f. If there were such a g, then to check that $g \circ f : X \to Y$ is $\mathscr{P}$-measurable, we would check that $f^{-1}[g^{-1}(B)]$ is $\mathscr{P}$-measurable for every $B \in \mathscr{B}_Y$. Since $g^{-1}(B) \in \mathscr{P}$, it suffices to check that $f^{-1}(P) \in \mathscr{P}$ for every $P \in \mathscr{P}$, which is the condition stated in Definition B.3. The stipulation in Definition B.3 that f have the same domain and range space is inconsequential as long as $\mathscr{P} = \mathscr{L}_X^\alpha$ for some $\alpha < \Omega$ or $\mathscr{P} = \mathscr{L}_X$ (see Proposition B.7). These are the only cases we consider. The closure of a σ-algebra under composition of mappings and the satisfaction of an equation like (31) are intimately related, as the following lemma shows.

Lemma B.8 Let X be a Borel space and let $\mathscr{P}$ be a σ-algebra on X. If $\mathscr{P}$ contains the analytic subsets of X and is closed under composition of functions, then

$$\mathscr{P} = \mathscr{S}(\mathscr{P}).$$

Proof If X is countable, the result is trivial, so we assume that X is uncountable. In light of Proposition 7.35(d), we need only prove that under the assumptions of the lemma we have $\mathscr{P} \supset \mathscr{S}(\mathscr{P})$. To do this, for an arbitrary Suslin scheme S for $\mathscr{P}$ we construct a $\mathscr{P}$-measurable function $f : X \to X$ and a set $P \in \mathscr{P}$ such that

$$f^{-1}(P) = N(S). \tag{32}$$

Let $\varphi : \mathscr{N} \to X$ be a Borel isomorphism from $\mathscr{N}$ onto X (Corollary 7.16.1), and let ψ be a one-to-one onto function from the set of positive integers to Σ. For $k = 1, 2, \ldots$, define $\bar{f}_k : \mathscr{N} \to \{1, 2\}$ by

$$\bar{f}_k(z) = \begin{cases} 1 & \text{if } \varphi(z) \in S[\psi(k)], \\ 2 & \text{otherwise,} \end{cases}$$

and define $\bar{f} : \mathscr{N} \to \mathscr{N}$ by

$$\bar{f}(z) = [\bar{f}_1(z), \bar{f}_2(z), \ldots].$$

Finally, let $f : X \to X$ be given by $f = \varphi \circ \bar{f} \circ \varphi^{-1}$. We show that f is $\mathscr{P}$-measurable. This is equivalent to showing that $\bar{f} \circ \varphi^{-1} : X \to \mathscr{N}$ is $\mathscr{P}$-measurable. But $\bar{f} \circ \varphi^{-1}$ takes values in $\{(\zeta_1, \zeta_2, \ldots) \in \mathscr{N} \mid \zeta_n \leq 2 \, \forall n\}$ which has as a subbase the collection of open sets $\{R(k), \tilde{R}(k) | k = 1, 2, \ldots\}$, where

$$R(k) = \{(\zeta_1, \zeta_2, \ldots) | \zeta_n \leq 2 \; \forall n \text{ and } \zeta_k = 1\}, \tag{33}$$
$$\tilde{R}(k) = \{(\zeta_1, \zeta_2, \ldots) | \zeta_n \leq 2 \; \forall n \text{ and } \zeta_k = 2\}.$$

By the remark following Definition 7.6, the $\mathscr{P}$-measurability of the sets

$$\varphi(\bar{f}^{-1}[R(k)]) = S[\psi(k)], \qquad k = 1, 2, \ldots,$$
$$\varphi(\bar{f}^{-1}[\tilde{R}(k)]) = X - S[\psi(k)], \qquad k = 1, 2, \ldots,$$

implies the $\mathscr{P}$-measurability of $\bar{f} \circ \varphi^{-1}$. It follows that f is $\mathscr{P}$-measurable.
Define $P \subset X$ by

$$P = \bigcup_{z \in \mathscr{N}} \bigcap_{s < z} \varphi(R[\psi^{-1}(s)]),$$

where $R(k)$ is given by (33). Then P is an analytic subset of X, so $P \in \mathscr{P}$.
We have

$$f^{-1}(P) = \varphi \left[\bigcup_{z \in \mathscr{N}} \bigcap_{s < z} \bar{f}^{-1}(R[\psi^{-1}(s)]) \right]$$

$$= \bigcup_{z \in \mathscr{N}} \bigcap_{s < z} S(s) = N(S),$$

so (32) holds.   Q.E.D.

Proposition B.11   Let X be a Borel space. The limit σ-algebra $\mathscr{L}_X$ is the smallest σ-algebra containing the analytic subsets of X which is closed under composition of functions.

Proof We show first that $\mathscr{L}_X$ is closed under composition of functions. It suffices to show that if $f : X \to X$ is $\mathscr{L}_X$-measurable, $\alpha < \Omega$, and $Q \in \mathscr{L}_X^\alpha$, then $f^{-1}(Q) \in \mathscr{L}_X$. If $\alpha = 0$, this is true by definition. Suppose that for some $\alpha < \Omega$ and for every $\beta < \alpha$ and $C \in \mathscr{L}_X^\beta$ we have $f^{-1}(C) \in \mathscr{L}_X$. We show that $f^{-1}(Q) \in \mathscr{L}_X$ for every $Q \in \mathscr{S}(\bigcup_{\beta < \alpha} \mathscr{L}_X^\beta)$, and this implies that $f^{-1}(Q) \in \mathscr{L}_X$ for every $Q \in \mathscr{L}_X^\alpha$. Choose $Q \in \mathscr{S}(\bigcup_{\beta < \alpha} \mathscr{L}_X^\beta)$ and let S be a Suslin scheme for $\bigcup_{\beta < \alpha} \mathscr{L}_X^\beta$ such that $Q = N(S)$. Then

$$f^{-1}(Q) = N(f^{-1} \circ S), \tag{34}$$

where $f^{-1} \circ S$ is the Suslin scheme defined by

$$(f^{-1} \circ S)(s) = f^{-1}[S(s)] \qquad \forall s \in \Sigma.$$

By the induction hypothesis, $f^{-1} \circ S$ is a Suslin scheme for $\mathscr{L}_X$, and we have from Proposition B.10 and (34) that $f^{-1}(Q) \in \mathscr{L}_X$.

The fact that $\mathscr{L}_X$ is the smallest σ-algebra containing the analytic subsets of X which is closed under composition of functions follows from Proposition B.10 and Lemma B.8.   Q.E.D.

Corollary B.11.1 Let X, Y, and Z be uncountable Borel spaces. If $f:X \to Y$ and $g:Y \to Z$ are limit-measurable, then $g \circ f:X \to Z$ is limit-measurable. In particular, if f and g are analytically measurable, then $g \circ f$ is limit-measurable. It is possible to choose f and g to be analytically measurable so that $g \circ f$ is not analytically measurable.

Proof Proposition B.9 implies that $\mathscr{A}_X$, $\mathscr{A}_Y$, and $\mathscr{A}_Z$ are properly contained in $\mathscr{L}_X$, $\mathscr{L}_Y$, and $\mathscr{L}_Z$, respectively. Apply Proposition B.7 to the results of Proposition B.11.   Q.E.D.

Using an argument similar to the first part of the proof of Proposition B.11, the reader may verify that if $f:X \to Y$ and $g:Y \to Z$ are analytically measurable, then $g \circ f$ is in fact $\mathscr{L}_X^2$-measurable. Indeed, one can show by induction that if f is $\mathscr{L}_X^m$-measurable and g is $\mathscr{L}_Y^n$-measurable, where m and n are integers, then $g \circ f$ is $\mathscr{L}_X^{m+n}$-measurable.

Let X be a Borel space, and for $Q \in \mathcal{U}_X$ define $\theta_Q:P(X) \to [0,1]$ by

$$\theta_Q(p) = p(Q). \tag{35}$$

Then θ_Q is universally measurable (Corollary 7.46.1). If Q is Borel-measurable, then θ_Q is Borel-measurable (Proposition 7.25), and if Q is analytically measurable, then θ_Q is analytically measurable (Proposition 7.43). We consider the case when Q is $\mathscr{L}_X^\alpha$-measurable.

Proposition B.12 Let X be a Borel space. If $Q \in \mathscr{L}_X$, then θ_Q defined by (35) is $\mathscr{L}_{P(X)}$-measurable. In fact if $\alpha < \Omega$ and $Q \in \mathscr{L}_X^\alpha$, then θ_Q is $\mathscr{L}_{P(X)}^\alpha$-measurable.

Proof The last statement is true when $\alpha = 0$. If it is true for every $\beta < \alpha$, where $\alpha < \Omega$, and S is a Suslin scheme for $\bigcup_{\beta < \alpha} \mathscr{L}_X^\beta$, then for any $c \in R$, (98) of Chapter 7 holds, where $A = N(S)$ and $K(s)$ is defined by (92) of Chapter 7. For each $s \in \Sigma$, $K(s) \in \bigcup_{\beta < \alpha} \mathscr{L}_X^\beta$, so by the induction hypothesis, the set $\{p \in P(X)|p[K(s)] \geq c - (1/n)\}$ is in $\bigcup_{\beta < \alpha} \mathscr{L}_{P(X)}^\beta$. It follows from (98) of Chapter 7 and Proposition 7.35(b) that

$$\{p \in P(X)|p[N(S)] \geq c\} \in \mathscr{S}\left(\bigcup_{\beta < \alpha} \mathscr{L}_{P(X)}^\beta\right) \subset \mathscr{L}_{P(X)}^\alpha.$$

Thus, if $Q \in \mathscr{S}(\bigcup_{\beta < \alpha} \mathscr{L}_{P(X)}^\beta)$, then θ_Q is $\mathscr{L}_{P(X)}^\alpha$-measurable. The collection of sets Q for which θ_Q is $\mathscr{L}_{P(X)}^\alpha$-measurable forms a Dynkin system, so by the Dynkin system theorem (Proposition 7.24), θ_Q is $\mathscr{L}_{P(X)}^\alpha$-measurable for every $Q \in \mathscr{L}_X^\alpha$. This completes the induction step.

If $Q \in \mathscr{L}_X$, then for some $\alpha < \Omega$, $Q \in \mathscr{L}_X^\alpha$, so θ_Q is $\mathscr{L}_{P(X)}^\alpha$-measurable, and therefore θ_Q is $\mathscr{L}_{P(X)}$-measurable.   Q.E.D.

B.5 Set Theoretic Aspects of Borel Spaces

The measurability properties of Borel spaces are closely linked to several issues in set theory which we have for the most part skirted. These issues are presented briefly here.

There is some controversy concerning the propriety of the axiom of choice and Cantor's continuum hypothesis in applied mathematics. The former is generally accepted and the latter is regarded with suspicion. The general axiom of choice says that given any index set A and a collection of nonempty sets $\{S_\alpha | \alpha \in A\}$, there is a function $f : A \to \bigcup_{\alpha \in A} S_\alpha$ such that $f(\alpha) \in S_\alpha$ for every $\alpha \in A$. We have used this axiom in Appendix A to construct examples. In particular, the set E of Example 1 of that appendix for which both E and E^c have p-outer measure one is constructed by means of the axiom of choice. We have also used this axiom to construct the set S in the proof of Lemma B.3, and this lemma was instrumental in proving that every uncountable Borel space is Borel-isomorphic to every other uncountable Borel space (Proposition B.3 and Corollary 7.16.1). However an alternative proof of Lemma B.3 which does not require the axiom of choice is possible, but is quite lengthy and will not be given.

The countable axiom of choice is the same as the general axiom except that the index set A is required to be countable. A paraphrase of this axiom is that given any countable collection of nonempty sets, one element can be chosen from each set. We have made extensive use of this axiom, such as in the choice, for each k, of a selector φ_k in the proof of Proposition 7.50(a). Indeed, much of real analysis and topology rests on the countable axiom of choice.

Solovay [S13] has shown that if the general axiom of choice is replaced by the weaker "principle of dependent choice," which is still stronger than the countable axiom of choice, then every subset of the real line may be assumed to be Lebesgue-measurable. A slight extension of this result shows that under these conditions every subset of any Borel space may be assumed to be universally measurable. Therefore, by choice of the proper axiom system, the measurability difficulties which are the subject of Part II can be made to disappear.

It is possible to show without the use of the axiom of choice that every uncountable Borel space X contains universally measurable sets which are not limit measurable. An unpublished proof of this is due to Richard Lockhart. If both the axiom of choice and the continuum hypothesis are adopted then it follows that $\mathcal{U}_X$ has a larger cardinality than $\mathcal{L}_X$. Since for each $\alpha < \Omega$, $\mathcal{B}_X \subset \mathcal{L}_X^\alpha$ and $\mathcal{B}_X$ has cardinality at least c, so does $\mathcal{L}_X^\alpha$. On the other hand, $\mathcal{L}_X^\alpha$ is contained in $\mathcal{P}(\mathcal{L}_X^\alpha)$ and there is a universal function for $\mathcal{P}(\mathcal{L}_X^\alpha)$, so the cardinality of $\mathcal{L}_X^\alpha$ is c. Now $\mathcal{L}_X = \bigcup_{\alpha < \Omega} \mathcal{L}_X^\alpha$, and the cardinality of the set of countable ordinals is less than or equal to c, so $\mathscr{L}_X$ has cardinality c. In contrast, under the assumption of the axiom of choice and Cantor's continuum hypothesis, $\mathscr{U}_X$ contains a set F of cardinality c which has measure zero with respect to every nonatomic probability measure [H5, Chapter III, Section 14]. Thus every subset of F is also in $\mathscr{U}_X$, and the cardinality of $\mathscr{U}_X$ is at least 2^c. It follows that $\mathscr{L}_X$ must be properly contained in $\mathscr{U}_X$.

Another relevant set theoretic work is that of Gödel [G1], who showed that it is consistent with the usual axioms of set theory to assume the existence of the complement of an analytic set in the unit square whose projection on an axis is not Lebesgue-measurable. This means that it is consistent with the usual axioms to assume the existence of an analytically measurable function $f : [0, 1][0, 1] \to R$ such that $f^*(x) = \inf_y f(x, y)$ is not Lebesgue measurable. This places a severe constraint on the types of strengthened versions of Proposition 7.47 which might be possible.

Appendix C

The Hausdorff Metric and the Exponential Topology

This appendix develops a metric topology on the collection of closed subsets (including the empty set $\varnothing$) of a compact metric space (X, d). We denote this collection of sets by 2^X. For $A \in 2^X$ and $x \in X$, define

$$d(x, A) = \min_{a \in A} d(x, a) \qquad \text{if} \quad A \neq \varnothing, \tag{1}$$

$$d(x, \varnothing) = \text{diam}(X) = \max_{y, z \in X} d(y, z). \tag{2}$$

Definition C.1 Let (X, d) be a compact metric space. The *Hausdorff metric* ρ on 2^X is defined by

$$\rho(A, B) = \max\left\{ \max_{a \in A} d(a, B), \max_{b \in B} d(b, A) \right\} \qquad \text{if} \quad A, B \neq \varnothing, \tag{3}$$

$$\rho(A, \varnothing) = \rho(\varnothing, A) = \text{diam}(X) \qquad \text{if} \quad A \neq \varnothing, \tag{4}$$

$$\rho(\varnothing, \varnothing) = 0. \tag{5}$$

We have written max in place of sup in (3), since every set in 2^X is compact and $d(x, A)$ is a continuous function of x for every $A \in 2^X$. To see this latter property, consider a set $A \in 2^X$. If $A = \varnothing$, then the function $d(x, A)$ is constant and hence continuous. If $A \neq \varnothing$, then for $x, y \in X$ and $a \in A$ we have

$$d(x, a) \leq d(x, y) + d(y, a).$$

By taking the infimum of both sides over $a \in A$, we obtain

$$d(x, A) - d(y, A) \leq d(x, y).$$

By reversing the roles of x and y, we have

$$|d(x, A) - d(y, A)| \leq d(x, y) \qquad \forall x, y \in X, \qquad (6)$$

which shows that $d(x, A)$ is a Lipschitz continuous function of x.

It is a tedious but straightforward task to verify that $(2^X, \rho)$ is a metric space, and this is left to the reader. We will prove that $(2^X, \rho)$ is a compact metric space. We first show some preliminary facts.

If A is a (not necessarily closed) subset of X, define

$$2^A = \{K \in 2^X | K \subset A\}.$$

We define two classes

$$\mathscr{G} = \{2^G | G \text{ is an open subset of } X\}, \qquad (7)$$

$$\mathscr{K} = \{2^X - 2^K | K \text{ is a closed subset of } X\}. \qquad (8)$$

To aid the reader, we will continue to denote points of X by lowercase Latin letters and subsets of X by uppercase Latin letters. Uppercase script letters will be used for subsets of 2^X, except for subsets of the form 2^A as defined above. In keeping with this practice, we denote open spheres in the two spaces as follows:

$$S_\varepsilon(x) = \{ y \in X | d(x, y) < \varepsilon \},$$

$$\mathscr{S}_\varepsilon(A) = \{ B \in 2^X | \rho(A, B) < \varepsilon \}.$$

Finally, classes of subsets of 2^X will be denoted by boldface script letters, as in the case of $\mathscr{G}$ and $\mathscr{K}$ defined above.

The topology obtained by taking $\mathscr{G} \cup \mathscr{K}$ as a subbase in 2^X is called the *exponential topology* and an extensive theory exists for it [K2, K3]. It can be developed for a nonmetrizable topological space X, but we are interested in it only when X is compact metric. In this case, the exponential topology is the topology generated by the Hausdorff metric, as we now show.

Proposition C.1 Let (X, d) be a compact metric space and ρ the Hausdorff metric on 2^X. The class $\mathscr{G} \cup \mathscr{K}$ as defined by (7) and (8) is a subbase for the topology on $(2^X, \rho)$.

Proof We first prove that when G is open and K is closed in X, then 2^G and $2^X - 2^K$ are open in $(2^X, \rho)$. If G or K is empty, then 2^G or $2^X - 2^K$, respectively, is easily seen to be open, so we assume G and K are nonempty. Suppose A is a nonempty closed subset of X and $A \in 2^G$. (The proof for $A = \varnothing$ is trivial.) Since A is compact, is a subset of G, and $X - G$ is closed, there exists ε with $0 < \varepsilon < \mathrm{diam}(X)$ such that

$$\min_{a \in A} d(a, X - G) \geq \varepsilon. \tag{9}$$

For $B \in \mathscr{S}_\varepsilon(A)$, we have $B \neq \varnothing$ and

$$\max_{b \in B} d(b, A) < \varepsilon. \tag{10}$$

From inequalities (9) and (10) we have that $B \subset G$. Hence $\mathscr{S}_\varepsilon(A) \subset 2^G$, and 2^G must be open. Turning to the case of $2^X - 2^K$ for K closed, we let $A \in 2^X - 2^K$ be nonempty. By definition, $A \notin 2^K$, so $A - K$ contains at least one point a_0. Since $X - K$ is open, we can find $\varepsilon > 0$ for which $S_\varepsilon(a_0) \subset X - K$. For $B \in \mathscr{S}_\varepsilon(A)$, we have

$$d(a_0, B) \leq \max_{a \in A} d(a, B) < \varepsilon,$$

which implies $B \cap S_\varepsilon(a_0) \neq \varnothing$ and $B \in 2^X - 2^K$. Therefore $\mathscr{S}_\varepsilon(A) \subset 2^X - 2^K$, and $2^X - 2^K$ is open.

Having thus shown that the sets 2^G and $2^X - 2^K$ are open in $(2^X, \rho)$ when G is open and K is closed, we must now show that given any open subset $\mathscr{G}$ of $(2^X, \rho)$ and any nonempty $A \in \mathscr{G}$, we can find open sets $G_1, G_2, \ldots, G_m$ and closed sets $K_1, K_2, \ldots, K_n$ in X for which

$$A \in 2^{G_1} \cap \cdots \cap 2^{G_m} \cap (2^X - 2^{K_1}) \cap \cdots \cap (2^X - 2^{K_n}) \subset \mathscr{G}.$$

Since $\mathscr{G}$ is open in $(2^X, \rho)$, there exists $\varepsilon > 0$ such that $\mathscr{S}_\varepsilon(A) \subset \mathscr{G}$. Since A is closed in the compact set X, there exist points $\{x_1, \ldots, x_n\}$ in A such that $A \subset \bigcup_{k=1}^n S_{\varepsilon/2}(x_k)$. Let

$$G_1 = \{x \in X \mid d(x, A) < \varepsilon\}$$

and

$$K_k = X - S_{\varepsilon/2}(x_k), \qquad k = 1, \ldots, n.$$

By construction, $A \in 2^{G_1}$ and, since for each k, $A \cap S_{\varepsilon/2}(x_k) \neq \varnothing$, we have $A \in 2^X - 2^{K_k}$. Therefore

$$A \in 2^{G_1} \cap (2^X - 2^{K_1}) \cap \cdots \cap (2^X - 2^{K_n}).$$

Suppose B is another set in $2^{G_1} \cap (2^X - 2^{K_1}) \cap \cdots \cap (2^X - 2^{K_n})$. The fact that $B \in 2^{G_1}$ implies

$$\max_{b \in B} d(b, A) < \varepsilon. \tag{11}$$

If for some $a_0 \in A$ we had $d(a_0, B) \geq \varepsilon$, then we would also have $S_\varepsilon(a_0) \subset X - B$. But for some $x_k \in A$, $a_0 \in S_{\varepsilon/2}(x_k)$ and this would imply in succession $S_{\varepsilon/2}(x_k) \subset X - B$, $B \subset K_k$, and $B \notin 2^X - 2^{K_k}$. This contradiction shows that

$$\max_{a \in A} d(a, B) < \varepsilon. \tag{12}$$

Inequalities (11) and (12) establish that $\rho(A, B) < \varepsilon$, and as a consequence

$$2^{G_1} \cap (2^X - 2^{K_1}) \cap \cdots \cap (2^X - 2^{K_n}) \subset S_\varepsilon(A) \subset \mathscr{G}. \qquad \text{Q.E.D.}$$

If a cover of a space contains no finite subcover, we say the cover is *essentially infinite*. To show that $(2^X, \rho)$ is compact when X is compact, we must show that no essentially infinite open cover of 2^X exists. As a consequence of the following lemma, this will be accomplished if we can show that the subbase $\mathscr{G} \cup \mathscr{K}$ contains no essentially infinite cover. We remind the reader that a topological space in which every open cover has a countable subcover is called Lindelöf, and in metrizable spaces this property is equivalent to separability.

Lemma C.1   Let Ω be a Lindelöf space and let $\mathscr{S}$ be a subbase for the topology on Ω. If there exists an essentially infinite open cover of Ω, then there exists one which is a subset of $\mathscr{S}$.

Proof   Let $\mathscr{B}$ be the base for the topology on Ω constructed by taking finite intersections of sets in $\mathscr{S}$ and let $\mathscr{C}$ be an essentially infinite open cover of Ω. Each $C \in \mathscr{C}$ has a representation $C = \bigcup_{\alpha \in A(C)} B_\alpha$, where $B_\alpha \in \mathscr{B}$ for every $\alpha \in A(C)$. The collection $\bigcup_{C \in \mathscr{C}} \{B_\alpha | \alpha \in A(C)\}$ is an essentially infinite open cover of Ω, and, by the Lindelöf property, it contains a countable, essentially infinite, open subcover $\mathscr{D} = \{B_1, B_2, \ldots\}$. Each B_k has a representation $B_k = \bigcap_{j=1}^{n(k)} S_{kj}$, where $S_{kj} \in \mathscr{S}, j = 1, \ldots, n(k)$. If for each j the cover $\mathscr{D}_j = \{S_{1j}, B_2, B_3, \ldots\}$ is not essentially infinite, then there exists a finite subcollection $\bar{\mathscr{D}}_j$ which also covers Ω. But then

$$\{B_1\} \cup \left[\bigcup_{j=1}^{n(1)} (\bar{\mathscr{D}}_j - \{S_{1j}\}) \right] \subset \mathscr{D}$$

is a finite subcover of Ω. This contradiction implies that for some index j_0, the cover $\mathscr{D}_{j_0}$ is essentially infinite. Denote $R_1 = S_{1j_0}$. In general, given $R_1, R_2, \ldots, R_n$ in $\mathscr{S}$ such that $B_k \subset R_k$, $k = 1, \ldots, n$, and $\{R_1, R_2, \ldots, R_n, B_{n+1}, B_{n+2}, \ldots\}$ is an essentially infinite open cover of Ω, we can use the preceding argument to construct $R_{n+1} \in \mathscr{S}$ for which $B_{n+1} \subset R_{n+1}$ and $\{R_1, R_2, \ldots, R_n, R_{n+1}, B_{n+2}, B_{n+3}, \ldots\}$ is an essentially infinite open cover of Ω. The collection $\{R_1, R_2, \ldots\}$ is an essentially infinite open cover contained in $\mathscr{S}$.   Q.E.D.

Proposition C.2   Let (X, d) be a compact metric space and ρ the Hausdorff metric on 2^X. The metric space $(2^X, \rho)$ is compact.

Proof   We first show that $(2^X, \rho)$ is separable. Since (X, d) is compact, it is separable. Let D be a countable dense subset of X and let

$$\mathscr{C} = \{\overline{S_{1/n}(x)} \mid x \in D, \, n = 1, 2, \ldots\}.$$

Let $\mathscr{D}$ consist of finite unions of sets in $\mathscr{C}$. Then $\mathscr{D}$ is countable and, as we now show, is dense in $(2^X, \rho)$. Given $A \in 2^X$ and $\varepsilon > 0$, choose a positive integer n satisfying $2/n < \varepsilon$. The collection of sets $\{S_{1/n}(x) \mid x \in D\}$ covers the compact set A, so there is a finite subcollection $\{S_{1/n}(x) \mid x \in F\}$ which also covers A and which satisfies $S_{1/n}(x) \cap A \neq \varnothing$ for every $x \in F$. The set $B = \bigcup_{x \in F} \overline{S_{1/n}(x)}$ is in $\mathscr{D}$ and satisfies $\rho(A, B) < \varepsilon$.

As a result of Proposition C.1, Lemma C.1, and the separability of $(2^X, \rho)$, to show that $(2^X, \rho)$ is compact we need only show that every open cover of 2^X which is a subset of $\mathscr{G} \cup \mathscr{H}$ contains a finite subcover of 2^X. Thus let $\{G_\alpha \mid \alpha \in A\}$ be a collection of open sets and $\{K_\beta \mid \beta \in B\}$ a collection of closed sets in X, and suppose

$$2^X = \left[\bigcup_{\alpha \in A} 2^{G_\alpha}\right] \cup \left[\bigcup_{\beta \in B} (2^X - 2^{K_\beta})\right].$$

Define the closed set $K_0 = \bigcap_{\beta \in B} K_\beta$. By definition, $K_0 \notin \bigcup_{\beta \in B}(2^X - 2^{K_\beta})$, so $K_0 \in \bigcup_{\alpha \in A} 2^{G_\alpha}$. Thus for some $\alpha_0 \in A$, we have $K_0 \in 2^{G_{\alpha_0}}$, i.e., $K_0 \subset G_{\alpha_0}$. This means that

$$X - G_{\alpha_0} \subset X - K_0 = \bigcup_{\beta \in B} (X - K_\beta),$$

and since $X - G_{\alpha_0}$ is compact, there exists a finite set $\{\beta_1, \beta_2, \ldots, \beta_n\} \subset B$ for which

$$X - G_{\alpha_0} \subset \bigcup_{k=1}^{n} (X - K_{\beta_k}). \tag{13}$$

To complete the proof, we show

$$2^X = 2^{G_{\alpha_0}} \cup \left[\bigcup_{k=1}^{n} (2^X - 2^{K_{\beta_k}})\right].$$

If $C \in 2^X$, then either $C \subset G_{\alpha_0}$, in which case $C \in 2^{G_{\alpha_0}}$, or else $C \cap (X - G_{\alpha_0}) \neq \varnothing$. In the latter case, (13) implies that for some k, $C \cap (X - K_{\beta_k}) \neq \varnothing$, i.e., $C \in 2^X - 2^{K_{\beta_k}}$.   Q.E.D.

We now develop some convergence notions in $(2^X, \rho)$. Let $\{A_n\}$ be a sequence of sets in 2^X. Define

$$\varlimsup_{n \to \infty} A_n = \left\{x \in X \,\middle|\, \liminf_{n \to \infty} d(x, A_n) = 0\right\}, \tag{14}$$

$$\varliminf_{n \to \infty} A_n = \left\{x \in X \,\middle|\, \limsup_{n \to \infty} d(x, A_n) = 0\right\}. \tag{15}$$

For example, if $X = [-1, 1]$ and $A_n = \{(-1)^n\}$, we have $\overline{\lim}_{n \to \infty} A_n = \{-1, 1\}$ and $\underline{\lim}_{n \to \infty} A_n = \varnothing$. If $X = [-1, 1]$ and $A_n = [-1/n, 1/n]$, we have

$$\overline{\lim_{n \to \infty}} A_n = \underline{\lim_{n \to \infty}} A_n = \{0\}.$$

Clearly we have $\underline{\lim}_{n \to \infty} A_n \subset \overline{\lim}_{n \to \infty} A_n$. It is also true that $\overline{\lim}_{n \to \infty} A_n$ and $\underline{\lim}_{n \to \infty} A_n$ are closed. To see this for $\overline{\lim}_{n \to \infty} A_n$, let $\{x_m\}$ be a sequence in $\overline{\lim}_{n \to \infty} A_n$ converging to x. Then from (6) we have for each m

$$\liminf_{n \to \infty} d(x, A_n) \le d(x, x_m) + \liminf_{n \to \infty} d(x_m, A_n) = d(x, x_m),$$

and since $d(x, x_m)$ can be made arbitrarily small by choosing m sufficiently large, we conclude that $x \in \overline{\lim}_{n \to \infty} A_n$. Replace $\liminf_{n \to \infty}$ by $\limsup_{n \to \infty}$ in the preceding argument to show that $\underline{\lim}_{n \to \infty} A_n$ is closed.

If $\overline{\lim}_{n \to \alpha} A_n = \underline{\lim}_{n \to \alpha} A_n$, we denote their common value by $\lim_{n \to \infty} A_n$. This notation is justified by the following proposition.

Proposition C.3 Let (X, d) be a compact metric space and ρ the Hausdorff metric on 2^X. Let $\{A_n\}$ be a sequence in 2^X. Then

$$\overline{\lim_{n \to \infty}} A_n = \underline{\lim_{n \to \infty}} A_n = A \qquad (16)$$

if and only if

$$\lim_{n \to \infty} \rho(A_n, A) = 0. \qquad (17)$$

Proof Assume for the moment that $A \ne \varnothing$ and suppose (16) holds. Then for each x in the compact set A, $d(x, A_n) \to 0$ as $n \to \infty$. Given $\varepsilon > 0$, let $\{x_1, \ldots, x_k\}$ be points of A such that the open spheres $S_{\varepsilon/2}(x_j), j = 1, \ldots, k$ cover A. Choose N large enough so that for all $n \ge N$

$$d(x_j, A_n) \le \varepsilon/2, \qquad j = 1, \ldots, k.$$

Now use the Lipschitz continuity [cf. (6)] of the function $x \to d(x, A_n)$ to conclude that

$$d(x, A_n) \le \varepsilon \qquad \forall x \in A.$$

This implies that

$$\lim_{n \to \infty} \max_{x \in A} d(x, A_n) = 0.$$

This equation and (3) imply that (17) will follow if we can show

$$\lim_{n \to \infty} \max_{y \in A_n} d(y, A) = 0. \qquad (18)$$

If (18) fails to hold, then for some $\varepsilon > 0$ there exists a sequence $y_k \in A_{n_k}$ such that $n_1 < n_2 < \cdots$ and

$$d(y_k, A) \geq \varepsilon \qquad \forall k. \tag{19}$$

The compactness of X implies that $\{y_k\}$ accumulates at some $y_0 \in X$ which, by (19) and the continuity of $x \to d(x, A)$, must satisfy $d(y_0, A) \geq \varepsilon$. But $y_0 \in \overline{\lim}_{n \to \infty} A_n$ by (14), and this contradicts (16). Hence (18) holds.

Still assuming $A \neq \varnothing$, we turn to the reverse implication of the proposition. If (17) holds, then

$$\lim_{n \to \infty} d(x, A_n) = 0 \qquad \forall x \in A, \tag{20}$$

and

$$\lim_{n \to \infty} \max_{y \in A_n} d(y, A) = 0. \tag{21}$$

Equation (20) implies that

$$A \subset \underline{\lim}_{n \to \infty} A_n \subset \overline{\lim}_{n \to \infty} A_n. \tag{22}$$

If $x \in \overline{\lim}_{n \to \infty} A_n$, then by definition there exists a sequence $y_k \in A_{n_k}$ such that $n_1 < n_2 < \cdots$ and

$$\lim_{k \to \infty} d(x, y_k) = 0. \tag{23}$$

We have from (6) that

$$d(x, A) \leq d(x, y_k) + d(y_k, A),$$

and, letting $k \to \infty$ and using (21) and (23), we conclude $d(x, A) = 0$. Since A is closed, this proves $x \in A$ and

$$\overline{\lim}_{n \to \infty} A_n \subset A. \tag{24}$$

Combine (22) and (24) to obtain (16).

Assume finally that $A = \varnothing$. If (16) holds, then all but finitely many of the sets A_n must be empty, for otherwise one could find $y_k \in A_{n_k}, n_1 < n_2 < \cdots$, and $\{y_k\}$ would accumulate at some $y_0 \in \overline{\lim}_{n \to \infty} A_n$. If all but finitely many of the sets A_n are empty, then (5) implies that (17) holds. Conversely, if (17) holds and $A = \varnothing$, then (4) implies that all but finitely many of the sets A_n are empty. Equation (16) follows from (2), (14), and (15).　　Q.E.D.

For the proof of Proposition 7.33 in Section 7.5 we need the concept of a function which is upper semicontinuous in the sense of Kuratowski, or in abbreviation, upper semicontinuous (K).

Definition C.2   Let Y be a metric space and X a compact metric space. A function $F: Y \to 2^X$ is *upper semicontinuous* (K) if for every convergent sequence $\{y_n\}$ in Y with limit y, we have $\overline{\lim}_{n \to \infty} F(y_n) \subset F(y)$.

The similarity of Definition C.2 to the idea of an upper semicontinuous real or extended real-valued function is apparent [Lemma 7.13(b)]. Although we will not discuss functions which are lower semicontinuous (K), it is interesting to note that such a concept exists and has the obvious definition, namely, that the function $F: Y \to 2^X$ is *lower semicontinuous* (K) if for every convergent sequence $\{y_n\}$ in Y with limit y, we have $\underline{\lim}_{n \to \infty} F(y_n) \supset F(y)$. It can be seen from Proposition C.3 that a function $F: Y \to 2^X$ is continuous in the usual sense (where 2^X has the exponential topology) if and only if it is both upper and lower semicontinuous (K). We carry the analogy with real-valued functions even farther by showing that an upper semicontinuous (K) function is Borel-measurable, and the remainder of the appendix is devoted to this.

Lemma C.2   Let Y be a metric space and X a compact metric space. If $F: Y \to 2^X$ is upper semicontinuous (K), then for each open set $G \subset X$, the set

$$\{y \in Y \mid F(y) \subset G\} = F^{-1}(2^G) \tag{25}$$

is open.

Proof   The openness of $F^{-1}(2^G)$ for every open G is in fact equivalent to upper semicontinuity (K), but we need only the weaker result stated. To prove it, we show that for G open, the set $F^{-1}(2^X - 2^G)$ is closed. If $\{y_n\}$ is a sequence in this set with limit $y \in Y$, then

$$F(y_n) \cap (X - G) \neq \varnothing, \qquad n = 1, 2, \ldots,$$

and so there exists a sequence $\{x_n\}$ in the compact set $X - G$ such that $x_n \in F(y_n)$, $n = 1, 2, \ldots$. This sequence has an accumulation point $x \in X - G$, and, by (14), $x \in \overline{\lim}_{n \to \infty} F(y_n)$. The upper semicontinuity (K) of F implies $x \in F(y)$, and so $F(y) \cap (X - G) \neq \varnothing$, i.e., $y \in F^{-1}(2^X - 2^G)$.   Q.E.D.

Proposition C.4   Let Y be a metric space, (X, d) a compact metric space, and let 2^X have the exponential topology. Let $F: Y \to 2^X$ be upper semicontinuous (K). Then F is Borel-measurable.

Proof   If $F: Y \to 2^X$ is upper semicontinuous (K) and G is an open subset of X, then $F^{-1}(2^G)$ is Borel-measurable in Y by Lemma C.2. If K is a closed subset of X, define open sets $G_n = \{x \mid d(x, K) < 1/n\}$. We have $K = \bigcap_{n=1}^{\infty} G_n$, and so a closed set A is a subset of K if and only if $A \subset G_n$, $n = 1, 2, \ldots$.

This implies $2^K = \bigcap_{n=1}^{\infty} 2^{G_n}$, and

$$F^{-1}(2^K) = \bigcap_{n=1}^{\infty} F^{-1}(2^{G_n})$$

is a G_δ-set, thus Borel-measurable in Y. It follows that for any set $\mathscr{G}$ in the subbase $\mathscr{G} \cup \mathscr{K}$ for the exponential topology on 2^X, $F^{-1}(\mathscr{G})$ is Borel-measurable in Y. By Proposition 7.1, any open set in 2^X can be represented as a countable union of finite intersections of sets in $\mathscr{G} \cup \mathscr{K}$ and so its inverse image under F is Borel-measurable.   Q.E.D.

References

[A1]   R. Ash, "Real Analysis and Probability." Academic Press, New York, 1972.

[A2]   K. J. Åström, Optimal control of Markov processes with incomplete state information, *J. Math. Anal. Appl.* **10** (1965), 174–205.

[B1]   R. Bellman, "Dynamic Programming." Princeton Univ. Press, Princeton, New Jersey, 1957.

[B2]   D. P. Bertsekas, Infinite-time reachability of state-space regions by using feedback control, *IEEE Trans. Automatic Control* **AC-17** (1972), 604–613.

[B3]   D. P. Bertsekas, On error bounds for successive approximation methods, *IEEE Trans. Automatic Control* **AC-21** (1976), 394–396.

[B4]   D. P. Bertsekas, "Dynamic Programming and Stochastic Control." Academic Press, New York, 1976.

[B5]   D. P. Bertsekas, Monotone mappings with application in dynamic programming, *SIAM J. Control Optimization* **15** (1977), 438–464.

[B6]   D. P. Bertsekas and S. Shreve, Existence of optimal stationary policies in deterministic optimal control, *J. Math. Anal. Appl.* (to appear).

[B7]   P. Billingsley, Invariance principle for dependent random variables, *Trans. Amer. Math. Soc.* **83** (1956), 250–282.

[B8]   D. Blackwell, Positive dynamic programming, *Proc. Fifth Berkeley Sympos. Math. Statist. and Probability*, *1965*, 415–418.

[B9]   D. Blackwell, Discounted dynamic programming, *Ann. Math. Statist.* **36** (1965), 226–235

[B10]   D. Blackwell, On stationary policies, *J. Roy. Statist. Soc.* **133A** (1970), 33–37.

[B11]   D. Blackwell, Borel-programmable functions, *Ann. Prob.* **6** (1978), 321–324.

[B12]   D. Blackwell, D. Freedman, and M. Orkin, The optimal reward operator in dynamic programming, *Ann. Probability* **2** (1974), 926–941.

[B13]   N. Bourbaki, "General Topology." Addison–Wesley, Reading, Massachusetts, 1966.

[B14]   D. W. Bressler and M. Sion, The current theory of analytic sets, *Canad. J. Math.* **16** (1964), 207–230.

[B15] L. D. Brown and R. Purves, Measurable selections of extrema, *Ann. Statist.* **1** (1973), 902–912.

[C1] D. Cenzer and R. D. Mauldin, Measurable parameterizations and selections, *Trans. Amer. Math. Soc.* (to appear).

[D1] C. Dellacherie, "Ensembles Analytiques, Capacités, Mesures de Hausdorff." Springer-Verlag, Berlin and New York, 1972.

[D2] E. V. Denardo, Contraction mappings in the theory underlying dynamic programming, *SIAM Rev.* **9** (1967), 165–177.

[D3] C. Derman, "Finite State Markovian Decision Processes." Academic Press, New York, 1970.

[D4] J. L. Doob, "Stochastic Processes." Wiley, New York, 1953.

[D5] L. Dubins and D. Freedman, Measurable sets of measures, *Pacific J. Math.* **14** (1964), 1211–1222.

[D6] L. Dubins and L. Savage, "Inequalities for Stochastic Processes (How to Gamble if you Must)." McGraw-Hill, New York, 1965. (Republished by Dover, New York, 1976.)

[D7] J. Dugundji, "Topology." Allyn & Bacon, Rockleigh, New Jersey, 1966.

[D8] E. B. Dynkin and A. A. Juskevic, "Controlled Markov Processes and their Applications." Moscow, 1975. (English translation to be published by Springer-Verlag.)

[F1] D. Freedman, The optimal reward operator in special classes of dynamic programming problems, *Ann. Probability.* **2** (1974), 942–949.

[F2] E. B. Frid, On a problem of D. Blackwell from the theory of dynamic programming, *Theor. Probability Appl.* **15** (1970), 719–722.

[F3] N. Furukawa, Markovian decision processes with compact action spaces, *Ann. Math. Statist.* **43** (1972) 1612–1622.

[F4] N. Furukawa and S. Iwamoto, Markovian decision processes and recursive reward functions, *Bull. Math. Statist.* **15** (1973), 79–91.

[F5] N. Furukawa and S. Iwamoto, Dynamic programming on recursive reward systems, *Bull. Math. Statist.* **17** (1976), 103–126.

[G1] K. Gödel, The consistency of the axiom of choice and of the generalized continuum-hypothesis, *Proc. Nat. Acad. Sci. U.S.A.* **24** (1938), 556–557.

[H1] P. R. Halmos, "Measure Theory." Van Nostrand–Reinhold, Princeton, New Jersey, 1950.

[H2] F. Hausdorff, "Set Theory." Chelsea, Bronx, New York, 1957.

[H3] C. J. Himmelberg, T. Parthasarathy, and F. S. Van Vleck, Optimal plans for dynamic programming problems, *Math. Operations Res.* **1** (1976), 390–394.

[H4] K. Hinderer, "Foundations of Nonstationary Dynamic Programming with Discrete Time Parameter." Springer-Verlag, Berlin and New York, 1970.

[H5] J. Hoffman–Jørgensen, "The Theory of Analytic Spaces." Aarhus Universitet, Aarhus, Denmark, 1970.

[H6] A. Hordijk, "Dynamic Programming and Markov Potential Theory." Mathematical Centre Tracts, Amsterdam, 1974.

[H7] R. Howard, "Dynamic Programming and Markov Processes." MIT Press, Cambridge, Massachusetts, 1960.

[J1] B. Jankov, On the uniformisation of A-sets, *Dokl. Akad. Nauk SSSR* **30** (1941), 591–592 (in Russian).

[J2] W. Jewell, Markov renewal programming I and II, *Operations Res.* **11** (1963), 938–971.

[J3] A. A. Juskevič (Yushkevich), Reduction of a controlled Markov model with incomplete data to a problem with complete information in the case of Borel state and control spaces, *Theor. Probability Appl.* **21** (1976), 153–158.

[K1] L. Kantorovich and B. Livenson, Memoir on analytical operations and projective sets, *Fund. Math.* **18** (1932), 214–279.

[K2]   K. Kuratowski, "Topology I." Academic Press, New York, 1966.

[K3]   K. Kuratowski, "Topology II." Academic Press, New York, 1968.

[K4]   K. Kuratowski and A. Mostowski, "Set Theory." North-Holland, Amsterdam, 1976.

[K5]   K. Kuratowski and C. Ryll-Nardzewski, A general theorem on selectors, *Bull. Polish Acad. Sci.* **13** (1965), 397–411.

[K6]   H. Kushner, "Introduction to Stochastic Control." Holt, New York, 1971.

[L1]   M. Loève, "Probability Theory." Van Nostrand-Reinhold, Princeton, New Jersey, 1963.

[L2]   N. Lusin, Sur les ensembles analytiques, *Fund. Math.* **10** (1927), 1–95.

[L3]   N. Lusin and W. Sierpinski, Sur quelques propriétés des ensembles (A), *Bull. Acad. Sci. Cracovie* (1918), 35–48.

[M1]   G. Mackey, Borel structure in groups and their duals, *Trans. Amer. Math. Soc.* **85** (1957), 134–165.

[M2]   A. Maitra, Discounted dynamic programming on compact metric spaces, *Sankhya* **30A** (1968), 211–216.

[M3]   J. McQueen, A modified dynamic programming method for Markovian decision problems, *J. Math. Anal. Appl.* **14** (1966), 38–43.

[M4]   P. A. Meyer, "Probability and Potentials." Ginn (Blaisdell), Boston, Massachusetts, 1966.

[M5]   P. A. Meyer and M. Traki, Reduites et jeux de hasard (Seminaire de Probabilites VII, Universite de Strasbourg, *in* "Lecture Notes in Mathematics," Vol. 321), pp. 155–171. Springer, Berlin, 1973.

[N1]   J. von Neumann, On rings of operators. Reduction theory, *Ann. of Math.* **50** (1949), 401–485.

[O1]   P. Olsen, Multistage stochastic programming with recourse: The equivalent deterministic problem, *SIAM J. Control Optimization* **14** (1976), 495–517.

[O2]   P. Olsen, When is a multistage stochastic programming problem well-defined?, *SIAM J. Control Optimization* **14** (1976), 518–527.

[O3]   P. Olsen, Multistage stochastic programming with recourse as mathematical programming in an L_p space, *SIAM J. Control Optimization* **14** (1976), 528–537.

[O4]   D. Ornstein, On the existence of stationary optimal strategies, *Proc. Amer. Math. Soc.* **20** (1969), 563–569.

[O5]   J. M. Ortega and W. C. Rheinboldt, "Iterative Solutions of Nonlinear Equations in Several Variables." Academic Press, New York, 1970.

[P1]   K. Parthasarathy, "Probability Measures on Metric Spaces." Academic Press, New York, 1967.

[P2]   Yu. V. Prohorov, Convergence of random processes and limit theorems in probability theory, *Theor. Probability Appl.* **1** (1956), 157–214.

[R1]   D. Rhenius, Incomplete information in Markovian decision models, *Ann. Statist.* **2** (1974), 1327–1334.

[R2]   R. T. Rockafellar, Integral functionals, normal integrands and measurable selections, *in* "Nonlinear Operators and the Calculus of Variations." Springer-Verlag, Berlin and New York, 1976.

[R3]   R. T. Rockafellar and R. Wets, Stochastic convex programming: relatively complete recourse and induced feasibility, *SIAM J. Control Optimization* **14** (1976), 574–589.

[R4]   R. T. Rockafellar and R. Wets, Stochastic convex programming: basic duality, *Pacific J. Math.* **62** (1976), 173–195.

[R5]   H. L. Royden, "Real Analysis." Macmillan, New York, 1968.

[S1]   S. Saks, "Theory of the Integral." Stechert, New York, 1937.

[S2]   Y. Sawaragi and T. Yoshikawa, Discrete-time Markovian decision processes with incomplete state information, *Ann. Math. Statist.* **41** (1970), 78–86.

[S3]   M. Schäl, On continuous dynamic programming with discrete time parameter, *Z. Wahrscheinlichkeitstheorie und Verw. Gebiete* **21** (1972), 279–288.

[S4]   M. Schäl, On dynamic programming: Compactness of the space of policies, *Stochastic Processes Appl.* **3** (1975), 345–364.

[S5]   M. Schäl, Conditions for optimality in dynamic programming and for the limit of *n*-stage optimal policies to be optimal, *Z. Wahrscheinlichkeitstheorie und Verw. Gebiete* **32** (1975), 179–196.

[S6]   E. Selivanovskij, Ob odnom klasse effektivnyh mnozestv (mnozestva C), *Mat. Sb.* **35** (1928), 379–413.

[S7]   S. Shreve, *A General Framework for Dynamic Programming with Specializations*, M. S. thesis (1977), Dept. of Elec. Eng., Univ. of Illinois, Urbana.

[S8]   S. Shreve, *Dynamic Programming in Complete Separable Spaces*, Ph.D. thesis (1977), Dept. of Math., Univ. of Illinois, Urbana.

[S9]   S. Shreve and D. P. Bertsekas, A new theoretical framework for finite horizon stochastic control, *Proc. Fourteenth Annual Allerton Conf. Circuit and System Theory, Allerton Park, Illinois, October, 1976*, 336–343.

[S10] S. Shreve and D. P. Bertsekas, Equivalent stochastic and deterministic optimal control problems, *Proc. 1976 IEEE Conf. Decision and Control, Clearwater Beach, Florida*, 705–709.

[S11] S. Shreve and D. P. Bertsekas, Alternative theoretical frameworks for finite horizon discrete-time stochastic optimal control, *SIAM J. Control Optimization* **16** (1978).

[S12] S. Shreve and D. P. Bertsekas, Universally measurable policies in dynamic programming *Mathematics of Operations Research* (to appear).

[S13] R. Solovay, A model of set-theory in which every set of reals is Lebesgue measurable, *Ann. Math.* **92** (1970), 1–56.

[S14] R. E. Strauch, Negative dynamic programming, *Ann Math. Statist.* **37** (1966), 871–890

[S15] C. Striebel, Sufficient statistics in the optimal control of stochastic systems, *J. Math. Anal. Appl.* **12** (1965), 576–592.

[S16] C. Striebel, "Optimal Control of Discrete Time Stochastic Systems." Springer-Verlag, Berlin and New York, 1975.

[S17] M. Suslin (Souslin), Sur une définition des ensembles measurables *B* sans nombres transfinis, *C. R. Acad. Sci. Paris* **164** (1917), 88–91.

[V1]   V. S. Varadarajan, Weak convergence of measures on separable metric spaces, *Sankhya* **19** (1958), 15–22.

[W1]   D. H. Wagner, Survey of measurable selection theorems, *SIAM J. Control Optimization* **15** (1977), 859–903.

[W2]   A. Wald, "Statistical Decision Functions." Wiley, New York, 1950.

[W3]   H. S. Witsenhausen, A standard form for sequential stochastic control, *Math. Systems Theory* **7** (1973), 5–11.

Table of Propositions, Lemmas, Definitions, and Assumptions

Chapter 2

Monotonicity Assumption ... 27

Chapter 3

Proposition 3.1 ... 40
Proposition 3.2 ... 43
Proposition 3.3 ... 44
Proposition 3.4 ... 46
Proposition 3.5 ... 47
Proposition 3.6 ... 50
Proposition 3.7 ... 51

Lemma 3.1 ... 45

Assumption F.1 ... 39
Assumption F.2 ... 40
Assumption F.3 ... 40

Chapter 4

Proposition 4.1 ... 53
Proposition 4.2 ... 55
Proposition 4.3 ... 56
Proposition 4.4 ... 57
Proposition 4.5 ... 59

Proposition 4.6 ... 60
Proposition 4.7 ... 62
Proposition 4.8 ... 64
Proposition 4.9 ... 64
Proposition 4.10 ... 68
Proposition 4.11 ... 69

Assumption C (Contraction
  Assumption) ... 52
Fixed Point
  Theorem ... 55

Chapter 5

Proposition 5.1 ... 71
Proposition 5.2 ... 73
Proposition 5.3 ... 75
Proposition 5.4 ... 78
Proposition 5.5 ... 78
Proposition 5.6 ... 79
Proposition 5.7 ... 80
Proposition 5.8 ... 81
Proposition 5.9 ... 84
Proposition 5.10 ... 86
Proposition 5.11 ... 87
Proposition 5.12 ... 88
Proposition 5.13 ... 89

318

Proposition 5.14   90
Proposition 5.15   90

Lemma 5.1   75
Lemma 5.2   82

Assumption I (Uniform
    Increase Assumption)   70
Assumption D (Uniform
    Decrease Assumption)   70
Assumption I.1   71
Assumption I.2   71
Assumption D.1   71
Assumption D.2   71

Chapter 6

Proposition 6.1   95
Proposition 6.2   95
Proposition 6.3   96
Proposition 6.4   97
Proposition 6.5   97

Assumption A.1   93
Assumption A.2   93
Assumption A.3   93
Assumption A.4   93
Assumption A.5   93
Assumption $\bar{F}$.2   94
Assumption $\bar{F}$.3   95
Exact Selection Assumption   95
Assumption C   96

Chapter 7

Proposition 7.1   106
Proposition 7.2 (Urysohn's
    Theorem)   106
Proposition 7.3
    (Alexandroff's Theorem)   107
Proposition 7.4   108
Proposition 7.5   109
Proposition 7.6   112
Proposition 7.7   113
Proposition 7.8   114
Proposition 7.9   116
Proposition 7.10   117
Proposition 7.11   118
Proposition 7.12   119
Proposition 7.13   119
Proposition 7.14   120
Proposition 7.15
    (Kuratowski's Theorem)   121

Proposition 7.16   121
Proposition 7.17   122
Proposition 7.18   124
Proposition 7.19   127
Proposition 7.20   127
Proposition 7.21   128
Proposition 7.22   130
Proposition 7.23   131
Proposition 7.24 (Dynkin
    System Theorem)   133
Proposition 7.25   133
Proposition 7.26   134
Proposition 7.27   135
Proposition 7.28   140
Proposition 7.29   144
Proposition 7.30   145
Proposition 7.31   148
Proposition 7.32   148
Proposition 7.33   153
Proposition 7.34   154
Proposition 7.35   158
Proposition 7.36   161
Proposition 7.37   164
Proposition 7.38   165
Proposition 7.39   165
Proposition 7.40   165
Proposition 7.41   166
Proposition 7.42 (Lusin's
    Theorem)   167
Proposition 7.43   169
Proposition 7.44   172
Proposition 7.45   175
Proposition 7.46   177
Proposition 7.47   179
Proposition 7.48   180
Proposition 7.49 (Jankov-von
    Neumann Theorem)   182
Proposition 7.50   184

Lemma 7.1 (Urysohn's Lemma)   105
Lemma 7.2   105
Lemma 7.3   116
Lemma 7.4   119
Lemma 7.5   119
Lemma 7.6   125
Lemma 7.7   125
Lemma 7.8   125
Lemma 7.9   127
Lemma 7.10   131
Lemma 7.11   139
Lemma 7.12   144
Lemma 7.13   146

Lemma 7.14	147	Lemma 8.3	196
Lemma 7.15	149	Lemma 8.4	197
Lemma 7.16	150	Lemma 8.5	202
Lemma 7.17	151	Lemma 8.6	205
Lemma 7.18	151	Lemma 8.7	206
Lemma 7.19	152		
Lemma 7.20	152	Definition 8.1	188
Lemma 7.21	154	Definition 8.2	190
Lemma 7.22	161	Definition 8.3	191
Lemma 7.23	162	Definition 8.4	194
Lemma 7.24	163	Definition 8.5	195
Lemma 7.25	164	Definition 8.6	206
Lemma 7.26	172	Definition 8.7	208
Lemma 7.27	173	Definition 8.8	210
Lemma 7.28	174		
Lemma 7.29	174	**Chapter 9**	
Lemma 7.30	177		
		Proposition 9.1	216
Definition 7.1	104	Proposition 9.2	219
Definition 7.2	105	Proposition 9.3	219
Definition 7.3	107	Proposition 9.4	220
Definition 7.4	112	Proposition 9.5	223
Definition 7.5	114	Proposition 9.6	224
Definition 7.6	117	Proposition 9.7	224
Definition 7.7	118	Proposition 9.8	225
Definition 7.8	120	Proposition 9.9	226
Definition 7.9	121	Proposition 9.10	226
Definition 7.10	122	Proposition 9.11	227
Definition 7.11	133	Proposition 9.12	227
Definition 7.12	134	Proposition 9.13	228
Definition 7.13	146	Proposition 9.14	231
Definition 7.14	157	Proposition 9.15	231
Definition 7.15	157	Proposition 9.16	232
Definition 7.16	160	Proposition 9.17	234
Definition 7.17	161	Proposition 9.18	236
Definition 7.18	167	Proposition 9.19	237
Definition 7.19	171	Proposition 9.20	239
Definition 7.20	171	Proposition 9.21	241
Definition 7.21	177		
		Lemma 9.1	220
		Lemma 9.2	221
Chapter 8		Lemma 9.3	230
Proposition 8.1	192	Definition 9.1	213
Proposition 8.2	198	Definition 9.2	214
Proposition 8.3	200	Definition 9.3	214
Proposition 8.4	203	Definition 9.4	216
Proposition 8.5	207	Definition 9.5	217
Proposition 8.6	209	Definition 9.6	217
Proposition 8.7	211	Definition 9.7	217
		Definition 9.8	217
Lemma 8.1	194	Definition 9.9	218
Lemma 8.2	196	Definition 9.10	229

Chapter 10

Proposition 10.1	246
Proposition 10.2	254
Proposition 10.3	256
Proposition 10.4	257
Proposition 10.5	262
Proposition 10.6	264
Lemma 10.1	253
Lemma 10.2	255
Lemma 10.3	260
Lemma 10.4	261
Definition 10.1	243
Definition 10.2	245
Definition 10.3	248
Definition 10.4	249
Definition 10.5	249
Definition 10.6	250
Definition 10.7	251
Definition 10.8	256

Chapter 11

Proposition 11.1	266
Proposition 11.2	267
Proposition 11.3	267
Proposition 11.4	268
Proposition 11.5	269
Proposition 11.6	270
Proposition 11.7	272

Appendix A

Proposition A.1	278
Lemma A.1	273
Lemma A.2	274
Lemma A.3	275
Definition A.1	273

Appendix B

Proposition B.1	282
Proposition B.2	285
Proposition B.3	289
Proposition B.4	290
Proposition B.5	291
Proposition B.6	292
Proposition B.7	293
Proposition B.8	295
Proposition B.9	296
Proposition B.10	297
Proposition B.11	299
Proposition B.12	300
Lemma B.1	285
Lemma B.2	285
Lemma B.3	286
Lemma B.4	287
Lemma B.5	288
Lemma B.6	294
Lemma B.7	295
Lemma B.8	298
Definition B.1	290
Definition B.2	293
Definition B.3	298

Appendix C

Proposition C.1	304
Proposition C.2	306
Proposition C.3	308
Proposition C.4	310
Lemma C.1	306
Lemma C.2	310
Definition C.1	303
Definition C.2	310

Index

A

Alexandroff's theorem, 107
Analytic measurability of a function, 171
Analytic set, 160
Analytic σ-algebra, 171
A posteriori distribution, 260ff
A priori distribution, 260ff
Axiom of choice, 301

B

Baire null space, 103, 109
Borel isomorphism, 121
Borel measurability of a function, 120
Borel programmable, 21
Borel σ-algebra, 117
Borel space, 118

C

Cantor's continuum hypothesis, 301
Completion
    of a metric space, 114
    of a σ-algebra, 167
Composition of measurable functions, 298
Contraction assumption, 52

Control
    constraint, 2, 26, 188, 216, 243, 245, 248,
        251, 271
    space, 2, 26, 188, 216, 243, 245, 248, 251,
        271
Cost
    corresponding to a policy, 2, 28, 191, 217,
        244, 249, 254
    one-stage, 2, 189, 216, 243, 245, 248, 251,
        271
    optimal, 2, 29, 191, 217, 244, 246, 250, 254
C-sets, 20

D

Disturbance kernel, 189, 243, 245, 271
Disturbance space, 189, 243, 245, 271
Dynamic programming (DP) algorithm, 3, 6,
        39, 57, 80, 198, 229, 259
Dynkin system, 133
Dynkin system theorem, 133

E

Epigraph, 82
Exact selection assumption, 95
Exponential topology, 304

F

Filtering, 261
Fixed point theorem (Banach), 55
F_σ-set, 102

G

G_δ-set, 102

H

Hausdorff metric, 303
Hilbert cube, 103
Homeomorphism, 104
Horizon
    finite, 28, 189, 243, 245, 248, 251, 271
    infinite, 70, 213, 216, 243, 245, 248, 251

I

Imperfect state information model, 248
Indicator function, 103
Information vector, 248
Isometry, 144

J

Jankov–von Neumann theorem, 182

K

Kuratowski's theorem, 121

L

Limit measurability, 298
Limit σ-algebra, 293
Lindelöf space, 106
Lower semianalytic function, 177
Lower semicontinuous function, 146
Lower semicontinuous model, 208
Lusin's theorem, 167

M

Metrizable space, 104
Monotonicity assumption, 27

N

Nonstationary model, 243

O

Observation kernel, 248
Observation space, 248
Optimality equation, 4, 57, 71, 73, 78ff, 225
    nonstationary, 246
Outer integral, 273
    monotone convergence theorem for, 278

P

Paved space, 157
Policy, 2, 6, 26, 91, 190, 214, 217, 243, 249
    analytically measurable, 190, 269ff
    Borel-measurable, 190
    ϵ-optimal, 29, 191, 215, 244
    $\{\epsilon_n\}$-dominated convergence to optimality, 29, 191, 245
    k-originating, 243
    limit-measurable, 190, 266ff
    Markov, 6, 190
    nonrandomized, 190, 249
    optimal, 29, 191, 215, 244
    p-ϵ-optimal, 12
    q-optimal, 256
    semi-Markov, 190
    stationary, 214
    uniformly N-stage optimal, 29, 206
    universally measurable, 190
    weakly q-ϵ-optimal, 256
p-outer measure, 166, 274
Projection mapping, 103

R

Regular probability measure, 122
Relative topology, 104
R-operator, 21

S

Second countable space, 106
Semi-Markov decision problems, 34
Separable space, 105
State space, 2, 26, 188, 216, 243, 245, 248, 251, 271

State transition kernel, 189, 243, 248, 251
Statistic sufficient for control, 250
  existence of, 259ff
Stochastic kernel, 134
Stochastic programming, 11ff
Suslin scheme, 157
  nucleus of, 157
  regular, 161
System function, 189, 216, 243, 245, 271

T

Topologically complete space, 107
Totally bounded space, 112

U

Uniform decrease assumption, 70
Uniform increase assumption, 70
Universal function, 290
Universal measurability of a function, 171
Universal σ-algebra, 167
Upper semicontinuous function, 146
Upper semicontinuous model, 210
Upper semicontinuous (K) function, 310
Urysohn's lemma, 105
Urysohn's theorem, 106

W

Weak topology on space of probability measures, 125